Essential Discrete Mathematics

Richard Johnsonbaugh

DEPAUL UNIVERSITY, CHICAGO

Essential Discrete Mathematics

MACMILLAN PUBLISHING COMPANY
NEW YORK
Collier Macmillan Publishers
LONDON

Macmillan Publishing Company
866 Third Avenue, New York, New York 10022

Collier Macmillan Canada, Inc.

LIBRARY OF CONGRESS CATALOGING-IN-PUBLICATION DATA
Johnsonbaugh, Richard,
    Essential discrete mathematics.
    Bibliography: p.
    Includes index.
    1. Mathematics—1961–   2. Electronic data
processing—Mathematics. I. Title.
QA39.2.J66   1987   510   86-18028
ISBN 0-02-360630-4
Printing: 1 2 3 4 5 6 7 8   Year: 7 8 9 0 1 2 3 4 5 6

ISBN   0-02-360630-4

Preface

TO THE INSTRUCTOR

A course in discrete mathematics has become a standard offering in many mathematics and computer science departments. This book for a one-term course in discrete mathematics represents a different philosophy and organization from those of most discrete mathematics books now available. This new approach will make discrete mathematics more accessible to students and will make the subject more inviting to students who are required to take a course in discrete mathematics.

The features in this book include the following:

1. Topics are introduced in the order of increasing sophistication.
2. A course in computer science or computer programming is not a prerequisite. (The recommended mathematics prerequisite is college algebra.)
3. A review and a self-test with answers in the back of the book conclude each chapter.

4. Many examples are drawn from familiar, nonmathematical, non-threatening settings.
5. Algorithms are described in ordinary English.
6. An optional *Student Study Guide* is available.
7. A number of bridges to previous work in mathematics, such as number systems, functions, and sequences, are provided.
8. Matrices and applications are included.
9. Logic is integrated into the main text.
10. Biographical sketches are included.

I will discuss some of these features in more detail.

So that students with a relatively modest background in mathematics can succeed in this course, topics are introduced in the order of increasing sophistication. As a prerequisite, I would recommend the equivalent of college algebra. Calculus and computer programming are not prerequisites.

The book begins with logic and circuits. Not only do these topics provide a gentle introduction to discrete mathematics, but also I have found that students enjoy analyzing and constructing circuits.

Chapter 2 deals with general mathematical concepts—sets, sequences, mathematical induction, functions, and matrices. Students will have seen some of this material before. I have taken a straightforward approach to these topics.

Chapters 3 and 4 deal with graphs and trees. Students enjoy this material and do not find it particularly abstract.

Beginning with Chapter 5 on algorithms, the level of abstraction begins to increase and more demands are made on the student for original thinking. At the minimum, one should cover Section 5.1, which discusses what an algorithm is, and Section 5.4, which deals with the time required to execute an algorithm. These topics are important in mathematics quite apart from their central role in computer science.

Permutations, combinations, and the pigeonhole principle form the subject matter of Chapter 6. A wealth of examples and discussion are provided to help the student master this tricky material.

Chapters 7 and 8 contain the most abstract topics in this book—recurrence relations, relations, and equivalence relations.

Algorithms are not formally treated until Chapter 5. (Algorithms are given prior to Chapter 5 but are not advertised as such.) Algorithms are stated in plain English and are also presented somewhat more formally as a stepwise process, but not in pseudocode. The latter description can be skipped if desired. Pseudocode is described in optional Section 5.2. Thereafter, pseudocode is presented only in optional sections flagged with †. Further, all computer science examples are given in a handful of optional sections marked with † that follow the regular sections. These optional sections are labeled Computer Notes. Exercises for Computer Notes sections are marked with the computer icon shown in the margin. Students who are taking a computer science course or who have programming experience should find the computer sections helpful.

Each chapter concludes with a chapter review section that lists the key ideas of the chapter and a self-test that includes questions from every section in the chapter. Answers to all self-test questions can be found in the Hints section at the end of the book. Students who have had programming experience may be interested in the computer exercises found at the end of each chapter.

For a taste of the examples that are drawn from familiar, nonmathematical, nonthreatening settings, sample the beginnings of Sections 1.1, 2.7, and 8.1; the beginning of Chapter 5; and Example 2.5.7.

Approximately one-third of the exercises have solutions in the Hints section at the back of the book. Exercises with solutions in the back of the book are marked with ''H.'' An optional *Student Study Guide* is available. This *Guide* contains solutions to another third of the exercises in the book. In addition, the *Guide* contains detailed explanations of examples and solutions to problems together with tips on how to solve problems. The style is informal.

The basic operations on matrices are introduced in Section 2.7. Section 3.4 treats the matrix representation of graphs. Matrices of relations are covered in Section 8.3.

Logic is integrated into the main text. The first two sections of the book introduce fundamental concepts in logic. Two additional sections on logic, Proofs and Arguments, and Categorical Propositions, Sections A.1 and A.2, can be covered at any time as needed.

Several optional sections can be omitted without loss of continuity. For example, if only a brief introduction to trees is desired, one would cover only Sections 4.1 (Introduction) and 4.2 (Terminology and Characterizations of Trees). On the other hand, by covering all of the sections, binary trees, isomorphic trees, and games trees would also be discussed. In this way it is possible to tailor the topics to local needs.

Because it is important for students to work a large number of exercises, over 1700 exercises are included. Although several more challenging than average exercises are marked with a star, there are an ample number of exercises that directly check students' understanding of the concepts and examples presented in the book. Ends of proofs are marked with the symbol ■.

An *Instructor's Guide* and a program diskette for IBM and IBM-compatible personal computers are available at no cost to adopters of this book. The *Instructor's Guide* contains a discussion about how to use this book, sample exams, sample syllabi, and solutions to most exercises whose solutions are not given in either the *Student Study Guide* or the book itself. The program diskette, which is coordinated with optional sections of the *Student Study Guide*, contains (Turbo) Pascal programs with which students can experiment to obtain hands-on experience with concepts in the book. There is no charge for making copies of this diskette to distribute to students.

TO THE STUDENT

A reasonable question is, "What is discrete mathematics?" Discrete mathematics deals mainly with the analysis of finite collections of objects, unlike continuous mathematics, which is concerned with infinite processes. Sorting is an example of a problem that uses discrete mathematics: How does one arrange a finite set of objects in order? Problems in classical physics belong to the world of continuous mathematics, calculus in particular. One example is: Find the force of water at *every* point against a dam. Since there are infinitely many points on the dam, this problem is studied by continuous methods. Computer science is finite in nature—computers have finite memories, instructions are executed at finite time intervals, programs are finite, and so on—thus computer science finds discrete mathematics extremely useful. However, the applications of discrete mathematics are by no means limited to computer science; operations research, business, engineering, economics, chemistry, political science, and biology, among other disciplines, find discrete mathematics an indispensable tool.

This course will build on concepts that you have already studied. You have undoubtedly studied sets and functions and perhaps logic and mathematical induction previously. These topics are not peculiar to discrete mathematics. However, in discrete mathematics there are fewer problems that require you to plug into a formula than in some of the other mathematics courses you may have taken. The way to succeed is to work many exercises. Practice really makes perfect in discrete mathematics.

About a third of the problems are marked with an "H," which indicates that the solution is given in the back of the book. Before looking in the back of the book, make an honest effort to solve the problem on your own. You will learn far more by working on a problem on your own, even if you do not obtain a complete solution, than by looking at a solution by someone else.

Each chapter concludes with a chapter review. You might try defining each term listed and illustrating its use in a problem. After reviewing the chapter, you can test yourself by taking the chapter self-test. Answers to all self-test questions can be found in the Hints section at the end of the book.

Should you desire other study material, a *Student Study Guide,* written especially to accompany this book, is available. This *Guide* contains solutions to another third of the exercises in the book. Written in a relaxed and informal style, the *Guide* contains additional explanations of examples and solutions to problems along with tips on how to solve problems.

You need not have had any experience with computers to study this book. However, if you know something about computers or about programming, you might be interested in the optional Computer Notes sections, which indicate connections between discrete mathematics and computer science. Exercises that require you to have studied the Computer Notes sections are flagged with the computer icon shown in the margin. Also, there are computer exercises at the end of each chapter.

ACKNOWLEDGMENTS

In writing this book, I have been assisted by many persons, including Henry S. Tropp, Humboldt State University; Joe Chan, DePaul University; Kam-Chan Lo, Pansophic Systems Inc.; Jerrold Grossman, Oakland University; Donald J. Albers, Menlo College; Gregory Bachelis, Wayne State University; Gary Phillips, Oakton Community College; Jane Edgard, Brevard Community College, Florida; Susan Forman, Bronx Community College; and Sadie Bragg, Borough of Manhattan Community College. Special thanks go to the students of DePaul University who used various drafts of this book.

I am indebted to the Department of Computer Science and Information Systems at DePaul University and its chairman, Helmut Epp, for providing time and encouragement for the development of this book.

I am fortunate to be able to work with some of the best people in the publishing industry at Macmillan. John Schultz, Picture Editor, located the photographs that accompany the biographical sketches. He also contributed to the biographies. I would like to thank Robert Freese, Book Designer, for his innovative design of the book. Elaine Wetterau, Production Supervisor, was instrumental in seeing that the publishing technicalities were executed properly. Bob Clark, Mathematics Editor, provided excellent reviewers and helped me keep track of myriad details. Gary Ostedt, Executive Editor, contributed many unique suggestions based on his wide experience in mathematics publishing and a great deal of warmly welcomed encouragement.

R. J.

Contents

1

Logic and Circuits 1

2

The Language of Mathematics 77

3

Graphs 153

4

Trees 217

†Can be omitted without loss of continuity.

5

Algorithms   *273*

6

Permutations, Combinations, and the Pigeonhole Principle   *317*

7

Recurrence Relations 377

8

Relations 425

Appendix: More on Logic 465

Essential Discrete Mathematics

1

But I proved beyond a shadow of a doubt and with geometric logic that a key did exist.

—from *The Caine Mutiny*

Logic and Circuits

Chapter 1 examines some concepts of **logic**, specifically, how we determine whether certain statements are true or false. Logical methods are used in mathematics to prove theorems and in computer science to prove that programs do what they are alleged to do. We will see that certain ideas in logic are closely related to the construction of circuits. The chapter concludes by showing how to design simple circuits like those used in digital computers.

1.1 Propositions

Which of the sentences are either true or false (but not both)?

(a) Benny Goodman has recorded classical music.
(b) The line ''Play it again, Sam'' occurs in the movie *Casablanca*.
(c) Earth is the only planet in the universe that has life.
(d) Buy two tickets to the Ungrateful Living concert for Friday.

Sentence (a) is true. Although Benny Goodman is best known for his jazz recordings, he recorded much classical music (e.g., the Weber Clarinet Concertos, numbers 1 and 2, with the Chicago Symphony Orchestra).

Although it is widely believed that the line ''Play it again, Sam'' occurs in *Casablanca*, it does not. The line that actually occurs is ''Play it, Sam. Play 'As Time Goes By.' '' Thus sentence (b) is false.

Sentence (c) is either true or false (but not both), but no one knows which at this time.

Sentence (d) is neither true nor false [(d) is a command].

A sentence that is either true or false, but not both, is called a **proposition**. Sentences (a)–(c) are propositions, whereas sentence (d) is not a proposition. Propositions are the basic building blocks of any theory of logic.

We will use lowercase letters, such as p, q, and r, to represent propositions. We will also use the notation

$$p: 1 + 1 = 3$$

to define p to be the proposition $1 + 1 = 3$.

In ordinary speech and writing, we combine propositions using connectives such as *and* and *or*. For example, the propositions "It is raining" and "I will take my umbrella" can be combined to form the single proposition "It is raining and I will take my umbrella." The formal definitions of *and* and *or* follow.

Definition 1.1.1.   Let p and q be propositions.

The *conjunction* of p and q, denoted $p \wedge q$, is the proposition

$$p \text{ and } q.$$

The *disjunction* of p and q, denoted $p \vee q$, is the proposition

$$p \text{ or } q.$$

Propositions such as $p \wedge q$ and $p \vee q$, which result from combining propositions, are called **compound propositions**.

Example 1.1.2.   If

$$p: 1 + 1 = 3,$$
$$q: \text{A decade is 10 years,}$$

then the conjunction of p and q is

$$p \wedge q: 1 + 1 = 3 \text{ and a decade is 10 years.}$$

The disjunction of p and q is

$$p \vee q: 1 + 1 = 3 \text{ or a decade is 10 years.}$$

The truth values of propositions such as conjunctions and disjunctions can be described by **truth tables**. The truth table of a proposition P made up of the individual propositions $p_1, \ldots, p_n$ lists all possible combinations of truth values for $p_1, \ldots, p_n$, T denoting true and F denoting false, and for each such combination lists the truth value of P.

Definition 1.1.3.   The truth value of the compound proposition $p \wedge q$ is defined by the truth table

p	q	$p \wedge q$
T	T	T
T	F	F
F	T	F
F	F	F

Notice that in the truth table in Definition 1.1.3 all four possible combinations of truth assignments for p and q are given.

Definition 1.1.3 states that the conjunction $p \wedge q$ is true only if p and q are both true; $p \wedge q$ is false otherwise.

Example 1.1.4.   Let p and q be as in Example 1.1.2. Then p is false, q is true, and the conjunction $p \wedge q$ is false.

Example 1.1.5.   Let

p: Benny Goodman has recorded classical music,
q: The Baltimore Orioles used to be the St. Louis Browns.

Then p and q are both true and the conjunction

$p \wedge q$: Benny Goodman has recorded classical music and the
Baltimore Orioles used to be the St. Louis Browns

is also true.

Example 1.1.6.   Let

p: $1 + 1 = 3$,
q: Chicago is in California.

Then p and q are both false and the conjunction

$p \wedge q$: $1 + 1 = 3$ and Chicago is in California

is false.

Definition 1.1.7. The truth value of the compound proposition $p \vee q$ is defined by the truth table

p	q	$p \vee q$
T	T	T
T	F	T
F	T	T
F	F	F

The *or* in the disjunction, $p \vee q$, is used in the *inclusive* sense; that is, $p \vee q$ is considered true if either p or q or *both* are true and $p \vee q$ is false only if both p and q are false. There is also an **exclusive-or** (see Exercise 20), in which p *exor* q is true if either p or q but *not* both is true.

Example 1.1.8. If p and q are as in Example 1.1.2, then p is false, q is true, and the disjunction $p \vee q$ is true.

Example 1.1.9. If p and q are as in Example 1.1.5, then p and q are both true and the disjunction

$p \vee q$: Benny Goodman has recorded classical music or the Baltimore Orioles used to be the St. Louis Browns

is also true.

Example 1.1.10. If p and q are as in Example 1.1.6, then p and q are both false and the disjunction

$p \vee q$: $1 + 1 = 3$ or Chicago is in California

is false.

If p is a proposition, we may form the **negation** $\bar{p}$ of p by reversing its truth values. The proposition $\bar{p}$ is sometimes read ''not p.''

Definition 1.1.11.   The *negation* of p, denoted $\bar{p}$, is the proposition

$$\text{not } p.$$

The truth value of the proposition $\bar{p}$ is defined by the truth table

p	$\bar{p}$
T	F
F	T

Example 1.1.12.   If

$$p: \text{Cary Grant starred in } \textit{Rear Window},$$

the negation of p is the proposition

$$\bar{p}: \text{Cary Grant did not star in } \textit{Rear Window}.$$

Since p is false, $\bar{p}$ is true.

Example 1.1.13.   Let

p: Beethoven lived in the eighteenth century,
q: The first all-electronic digital computer was constructed in the twentieth century,
r: Audrey Meadows was the original ''Alice'' in ''The Honeymooners.''

Represent the proposition

Either Beethoven lived in the eighteenth century and it is not the case that the first all-electronic digital computer was constructed in the twentieth century; or Audrey Meadows was the original ''Alice'' in ''The Honeymooners''

symbolically and determine whether it is true or false.

The proposition may be written symbolically as

$$(p \wedge \bar{q}) \vee r.$$

If we replace each symbol by its truth value, we find

$$(p \wedge \overline{q}) \vee r = (\text{T} \wedge \overline{\text{T}}) \vee \text{F}$$
$$= (\text{T} \wedge \text{F}) \vee \text{F}$$
$$= \text{F} \vee \text{F}$$
$$= \text{F}.$$

Therefore, the given proposition is false.

†COMPUTER NOTES FOR SECTION 1.1

The execution of statements within a control structure in a computer program is governed by evaluating the truth of propositions.

For example, in the **if-then-else** structure

> **if** p **then**
>> *action* 1
>
> **else**
>> *action* 2

if the proposition p is true, *action* 1 (but not *action* 2) is executed and if the proposition p is false, *action* 2 (but not *action* 1) is executed.

In the **while loop** structure

> **while** p **do**
>> *action*

action will be repeatedly executed as long as p is true.

Example 1.1.14.   When the following program segment is executed,

$$i := 2$$
$$j := 3$$

if $i = 1$ **or** $j + 1 = 4$ **then**   (1.1.1)
> print "FIRST"

else
> print "SECOND"

†Any Computer Notes section can be omitted without loss of continuity.

we will print FIRST. (We denote the assignment operator by $:=$. Thus, when $x := y$ is executed, the value of y is assigned to x or, equivalently, the value of x is replaced by the value of y. We denote the equality relation by $=$. Thus given values of x and y, $x = y$ is a proposition—the statement $x = y$ is either true or false.)

If we let

$$p: i = 1, \qquad q: j + 1 = 4,$$

the proposition that appears as the condition in the if-then-else structure (1.1.1) can be written $p \lor q$. Since p is false and q is true, $p \lor q$ is true. Therefore, we execute the instruction between the **if** and the **else**.

Example 1.1.15.   When the following program segment is executed,

$$
\begin{aligned}
&i := 2 \\
&j := 3 \\
&\textbf{if } i = 2 \textbf{ and not } (j - 1 = 2) \textbf{ then} \qquad\qquad (1.1.2)\\
&\qquad \text{print ``FIRST''} \\
&\textbf{else} \\
&\qquad \text{print ``SECOND''}
\end{aligned}
$$

we will print SECOND. If we let

$$p: i = 2, \qquad q: j - 1 = 2,$$

the proposition that appears as the condition in the if-then-else structure (1.1.2) can be written $p \land \overline{q}$. The proposition p is true. Since q is true, $\overline{q}$ is false. Now $p \land \overline{q}$ is false because $\overline{q}$ is false. Therefore, we execute the instruction following **else**.

Example 1.1.16.   In the following program segment,

$$
\begin{aligned}
&i := 1 \\
&j := 1 \\
&\textbf{while } (i < 2 \textbf{ and } j < 5) \textbf{ or } i + j = 5 \textbf{ do} \\
&\qquad \textbf{begin} \qquad\qquad\qquad\qquad\qquad\qquad\qquad (1.1.3)\\
&\qquad i := i + 2 \\
&\qquad j := j + 1 \\
&\qquad \textbf{end}
\end{aligned}
$$

the while loop will be executed twice. (In case the action to be taken consists of multiple statements, we use **begin** and **end**, as well as indentation, to clearly delimit the action.)

If we let

$$p: i < 2, \qquad q: j < 5, \qquad r: i + j = 5$$

the proposition that appears as the condition in the while loop (1.1.3) can be written $(p \wedge q) \vee r$. Initially, p is true, q is true, and r is false. In this case, $(p \wedge q) \vee r$ is true and the while loop is executed. At this point, i is 3 and j is 2, so this time p is false, q is true, and r is true. Therefore, $(p \wedge q) \vee r$ is again true and the while loop is executed a second time. At this point i is 5 and j is 3, so this time p is false, q is true, and r is false. Therefore, $(p \wedge q) \vee r$ is false and the while loop ceases to execute.

EXERCISES

Evaluate each proposition in Exercises 1–6 for the truth values

$$p = F, \qquad q = T, \qquad r = F.$$

1H.† $p \vee q$ **2.** $\bar{p} \vee \bar{q}$

3. $\bar{p} \vee q$ **4H.** $\bar{p} \vee \overline{(q \wedge r)}$

5. $\overline{(p \vee q)} \wedge (\bar{p} \vee r)$ **6.** $(p \vee \bar{r}) \wedge (q \vee r) \vee \overline{(r \vee p)}$

Write the truth table of each proposition in Exercises 7–14.

7H. $p \wedge \bar{q}$ **8.** $(\bar{p} \vee \bar{q}) \vee p$

9. $(p \vee q) \wedge \bar{p}$ **10H.** $(p \wedge q) \wedge \bar{p}$

11. $(p \wedge q) \vee (\bar{p} \vee q)$ **12.** $\overline{(p \wedge q)} \vee (r \wedge \bar{p})$

13H. $(p \vee q) \wedge (\bar{p} \vee q) \wedge (p \vee \bar{q}) \wedge (\bar{p} \vee \bar{q})$

14. $\overline{(p \wedge q)} \vee (\bar{q} \vee r)$

Assume that a, b, and c are real numbers. In Exercises 15–19, represent the given statement symbolically by letting

$$p: a < b, \qquad q: b < c, \qquad r: a < c.$$

15H. $a < b$ and $b < c$

16. $(a \geq b$ and $b < c)$ or $a \geq c$.

†An exercise marked with "H" has a hint or solution in the back of the book.

17. It is not the case that ($a < b$ and $b < c$).

18H. $a < b$ or it is not the case that ($b < c$ and $a < c$).

19. (It is not the case that ($a < b$ and ($a < c$ or $b < c$))) or ($a \geq b$ and $a < c$).

20. Give the truth table for the exclusive-or of p and q in which p *exor* q is true if either p or q but not both is true.

In Exercises 21–26, formulate the symbolic expression in words using

p: Today is Monday,
q: It is raining,
r: It is hot.

21H. $p \vee q$ **22.** $\bar{p} \wedge (q \vee r)$ **23.** $\overline{p \vee q} \wedge r$

24H. $(p \wedge q) \wedge \overline{(r \vee p)}$

25. $(p \wedge (q \vee r)) \wedge (r \vee (q \vee p))$

26. $(p \vee (\bar{p} \wedge \overline{(q \vee r)})) \wedge (p \vee \overline{(r \vee q)})$

For each program segment in Exercises 27–31, determine the number of times the statement $x := x + 1$ will be executed.

27H. $i := 1$
    if $i < 2$ **or** $i > 0$ **then**
      $x := x + 1$
    else
      $x := x + 2$

28. $i := 2$
    if ($i < 0$ **and** $i > 1$) **or** $i = 3$ **then**
      $x := x + 1$
    else
      $x := x + 2$

29. $i := 1$
    while $i < 3$ **do**
      begin
      $x := x + 1$
      $i := i + 1$
      end

30H. $i := 1$
    while ($i > 0$ **and** $i < 3$) **or** $i = 3$ **do**
      begin

$$x := x + 1$$
$$i := i + 1$$
end

31. $i := 1$
     while $i > 0$ **and** $i < 4$ **do**
     begin
     if $i < 2$ **then**
       $i := i + 1$
     else
       $i := i + 2$
     $x := x + 1$
     end

1.2 Conditional Propositions and Logical Equivalence

Consider the statement

> If the Mathematics Department gets an additional \$20,000, then it will hire one new faculty member.   (1.2.1)

Statement (1.2.1) states that on the condition that the Mathematics Department gets an additional \$20,000, then the Mathematics Department will hire one new faculty member. A proposition such as (1.2.1) is called a **conditional proposition**.

Definition 1.2.1.   If p and q are propositions, the compound proposition

$$\text{if } p \text{ then } q \qquad (1.2.2)$$

is called a *conditional proposition* and is denoted

$$p \rightarrow q.$$

The proposition p is called the *hypothesis* (or *antecedent*) and the proposition q is called the *conclusion* (or *consequent*).

Example 1.2.2.   If we define

> p: The Mathematics Department gets an additional $20,000,
> q: The Mathematics Department hires one new faculty member,

then statement (1.2.1) assumes the form (1.2.2). The hypothesis is the statement "The Mathematics Department gets an additional $20,000," and the conclusion is the statement "The Mathematics Department hires one new faculty member."

Some statements not of the form (1.2.2) may be rephrased as conditional propositions, as the next example illustrates.

Example 1.2.3.   Restate each proposition in the form (1.2.2) of a conditional proposition.

(a) Mary will be a good student if she studies hard.

(b) John may take calculus only if he has sophomore, junior, or senior standing.

(c) When you sing, my ears hurt.

(d) A necessary condition for the triangle t to be equilateral is that t be equiangular.

(e) A sufficient condition for Ralph to visit California is that he goes to Disneyland.

(a) The hypothesis is the clause following *if*; thus an equivalent formulation is

> If Mary studies hard, then she will be a good student.

(b) The *only if* clause is the conclusion; that is,

$$\text{if } p, \text{ then } q$$

is considered logically the same as

$$p \text{ only if } q.$$

An equivalent formulation is

> If John takes calculus, then he has sophomore,
> junior, or senior standing.

The "if p, then q" formulation emphasizes the hypothesis, whereas the "p only if q" formulation emphasizes the conclusion; the difference is only stylistic.

(c) *When* means the same as *if*; thus an equivalent formulation is

If you sing, then my ears hurt.

(d) A **necessary condition** is another name for the conclusion; thus an equivalent formulation is

If the triangle t is equilateral, then t is equiangular.

(e) A **sufficient condition** is another name for the hypothesis; thus an equivalent formulation is

If Ralph goes to Disneyland, then he visits California.

Consider the problem of assigning a truth value to the conditional proposition (1.2.2). In Example 1.2.2, if it is true that the Mathematics Department gets an additional $20,000 and it is also true that the Mathematics Department hires one new faculty member, we would regard statement (1.2.1) as true. On the other hand, if it is true that the Mathematics Department gets an additional $20,000, but it is false that the Mathematics Department hires one new faculty member, we would regard statement (1.2.1) as false. In general, the conditional proposition $p \rightarrow q$ is true provided that the hypothesis p is true and the conclusion q is also true. On the other hand, $p \rightarrow q$ is false provided that the hypothesis p is true and the conclusion q is false. (You should not be able to deduce a false conclusion from a true hypothesis!) The situation becomes murkier when the hypothesis p is false. Suppose, for example, it is false that the Mathematics Department gets an additional $20,000. Then all bets are off as to whether the Department will hire one new faculty member. When the hypothesis is false, the conclusion plays no role in determining whether (1.2.1) is true or false. In case the hypothesis p is false, the standard definition declares (1.2.2) to be true whether the conclusion q is true or false. The preceding discussion is summarized in the next definition.

Definition 1.2.4. The truth value of the conditional proposition $p \rightarrow q$ is defined by the following truth table:

p	q	$p \rightarrow q$
T	T	T
T	F	F
F	T	T
F	F	T

Example 1.2.5.   Let

$$p: 1 > 2, \qquad q: 4 < 8.$$

Then p is false and q is true. Therefore,

$$p \to q \text{ is true}, \qquad q \to p \text{ is false}.$$

Example 1.2.6.   Assuming that p is true, q is false, and r is true, find the truth value of each proposition.

(a) $(p \wedge q) \to r$
(b) $(p \vee q) \to \bar{r}$
(c) $p \wedge (q \to r)$
(d) $p \to (q \to r)$

We replace each symbol p, q, and r by its truth value to obtain the truth value of the proposition:

(a) $(T \wedge F) \to T = F \to T = \text{true}$
(b) $(T \vee F) \to \bar{T} = T \to F = \text{false}$
(c) $T \wedge (F \to T) = T \wedge T = \text{true}$
(d) $T \to (F \to T) = T \to T = \text{true}$

In ordinary language, the hypothesis and conclusion in a conditional proposition are normally related, but in logic, the hypothesis and conclusion in a conditional proposition are not required to refer to the same subject matter. For example, in logic, we permit propositions such as

If $5 < 3$, then Alfred Hitchcock directed *The Third Man*.

(In fact, since the hypothesis is false, this proposition is true.) Logic is concerned with the form of propositions and the relation of propositions to each other and not with the subject matter itself.

Example 1.2.5 shows that the proposition $p \to q$ can be true while the proposition $q \to p$ is false. We call the proposition $q \to p$ the **converse** of the proposition $p \to q$. Thus a conditional proposition can be true while its converse is false.

Example 1.2.7.   Write each conditional proposition symbolically. Write the converse of each statement symbolically and in words. Also, find the truth value of each conditional proposition and its converse.

(a) If $1 < 2$, then $3 < 6$.
(b) If $1 > 2$, then $3 < 6$.

(a) Let

$$p: 1 < 2, \qquad q: 3 < 6.$$

The given statement may be written symbolically as

$$p \rightarrow q.$$

Since p and q are both true, this statement is true. The converse may be written symbolically as

$$q \rightarrow p$$

and in words as

$$\text{If } 3 < 6, \text{ then } 1 < 2.$$

Since p and q are both true, the converse $q \rightarrow p$ is true.
  (b) Let

$$p: 1 > 2, \qquad q: 3 < 6.$$

The given statement may be written symbolically as

$$p \rightarrow q.$$

Since p is false and q is true, this statement is true. The converse may be written symbolically as

$$q \rightarrow p$$

and in words as

$$\text{If } 3 < 6, \text{ then } 1 > 2.$$

Since q is true and p is false, the converse $q \rightarrow p$ is false.

Another useful compound proposition is

$$p \text{ if and only if } q. \qquad (1.2.3)$$

Such a statement is interpreted to mean

$$(\text{if } p \text{ then } q) \text{ and } (\text{if } q \text{ then } p). \qquad (1.2.4)$$

Let us determine the truth value of the proposition (1.2.3). Suppose that p and q are both true. Then both conditional propositions in (1.2.4) are true. *And*ing true values produces a true value (Definition 1.1.3); hence (1.2.4) is true. Since (1.2.3) is interpreted as (1.2.4), we regard (1.2.3) as true if both p and q are true. If both p and q are false, again both

conditional propositions in (1.2.4) are true and so (1.2.4) is true. Therefore, if both p and q are false, we regard (1.2.3) as true. If p is false
and q is true, then the second conditional proposition in (1.2.4) is false.
When one value in an *and* operation is false, the result is false. Thus we
regard (1.2.3) as false if p is false and q is true. Similarly, if p is true
and q is false, we regard (1.2.3) as false. This discussion motivates the
following definition.

Definition 1.2.8.   If p and q are propositions, the compound proposition

$$p \text{ if and only if } q$$

is called a *biconditional proposition* and is denoted

$$p \leftrightarrow q.$$

The truth value of the proposition $p \leftrightarrow q$ is defined by the following
truth table:

p	q	$p \leftrightarrow q$
T	T	T
T	F	F
F	T	F
F	F	T

Notice that $p \leftrightarrow q$ is true precisely when either p and q are both true
or p and q are both false. An alternative way to state "p if and only if
q" is "p is a necessary and sufficient condition for q." "p if and only
if q" is sometimes written "p iff q."

Example 1.2.9.   Let a, b, and c be the lengths of the sides of the triangle T with c the
longest length. The statement

$$T \text{ is a right triangle if and only if } a^2 + b^2 = c^2 \quad (1.2.5)$$

can be written symbolically as

$$p \leftrightarrow q$$

if we define

$$p: T \text{ is a right triangle,} \qquad q: a^2 + b^2 = c^2.$$

Statement (1.2.5) asserts that

$$\text{If } T \text{ is a right triangle, then } a^2 + b^2 = c^2 \qquad (1.2.6)$$

and

$$\text{If } a^2 + b^2 = c^2, \text{ then } T \text{ is a right triangle.} \qquad (1.2.7)$$

An alternative way to state (1.2.5) is: A necessary and sufficient condition for the triangle T to be a right triangle is that the sides satisfy $a^2 + b^2 = c^2$.

In some cases, two different compound propositions have the same truth values no matter what truth values their constituent propositions have. Such propositions are said to be **logically equivalent**.

Definition 1.2.10.   Suppose that the compound propositions P and Q are made up of the propositions $p_1, \ldots, p_n$. We say that P and Q are *logically equivalent* and write

$$P \equiv Q,$$

provided that given any truth values of $p_1, \ldots, p_n$, either P and Q are both true or P and Q are both false.

Example 1.2.11 De Morgan's Laws for Logic.   We will verify the first of **De Morgan's laws**

$$\overline{p \vee q} \equiv \bar{p} \wedge \bar{q}, \qquad \overline{p \wedge q} \equiv \bar{p} \vee \bar{q}$$

leaving the second as an exercise (see Exercise 47).

By writing the truth tables for $P = \overline{p \vee q}$ and $Q = \bar{p} \wedge \bar{q}$, we can verify that given any truth values of p and q, either P and Q are both true or P and Q are both false:

p	q	$\overline{p \vee q}$	$\bar{p} \wedge \bar{q}$
T	T	F	F
T	F	F	F
F	T	F	F
F	F	T	T

Thus P and Q are logically equivalent.

Augustus De Morgan   (1806–1871)

Augustus De Morgan was born in India but raised and educated in England. He became professor of mathematics at University College, London, in 1828. He resigned his position three years later in support of a professor of anatomy who was fired by the college council without giving reasons. Mathematical induction had been used prior to De Morgan, but he was the first to carefully describe the process. (He was the first to use the term "mathematical induction.") He was interested in logic, foundations of algebra, and teaching. [*Courtesy of the Ann Ronan Picture Library*]

Our next example gives a logically equivalent form of the negation of $p \to q$.

Example 1.2.12.   Show that the negation of $p \to q$ is logically equivalent to $p \land \bar{q}$.
We must show that

$$\overline{p \to q} \equiv p \land \bar{q}.$$

By writing the truth tables for $P = \overline{p \to q}$ and $Q = p \land \bar{q}$, we can verify that given any truth values of p and q, either P and Q are both true or P and Q are both false:

p	q	$\overline{p \to q}$	$p \land \bar{q}$
T	T	F	F
T	F	T	T
F	T	F	F
F	F	F	F

Thus P and Q are logically equivalent.

We defined $p \leftrightarrow q$ so that it would mean the same as $p \to q$ *and* $q \to p$. We can now show that, according to our definitions, $p \leftrightarrow q$ is logically equivalent to $p \to q$ *and* $q \to p$.

Example 1.2.13.   The truth table shows that

$$p \leftrightarrow q \equiv (p \to q) \land (q \to p).$$

p	q	$p \leftrightarrow q$	$p \to q$	$q \to p$	$(p \to q) \land (q \to p)$
T	T	T	T	T	T
T	F	F	F	T	F
F	T	F	T	F	F
F	F	T	T	T	T

We conclude this section by defining the **contrapositive** of a conditional proposition. We will see (Theorem 1.2.16) that the contrapositive is an alternative, logically equivalent form of the conditional proposition. Exercise 48 gives another logically equivalent form of the conditional proposition.

Definition 1.2.14.   The *contrapositive* (or *transposition*) of the conditional proposition $p \rightarrow q$ is the proposition $\bar{q} \rightarrow \bar{p}$.

Notice the difference between the contrapositive and the converse. The converse of a conditional proposition merely reverses the roles of p and q, whereas the contrapositive reverses the roles of p and q *and* negates each of them.

Example 1.2.15.   Write the proposition

$$\text{If } 1 < 4, \text{ then } 5 > 8$$

symbolically. Write the converse and the contrapositive both symbolically and in words. Find the truth value of each proposition.
  If we define

$$p: 1 < 4, \qquad q: 5 > 8,$$

then the given proposition may be written symbolically as

$$p \rightarrow q.$$

The converse is

$$q \rightarrow p,$$

or, in words,

$$\text{If } 5 > 8, \text{ then } 1 < 4.$$

The contrapositive is

$$\bar{q} \rightarrow \bar{p},$$

or, in words,

$$\text{If } 5 \leq 8, \text{ then } 1 \geq 4.$$

We see that $p \rightarrow q$ is false, $q \rightarrow p$ is true, and $\bar{q} \rightarrow \bar{p}$ is false.

An important fact is that a conditional proposition and its contrapositive are logically equivalent.

Theorem 1.2.16.   *The conditional proposition $p \rightarrow q$ and its contrapositive $\bar{q} \rightarrow \bar{p}$ are logically equivalent.*

***PROOF*.**   The truth table

p	q	$p \rightarrow q$	$\bar{q} \rightarrow \bar{p}$
T	T	T	T
T	F	F	F
F	T	T	T
F	F	T	T

shows that $p \rightarrow q$ and $\bar{q} \rightarrow \bar{p}$ are logically equivalent.   ■

EXERCISES

In Exercises 1–8, restate each proposition in the form (1.2.2) of a conditional proposition.

1H.   All Cubs are great baseball players.

2.   For the real number x, $|x| < 2$ provided that $0 < x < 2$.

3.   When better cars are built, Buick will build them.

4H.   A sufficient condition for the function f to be integrable is that f be continuous.

5.   The Cubs will win the World Series when they get a left-handed relief pitcher.

6.   A necessary condition for the graph G to have a Hamiltonian cycle is that G be connected.

7H.   The audience will go to sleep if the chairperson gives the lecture.

8.   The program is readable only if it is well structured.

9H.   Write the converse of each proposition in Exercises 1–8.

10.   Write the contrapositive of each proposition in Exercises 1–8.

Assuming that p and r are false and that q and s are true, find the truth value of each proposition in Exercises 11–18.

11H. $p \rightarrow q$ **12.** $\bar{p} \rightarrow \bar{q}$ **13.** $\overline{p \rightarrow q}$

14H. $(p \rightarrow q) \wedge (q \rightarrow r)$

15. $(p \rightarrow q) \rightarrow r$

16. $p \rightarrow (q \rightarrow r)$

17H. $(s \rightarrow (p \wedge \bar{r})) \wedge ((p \rightarrow (r \vee q)) \wedge s)$

18. $((p \wedge \bar{q}) \rightarrow (q \wedge r)) \rightarrow (s \vee \bar{q})$

Assume that a, b, and c are real numbers. In Exercises 19–24, represent the given statement symbolically by letting

$$p: a < b, \qquad q: b < c, \qquad r: a < c.$$

19H. If $a < b$, then $b \geq c$.

20. If ($a < b$ and $b < c$), then $a < c$.

21. If ($a \geq b$ and $b < c$), then $a \geq c$.

22H. If it is not the case that ($a < c$ and $b < c$), then $a \geq c$.

23. $a < b$ if and only if ($b < c$ and $a < c$).

24. If it is not the case that ($a < b$ and (either $a < b$ or $b < c$)), then (if $a \geq b$, then $a < c$).

In Exercises 25–30, formulate the symbolic expression in words using

p: Today is Monday,
q: It is raining,
r: It is hot.

25H. $p \rightarrow q$ **26.** $\bar{q} \rightarrow (r \wedge p)$

27. $\bar{p} \rightarrow (q \vee r)$ **28H.** $p \vee q \leftrightarrow r$

29. $(p \wedge (q \vee r)) \rightarrow (r \vee (q \vee p))$

30. $(p \vee (\bar{p} \wedge \overline{(q \vee r)})) \rightarrow (p \vee \overline{(r \vee q)})$

In Exercises 31–34, write each conditional proposition symbolically. Write the converse and contrapositive of each statement symbolically and in words. Also, find the truth value of each conditional proposition, its converse, and its contrapositive.

31H. If $4 < 6$, then $9 > 12$.

32. If $4 > 6$, then $9 > 12$.

33. $|1| < 3$ if $-3 < 1 < 3$.

34H. $|4| < 3$ if $-3 < 4 < 3$.

For each pair of propositions P and Q in Exercises 35–44, state whether or not $P \equiv Q$.

35H. $P = p$, $Q = p \vee q$

36. $P = p \wedge q$, $Q = \bar{p} \vee \bar{q}$

37. $P = p \to q, Q = \bar{p} \lor q$

38H. $P = p \land (\bar{q} \lor r), Q = p \lor (q \land \bar{r})$

39. $P = p \land (q \lor r), Q = (p \lor q) \land (p \lor r)$

40. $P = p \to q, Q = \bar{q} \to \bar{p}$

41H. $P = p \to q, Q = p \leftrightarrow q$

42. $P = (p \to q) \land (q \to r), Q = p \to r$

43. $P = (p \to q) \to r, Q = p \to (q \to r)$

44H. $P = (s \to (p \land \bar{r})) \land ((p \to (r \lor q)) \land s), Q = p \lor t$

45H. Define the truth table for *imp1* by

p	q	p *imp1* q
T	T	T
T	F	F
F	T	F
F	F	T

Show that

$$p \text{ } imp1 \text{ } q \equiv q \text{ } imp1 \text{ } p.$$

46.   Define the truth table for *imp2* by

p	q	p *imp2* q
T	T	T
T	F	F
F	T	T
F	F	F

(a) Show that

$$(p \text{ } imp2 \text{ } q) \land (q \text{ } imp2 \text{ } p) \not\equiv p \leftrightarrow q. \qquad (1.2.8)$$

(b) Show that (1.2.8) remains true if we alter *imp2* so that if p is false and q is true, then p *imp2* q is false.

47.   Verify the second De Morgan law, $\overline{p \land q} \equiv \bar{p} \lor \bar{q}$.

48H. Show that $(p \to q) \equiv (\bar{p} \lor q)$.

George Boole   (1815–1864)

George Boole was a professor of mathematics in Ireland when he wrote *An Investigation of The Laws of Thought* which suggested an algebraic approach to formal logic. He died of pneumonia at a relatively young age, allegedly because of a chill that developed after he was caught in the rain prior to giving a lecture. [*Courtesy of the Mathematics Library, Columbia University*]

1.3 Combinatorial Circuits

Logic and digital computers are intimately related. We saw in the preceding sections that a proposition assumes one of two states—true or false. In a digital computer at the most fundamental level, all programs and data are ultimately reducible to two states, typically denoted 0 and 1. In this context, 0 and 1 are called **bits**. (Bit is an abbreviation for "binary digit.") The development of a system of logic incorporating the concepts of the previous sections was begun in the middle of the nineteenth century by the English mathematician George Boole. Almost a century after Boole's work, it was observed, especially by Claude E. Shannon, that the methods developed by Boole to handle logic could also be used to analyze electrical circuits. In this section we explore the analysis of digital circuits.

A variety of devices have been used throughout the years in digital computers to store bits. Electronic circuits allow these storage devices to communicate with each other. A bit in one part of a circuit is transmitted to another part of the circuit as a voltage. Thus two voltage levels are needed—for example, a high voltage can communicate 1 and a low voltage can communicate 0.

In this section we discuss **combinatorial circuits**. A combinatorial circuit is a circuit in which the output is uniquely defined for every combination of inputs. A combinatorial circuit has no memory; previous inputs and the state of the system do not affect the output of a combinatorial circuit.

Combinatorial circuits can be constructed using solid-state devices, called **gates**, which are capable of switching voltage levels (bits). We begin by discussing AND, OR, and NOT gates.

Definition 1.3.1. An *AND gate* receives inputs x_1 and x_2, where x_1 and x_2 are bits, and produces output denoted $x_1 \wedge x_2$, where

$$x_1 \wedge x_2 = \begin{cases} 1 & \text{if } x_1 = 1 \text{ and } x_2 = 1 \\ 0 & \text{otherwise.} \end{cases}$$

An AND gate is drawn as shown in Figure 1.3.1.

Claude E. Shannon *(1916–   )*

Claude E. Shannon is emeritus professor of electrical engineering at Massachusetts Institute of Technology. He developed his theory of computer circuits in his masters thesis at MIT in 1937. He is the founder of *information theory*, which addresses the problem of transmitting information in the most efficient way. He was an early contributor to computer chess. A game and puzzle enthusiast, he found a sentence, "Squdgy fez, blank jimp crwth vox," that uses each letter of the alphabet exactly once. (Each of the words is in the dictionary!) [*Courtesy of The MIT Museum*]

Figure 1.3.1

Notice that if we interpret 1 as true and 0 as false, Definition 1.3.1 (of the AND gate) is the same as Definition 1.1.3 (of the logical *and* $\wedge$).

Definition 1.3.2. An *OR gate* receives inputs x_1 and x_2, where x_1 and x_2 are bits, and produces output denoted $x_1 \vee x_2$, where

$$x_1 \vee x_2 = \begin{cases} 1 & \text{if } x_1 = 1 \text{ or } x_2 = 1 \\ 0 & \text{otherwise.} \end{cases}$$

An OR gate is drawn as shown in Figure 1.3.2.

Figure 1.3.2

Notice that if we interpret 1 as true and 0 as false, Definition 1.3.2 (of the OR gate) is the same as Definition 1.1.7 (of the logical *or* $\vee$).

Definition 1.3.3. A *NOT gate* (or *inverter*) receives input x, where x is a bit, and produces output denoted $\bar{x}$, where

$$\bar{x} = \begin{cases} 1 & \text{if } x = 0 \\ 0 & \text{if } x = 1. \end{cases}$$

A NOT gate is drawn as shown in Figure 1.3.3.

Figure 1.3.3

Notice that if we interpret 1 as true and 0 as false, Definition 1.3.3 (of the NOT gate) is the same as Definition 1.1.11 (of the logical *not* ⁻).

A **logic table** is similar to a truth table. A logic table of a combinatorial circuit lists all possible inputs together with the resulting outputs.

Example 1.3.4. Following are the logic tables for the basic AND, OR, and NOT circuits (Figures 1.3.1 to 1.3.3).

x_1	x_2	$x_1 \wedge x_2$
1	1	1
1	0	0
0	1	0
0	0	0

x_1	x_2	$x_1 \vee x_2$
1	1	1
1	0	1
0	1	1
0	0	0

x	$\bar{x}$
1	0
0	1

We note that performing the operation AND (OR) is the same as taking the minimum (maximum) of the two bits x_1 and x_2.

Example 1.3.5. The circuit of Figure 1.3.4 is an example of a combinatorial circuit since the output y_1 is uniquely defined for each combination of inputs x_1, x_2, and x_3. The logic table for this combinatorial circuit follows.

x_1	x_2	x_3	y_1
1	1	1	0
1	1	0	0
1	0	1	0
1	0	0	1
0	1	1	0
0	1	0	1
0	0	1	0
0	0	0	1

Notice that all possible combinations of values for the inputs x_1, x_2, and x_3 are listed. For a given set of inputs, we can compute the value of the output y_1 by tracing the flow through the circuit. For example, the fourth line of the table gives the value of the output y_1 for the input values

$$x_1 = 1, \qquad x_2 = 0, \qquad x_3 = 0.$$

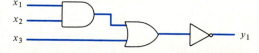

Figure 1.3.4

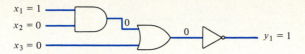

Figure 1.3.5

If $x_1 = 1$ and $x_2 = 0$, the output from the AND gate is 0 (see Figure 1.3.5). Since $x_3 = 0$, the inputs to the OR gate are both 0. Therefore, the output of the OR gate is 0. Since the input to the NOT gate is 0, it produces output $y_1 = 1$.

Example 1.3.6. The circuit of Figure 1.3.6 is not a combinatorial circuit because the output y is not uniquely defined for each combination of inputs x_1 and x_2. For example, suppose that $x_1 = 1$ and $x_2 = 0$. If the output of the AND gate is 0, then $y = 0$. On the other hand, if the output of the AND gate is 1, then $y = 1$.

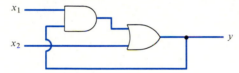

Figure 1.3.6

Example 1.3.7. Individual combinatorial circuits may be interconnected to form a new combinatorial circuit. The combinatorial circuits C_1, C_2, and C_3 of Figure 1.3.7 may be combined, as shown, to obtain the combinatorial circuit C. To assure that the result of interconnecting combinatorial circuits is itself a combinatorial circuit, when adding a new combinatorial circuit C' to an existing combinatorial circuit C'', we should connect outputs of C' to inputs of C'' or inputs of C' to outputs of C'', but not both. Doing both may give a noncombinatorial circuit (see Figure 1.3.6).

Example 1.3.8. A combinatorial circuit with one output, like that in Figure 1.3.4, can be represented by an expression using the symbols $\wedge$, $\vee$, and $^{-}$. We follow the flow of the circuit symbolically. First, x_1 and x_2 are ANDed (see Figure 1.3.8), which produces output $x_1 \wedge x_2$. This output is then

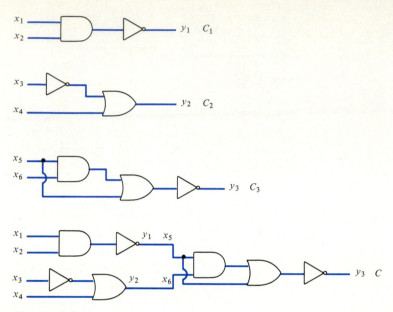

Figure 1.3.7

ORed with x_3 to produce output $(x_1 \wedge x_2) \vee x_3$. This output is then NOTed. Thus the output y_1 may be written

$$y_1 = \overline{(x_1 \wedge x_1) \vee x_3}. \tag{1.3.1}$$

Expressions involving variables and the symbols $\wedge$, $\vee$, and $^-$ are called **Boolean expressions**.

In a Boolean expression in which parentheses are not used to specify the order of operations, we assume that $\wedge$ is evaluated before $\vee$. This is the same assumption that most programming languages make.

Figure 1.3.8

Example 1.3.9.   For $x_1 = 0$, and $x_2 = 0$, and $x_3 = 1$, the value of the Boolean expression $x_1 \wedge x_2 \vee x_3$ is

$$x_1 \wedge x_2 \vee x_3 = 0 \wedge 0 \vee 1 = 0 \vee 1 = 1.$$

Example 1.3.8 showed how to represent a combinatorial circuit with one output as a Boolean expression. The following example shows how to construct a combinatorial circuit that represents a Boolean expression.

Example 1.3.10. Find the combinatorial circuit corresponding to the Boolean expression

$$(x_1 \wedge (\overline{x_2} \vee x_3)) \vee x_2$$

and write the logic table for the circuit obtained.

We begin with the expression $\overline{x_2} \vee x_3$ in the innermost parentheses. This expression is converted to a combinatorial circuit as shown in Figure 1.3.9. The output of this circuit is ANDed with x_1 to produce the circuit drawn in Figure 1.3.10. Finally, the output of this circuit is ORed with x_2 to give the desired circuit drawn in Figure 1.3.11.

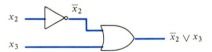

Figure 1.3.9

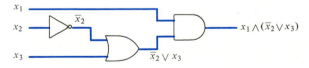

Figure 1.3.10

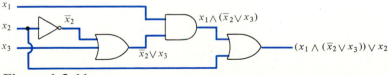

Figure 1.3.11

The logic table is as follows:

x_1	x_2	x_3	$(x_1 \wedge (\overline{x_2} \vee x_3)) \vee x_2$
1	1	1	1
1	1	0	1
1	0	1	1
1	0	0	1
0	1	1	1
0	1	0	1
0	0	1	0
0	0	0	0

EXERCISES

In Exercises 1–6, write a Boolean expression that represents the given combinatorial circuit, write the logic table, and write the output of each gate symbolically as in Figure 1.3.8.

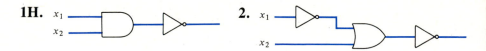

1H.

2.

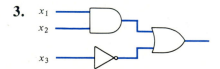

3.

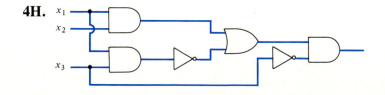

4H.

5.

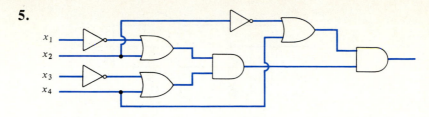

6. The circuit C at the bottom of Figure 1.3.7

In Exercises 7–11, find the value of the Boolean expressions for

$$x_1 = 1, \qquad x_2 = 1, \qquad x_3 = 0, \qquad x_4 = 1.$$

7H. $\overline{x_1 \wedge x_2}$

8. $x_1 \vee (\overline{x_2} \wedge x_3)$

9. $(x_1 \wedge \overline{x_2}) \vee (x_1 \vee \overline{x_3})$

10H. $(x_1 \wedge (x_2 \vee (x_1 \wedge \overline{x_2}))) \vee \overline{((x_1 \wedge \overline{x_2}) \vee \overline{(x_1 \wedge \overline{x_3})})}$

11. $(((x_1 \wedge x_2) \vee (x_3 \wedge \overline{x_4})) \vee \overline{((x_1 \vee x_3) \wedge (\overline{x_2} \vee x_3)))}$
$\vee (x_1 \wedge \overline{x_3})$

12. Find a combinatorial circuit corresponding to each Boolean expression in Exercises 7–11 and write the logic table.

Exercises 13–15 refer to the circuit

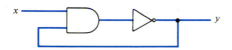

13H. Show that this circuit is not a combinatorial circuit.

14. Show that if $x = 0$, the output y is uniquely determined.

15. Show that if $x = 1$, the output y is undetermined.

A **switching circuit** is an electrical network consisting of switches each of which is open or closed. An example is given in Figure 1.3.12. If switch X is open, we write $X = 0$. If switch X is closed, we write $X = 1$. Switches labeled with the same letter, such as B in Figure 1.3.12, are either all open or all closed. Switch X, like A in Figure

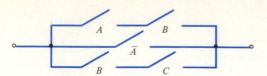

Figure 1.3.12

1.3.12, is open if and only if switch $\overline{X}$, like $\overline{A}$, is closed. If current can flow between the extreme left and right ends of the circuit, we say that the output of the circuit is 1; otherwise, we say that the output of the circuit is 0. A **switching table** gives the output of the circuit for all values of the switches. The switching table for Figure 1.3.12 is as follows:

A	B	C	Circuit Output
1	1	1	1
1	1	0	1
1	0	1	0
1	0	0	0
0	1	1	1
0	1	0	1
0	0	1	1
0	0	0	1

16H. Draw a circuit with two switches A and B having the property that the circuit output is 1 precisely when both A and B are closed. This configuration is labeled $A \wedge B$ and is called a **series circuit.**

17. Draw a circuit with two switches A and B having the property that the circuit output is 1 precisely when either A or B is closed. This configuration is labeled $A \vee B$ and is called a **parallel circuit.**

18. Verify that the circuit of Figure 1.3.12 can be represented symbolically as

$$(A \wedge B) \vee \overline{A} \vee (B \wedge C).$$

Represent each circuit in Exercises 19–24 symbolically and give its switching table.

19H.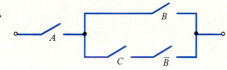

20.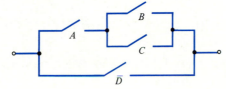

21.

22H.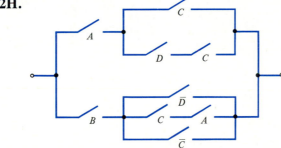

23.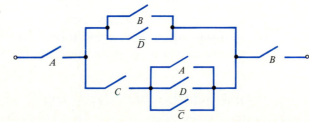

24.

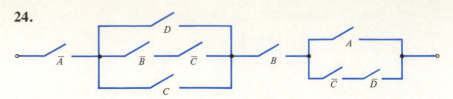

Represent the expressions in Exercises 25–29 as switching circuits and write the switching tables.

25H. $(A \lor \bar{B}) \land A$ **26.** $A \lor (\bar{B} \land C)$

27. $(\bar{A} \land B) \lor (C \land A)$

28H. $(A \land ((B \land \bar{C}) \lor (\bar{B} \land C))) \lor (\bar{A} \land B \land C)$

29. $A \land ((B \land C \land \bar{D}) \lor ((\bar{B} \land C) \lor D) \lor (\bar{B} \land \bar{C} \land D))$
$\land (B \lor \bar{D})$

30. Does every Boolean expression represent a switching circuit? Why or why not?

1.4 Synthesis of Circuits

In Section 1.3 we introduced combinatorial circuits and showed how combinatorial circuits and Boolean expressions are related. A typical application of this theory is to construct a combinatorial circuit to carry out some specified task. For example, we might want to design a combinatorial circuit to add two numbers. In this section we provide an introduction to the design of combinatorial circuits. In Section 1.6 we construct some circuits that might be used in a digital computer.

Suppose that we want to construct a particular combinatorial circuit. We can specify the properties the combinatorial circuit is supposed to have by listing the desired outputs for each choice of inputs. We illustrate such a specification by considering the **exclusive-OR**.

Example 1.4.1 Exclusive-OR.   The *exclusive-OR* $x_1 \oplus x_2$ of x_1 and x_2 is defined to be 1 when either x_1 or x_2, but not both, is 1. List the inputs and outputs that define the exclusive-OR.

The solution is given by the following table:

x_1	x_2	$x_1 \oplus x_2$
1	1	0
1	0	1
0	1	1
0	0	0

To construct a combinatorial circuit that computes $x_1 \oplus x_2$, we must find a Boolean expression X with a logic table identical to the table of Example 1.4.1 and then simply construct the circuit corresponding to X. Our next example provides a Boolean expression with a logic table identical to the table of Example 1.4.1. In the remainder of this section we show how to find these Boolean expressions.

Example 1.4.2. Verify that the Boolean expression

$$(x_1 \wedge \bar{x}_2) \vee (\bar{x}_1 \wedge x_2)$$

has the same logic table as $x_1 \oplus x_2$.

The logic table for $(x_1 \wedge \bar{x}_2) \vee (\bar{x}_1 \wedge x_2)$ follows.

x_1	x_2	$(x_1 \wedge \bar{x}_2) \vee (\bar{x}_1 \wedge x_2)$
1	1	0
1	0	1
0	1	1
0	0	0

We see that $(x_1 \wedge \bar{x}_2) \vee (\bar{x}_1 \wedge x_2)$ has the same table as $x_1 \oplus x_2$.

Example 1.4.3. The combinatorial circuit corresponding to the Boolean expression $(x_1 \wedge \bar{x}_2) \vee (\bar{x}_1 \wedge x_2)$ is shown in Figure 1.4.1. Since the tables for $(x_1 \wedge \bar{x}_2) \vee (\bar{x}_1 \wedge x_2)$ and $x_1 \oplus x_2$ are the same, the circuit of Figure 1.4.1 computes the exclusive-OR of x_1 and x_2.

Examples 1.4.1 to 1.4.3 show that the problem of constructing a combinatorial circuit for a given table T can be reduced to the problem of finding a Boolean expression whose logic table coincides with T. Our

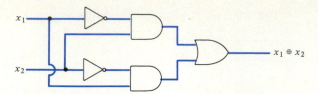

Figure 1.4.1

next example illustrates a general method that always produces the desired Boolean expression.

Example 1.4.4.   Find a Boolean expression whose logic table is shown in Table 1.4.1.

Table 1.4.1

x_1	x_2	x_3	y
1	1	1	1
1	1	0	0
1	0	1	0
1	0	0	1
0	1	1	0
0	1	0	1
0	0	1	0
0	0	0	0

Consider the first row of the table and the combination†

$$x_1 \wedge x_2 \wedge x_3. \tag{1.4.1}$$

Notice that if $x_1 = x_2 = x_3 = 1$, as indicated in the first row of the table, then (1.4.1) is 1. The values of x_i given by any other row of the table give (1.4.1) the value 0. Similarly, for the fourth row of the table we may construct the combination

$$x_1 \wedge \overline{x_2} \wedge \overline{x_3}. \tag{1.4.2}$$

Expression (1.4.2) has the value 1 for the values of x_i given by the fourth row of the table, while the values of x_i given by any other row of the table give (1.4.2) the value 0.

---

†We omit parentheses in expressions such as (1.4.1) since $(x_1 \wedge x_2) \wedge x_3 = x_1 \wedge (x_2 \wedge x_3)$. In general, $x_1 \wedge x_2 \wedge \cdots \wedge x_n$ and $x_1 \vee x_2 \vee \cdots \vee x_n$ are unambiguous.

In general, we consider a row R of the table where the output is 1. We then form the combination $x_1 \wedge x_2 \wedge x_3$ and place a bar over each x_i whose value is 0 in row R. The combination formed is 1 if and only if the x_i have the values given in row R. Thus, for row 6, we obtain the combination

$$\overline{x_1} \wedge x_2 \wedge \overline{x_3}. \tag{1.4.3}$$

Next, we OR the terms (1.4.1)–(1.4.3) to obtain the Boolean expression

$$(x_1 \wedge x_2 \wedge x_3) \vee (x_1 \wedge \overline{x_2} \wedge \overline{x_3}) \vee (\overline{x_1} \wedge x_2 \wedge \overline{x_3}). \tag{1.4.4}$$

We claim that for any choice of values of x_1, x_2 and x_3, expression (1.4.4) and y in Table 1.4.1 are equal. To verify this, first suppose that x_1, x_2, x_3 have values given by a row of Table 1.4.1 for which $y = 1$. Then one of (1.4.1)–(1.4.3) is 1, so the value of (1.4.4) is 1. On the other hand, if x_1, x_2, x_3 have values given by a row of Table 1.4.1 for which $y = 0$, then all of (1.4.1)–(1.4.3) are 0, so the value of (1.4.4) is 0. Thus, for any choice of values of x_1, x_2, and x_3, expression (1.4.4) and y in Table 1.4.1 are equal.

The representation (1.4.4) is called the **disjunctive normal form** of a Boolean expression. Each term $x_1 \wedge x_2 \wedge x_3$ in the disjunctive normal form is called a **minterm**.

Example 1.4.5. Find a combinatorial circuit corresponding to Table 1.4.1.

In Example 1.4.4 we found the Boolean expression (1.4.4) whose logic table is Table 1.4.1. The combinatorial circuit corresponding to expression (1.4.4) provides the solution (see Figure 1.4.2).

The method of finding a combinatorial circuit corresponding to a given table T can be described as follows:

1. Find the disjunctive normal form D of a Boolean expression whose logic table is identical to T.
2. Draw the combinatorial circuit that represents D.

We will illustrate the method with an additional example.

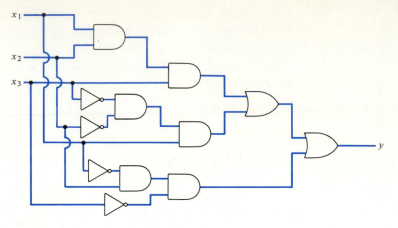

Figure 1.4.2

Example 1.4.6.   Find a combinatorial circuit corresponding to the table

x_1	x_2	x_3	y
1	1	1	0
1	1	0	0
1	0	1	0
1	0	0	1
0	1	1	1
0	1	0	1
0	0	1	0
0	0	0	1

The disjunctive normal form of a Boolean expression having the same table as the given table is

$$(x_1 \wedge \overline{x_2} \wedge \overline{x_3}) \vee (\overline{x_1} \wedge x_2 \wedge x_3) \vee (\overline{x_1} \wedge x_2 \wedge \overline{x_3})$$
$$\vee (\overline{x_1} \wedge \overline{x_2} \wedge \overline{x_3}).$$

The combinatorial circuit that corresponds to this Boolean expression is shown in Figure 1.4.3.

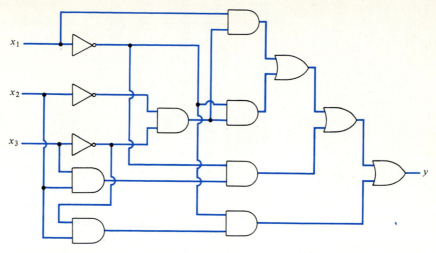

Figure 1.4.3

EXERCISES

In Exercises 1–10, find the disjunctive normal form of a Boolean expression having a logic table the same as the given table and draw the combinatorial circuit corresponding to the disjunctive normal form.

1H.

x_1	x_2	y
1	1	0
1	0	1
0	1	0
0	0	0

2.

x_1	x_2	y
1	1	1
1	0	1
0	1	0
0	0	0

3.

x_1	x_2	y
1	1	1
1	0	0
0	1	1
0	0	1

4H.

x_1	x_2	y
1	1	0
1	0	1
0	1	0
0	0	1

5.

x_1	x_2	x_3	y
1	1	1	1
1	1	0	1
1	0	1	0
1	0	0	1
0	1	1	0
0	1	0	0
0	0	1	1
0	0	0	1

6.

x_1	x_2	x_3	y
1	1	1	1
1	1	0	1
1	0	1	0
1	0	0	1
0	1	1	1
0	1	0	1
0	0	1	0
0	0	0	0

7H.

x_1	x_2	x_3	y
1	1	1	1
1	1	0	1
1	0	1	1
1	0	0	0
0	1	1	0
0	1	0	1
0	0	1	1
0	0	0	1

8.

x_1	x_2	x_3	y
1	1	1	0
1	1	0	1
1	0	1	1
1	0	0	1
0	1	1	1
0	1	0	1
0	0	1	1
0	0	0	0

9.

x_1	x_2	x_3	y
1	1	1	1
1	1	0	0
1	0	1	0
1	0	0	1
0	1	1	0
0	1	0	0
0	0	1	0
0	0	0	1

10H.

x_1	x_2	x_3	y
1	1	1	0
1	1	0	0
1	0	1	0
1	0	0	1
0	1	1	1
0	1	0	1
0	0	1	1
0	0	0	0

The **conjunctive normal form** of a Boolean expression is obtained from a table with inputs $x_1, x_2, \ldots, x_n$ by the following method. For each row with output 0, form the expression

$$x_1 \vee x_2 \vee \cdots \vee x_n,$$

and place a bar over x_i if x_i is 1 in that row. Such an expression is called a **maxterm**. Then AND all of the resulting maxterms. For example, the conjunctive normal form for the table of Example 1.4.6 is

$$(\overline{x_1} \vee \overline{x_2} \vee \overline{x_3}) \wedge (\overline{x_1} \vee \overline{x_2} \vee x_3) \wedge (\overline{x_1} \vee x_2 \vee \overline{x_3})$$
$$\wedge (x_1 \vee x_2 \vee \overline{x_3}).$$

11.   Give an argument to show that the logic table of the preceding expression is identical to the table of Example 1.4.6.

12H. Find the conjunctive normal forms of the Boolean expressions having logic tables the same as the tables of Exercises 1–10 and draw the combinatorial circuits corresponding to the conjunctive normal forms.

1.5 Simplification of Boolean Expressions

The logic table of the Boolean expression

$$x \vee \overline{y} \qquad\qquad (1.5.1)$$

is

x	y	$x \vee \overline{y}$
1	1	1
1	0	1
0	1	0
0	0	1

and the logic table of the Boolean expression

$$(x \wedge y) \vee (x \wedge \overline{y}) \vee (\overline{x} \wedge \overline{y}) \qquad\qquad (1.5.2)$$

is

x	y	$(x \wedge y) \vee (x \wedge \overline{y}) \vee (\overline{x} \wedge \overline{y})$
1	1	1
1	0	1
0	1	0
0	0	1

We see that for each assignment of bits to x and y, the expressions are equal. In this case, we say that the Boolean expressions are **equal** and we write

$$x \vee \bar{y} = (x \wedge y) \vee (x \wedge \bar{y}) \vee (\bar{x} \wedge \bar{y}).$$

Just as in algebra we consider

$$x + y$$

a simplification of

$$\frac{x^2 + xy}{x},$$

we consider

$$x \vee \bar{y}$$

a simplification of

$$(x \wedge y) \vee (x \wedge \bar{y}) \vee (\bar{x} \wedge \bar{y}).$$

The combinatorial circuit corresponding to expression (1.5.1) is shown in Figure 1.5.1 and the combinatorial circuit corresponding to expression (1.5.2) is shown in Figure 1.5.2. These circuits are said to be **equivalent** because, for each assignment of inputs to these circuits, the outputs are identical. (We see that two combinatorial circuits are equivalent if and only if their corresponding Boolean expressions are equal.) Even though the circuits of Figures 1.5.1 and 1.5.2 are equivalent, we would prefer

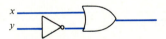

Figure 1.5.1

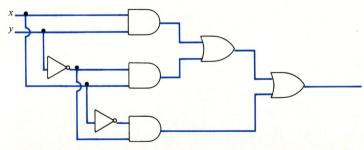

Figure 1.5.2

the circuit of Figure 1.5.1 to that of Figure 1.5.2 since the circuit of Figure 1.5.1 uses fewer gates than the circuit of Figure 1.5.2. It is apparent that the problem of simplifying a circuit can be solved by simplifying the representing Boolean expression. **Karnaugh maps** allow us to simplify Boolean expressions given in disjunctive normal form. We illustrate with an example.

Example 1.5.1.   Simplify the Boolean expression

$$(x \wedge y) \vee (x \wedge \bar{y}) \vee (\bar{x} \wedge \bar{y}). \tag{1.5.3}$$

Figure 1.5.3

We begin with Figure 1.5.3. Notice that each interior square represents one of the four possible minterms,

$$x \wedge y, \bar{x} \wedge y, x \wedge \bar{y}, \bar{x} \wedge \bar{y},$$

that can occur in the disjunctive normal form of a Boolean expression in two variables x and y. For example, the upper left square represents $x \wedge y$ and the lower left square represents $x \wedge \bar{y}$.

We place a star in each interior square that represents a minterm in expression (1.5.3) (see Figure 1.5.4). We next place each star in at least one circle. We must use as few circles as possible. Among all choices that minimize the number of circles, we select one in which the circles are as large as possible. The rules for drawing circles are

1. A single square that contains a star may be circled.

Figure 1.5.4

2. Adjacent squares that contain stars may be circled. (Squares are *adjacent* if they share a common side.)
3. If all squares contain stars, the entire figure may be circled.

In our case, we use only rule 2. Since the bottom two squares contain stars and they are adjacent, they may be circled. Since the left two squares contain stars and they are adjacent, they may also be circled. The resulting Karnaugh map is shown in Figure 1.5.5.

A simplified Boolean expression can be obtained from the Karnaugh map. The rules for obtaining the terms of a simplified Boolean expression are

1. A single circled square corresponds to the minterm given by its row and column.
2. A circled column corresponds to the term that identifies the column.
3. A circled row corresponds to the term that identifies the row.†
4. A circle that encloses all four squares corresponds to 1.

In our case, the circled column corresponds to x and the circled row corresponds to $\bar{y}$. To obtain the simplified Boolean expression, we *or* the terms. In our case, we obtain the simplified expression $x \vee \bar{y}$.

Figure 1.5.5

Example 1.5.2.   The Karnaugh map of the expression

$$(x \wedge y) \vee (\bar{x} \wedge y) \vee (x \wedge \bar{y})$$

is shown in Figure 1.5.6. The simplified expression is

$$x \vee y.$$

---

†Rules 2 and 3 result from the equality of the Boolean expressions a and $(a \wedge b) \vee (a \wedge \bar{b})$. This fact can be verified by writing the logic tables for the two expressions.

Figure 1.5.6

Example 1.5.3.   The Karnaugh map of the expression

$$(\bar{x} \wedge y) \vee (x \wedge \bar{y})$$

is shown in Figure 1.5.7. Since no squares are adjacent, only rule 1 applies and we circle individual squares. Since the individual squares correspond to the minterms given by the rows and columns, the resulting expression is

$$(\bar{x} \wedge y) \vee (x \wedge \bar{y}).$$

Notice that the Karnaugh map did not lead to any simplification of the given expression.

Figure 1.5.7

Example 1.5.4.   The Karnaugh map of the expression

$$(x \wedge y) \vee (\bar{x} \wedge y) \vee (x \wedge \bar{y}) \vee (\bar{x} \wedge \bar{y})$$

is shown in Figure 1.5.8. Since all squares contain stars, the entire figure may be circled. The given expression simplifies to 1; that is, the given expression is equal to 1 for each assignment of bits to x and y.

Karnaugh maps can be used to simplify Boolean expressions in disjunctive normal form when the number of variables is not too large. The remainder of this section is devoted to showing how to use Karnaugh

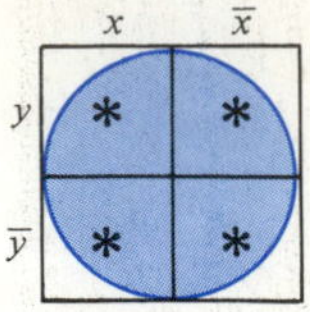

Figure 1.5.8

maps to simplify three-variable Boolean expressions in disjunctive normal form.

 We begin with Figure 1.5.9. Notice that each interior square represents one of the eight possible minterms

$$x \wedge y \wedge z, \bar{x} \wedge y \wedge z, x \wedge \bar{y} \wedge z, x \wedge y \wedge \bar{z},$$
$$x \wedge \bar{y} \wedge \bar{z}, \bar{x} \wedge y \wedge \bar{z}, \bar{x} \wedge \bar{y} \wedge z, \bar{x} \wedge \bar{y} \wedge \bar{z}$$

that can occur in the disjunctive normal form of a Boolean expression in the variables x, y, and z. For example, the square in row one and column two represents $x \wedge \bar{y} \wedge z$ and the square in row two and column four represents $\bar{x} \wedge y \wedge \bar{z}$. As in the case of a two-variable Karnaugh map, we begin by placing a star in each interior square that represents a minterm in the given expression. We illustrate with a couple of examples.

	$x \wedge y$	$x \wedge \bar{y}$	$\bar{x} \wedge \bar{y}$	$\bar{x} \wedge y$
z		*	*	*
$\bar{z}$			*	*

Figure 1.5.9

Example 1.5.5.   Figure 1.5.9 shows how the Boolean expression

$$(x \wedge \bar{y} \wedge z) \vee (\bar{x} \wedge \bar{y} \wedge z) \vee (\bar{x} \wedge y \wedge z)$$
$$\vee (\bar{x} \wedge \bar{y} \wedge \bar{z}) \vee (\bar{x} \wedge y \wedge \bar{z})$$

is represented.

Example 1.5.6.   Figure 1.5.10 shows how the Boolean expression

$$(x \wedge y \wedge z) \vee (x \wedge \bar{y} \wedge \bar{z}) \vee (\bar{x} \wedge \bar{y} \wedge z) \vee (\bar{x} \wedge y \wedge z)$$

is represented.

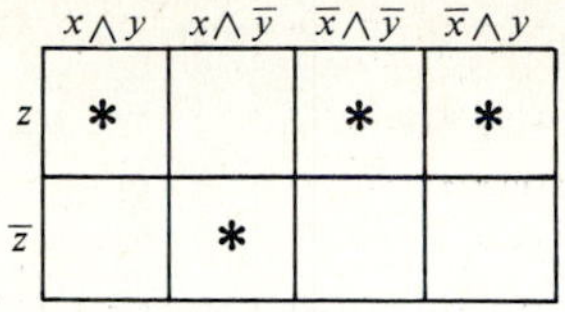

	$x \wedge y$	$x \wedge \overline{y}$	$\overline{x} \wedge \overline{y}$	$\overline{x} \wedge y$
z	*		*	*
$\overline{z}$		*		

Figure 1.5.10

As in the case of a two-variable Karnaugh map, we place each star in at least one circle. We must use as few circles as possible. Among all choices that minimize the number of circles, we select one in which the circles are as large as possible. The rules for drawing circles are

1. A single square that contains a star may be circled.
2. Two adjacent squares that contain stars may be circled. (Squares are *adjacent* if they share a common side. Furthermore, the squares at the extreme left and right of a single row as in Figure 1.5.11 are considered adjacent.)

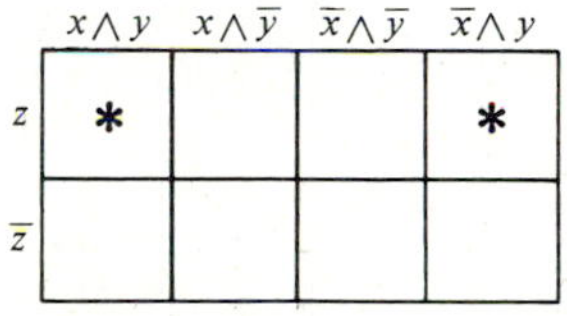

	$x \wedge y$	$x \wedge \overline{y}$	$\overline{x} \wedge \overline{y}$	$\overline{x} \wedge y$
z	*			*
$\overline{z}$				

Figure 1.5.11

3. Four squares that contain stars may be circled provided each is adjacent to exactly two of the others. (Examples are shown in Figures 1.5.12 and 1.5.13.)
4. Four squares in a row that contain stars may be circled. (An example is shown in Figure 1.5.14.)
5. If all squares contain stars, the entire figure may be circled.

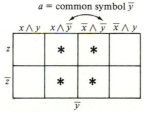

Figure 1.5.12

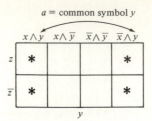

Figure 1.5.13

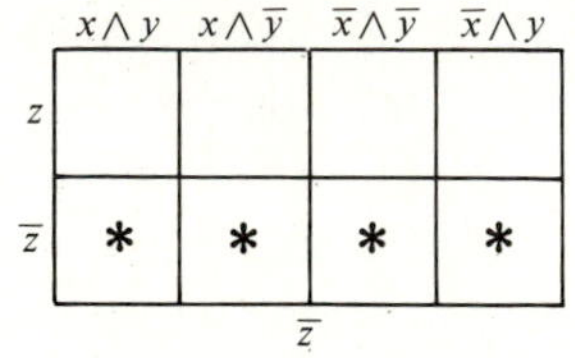

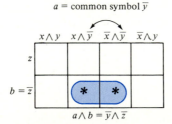

Figure 1.5.14

A simplified Boolean expression can be obtained from the Karnaugh map. The rules for obtaining the terms of a simplified Boolean expression are

1. A single circled square corresponds to the minterm given by its row and column.
2. A circled column corresponds to the expression that identifies the column.
3. Two stars in a row that are circled correspond to $a \wedge b$ where $a \wedge b$ is obtained as follows. One of x, $\bar{x}$, y, or $\bar{y}$ will appear in both column labels; a is equal to this common symbol. (For example, in Figure 1.5.15, $\bar{y}$ appears in both column labels.) b is

Figure 1.5.15

equal to the row label. (In Figure 1.5.15, b is equal to $\bar{z}$; thus, Figure 1.5.15 represents $\bar{y} \wedge \bar{z}$.)

4. A circle that contains four squares in two rows each adjacent to exactly two of others corresponds to a, where a appears in both column labels. (For example, Figure 1.5.12 represents $\bar{y}$ since $\bar{y}$ appears in both column labels. Figure 1.5.13 represents y since y appears in both column labels.)

5. A circle that contains four squares in a row corresponds to the row label. (For example, Figure 1.5.14 represents $\bar{z}$.)

6. A circle that encloses all eight squares corresponds to 1.

The simplified Boolean expression is obtained by *or*ing the terms.

Example 1.5.7. Simplify the Boolean expression

$$(x \wedge y \wedge z) \vee (x \wedge y \wedge \bar{z}) \vee (\bar{x} \wedge \bar{y} \wedge z)$$
$$\vee (\bar{x} \wedge y \wedge z) \vee (\bar{x} \wedge y \wedge \bar{z}).$$

The Karnaugh map is shown in Figure 1.5.16. The circle that contains four stars represents y and the circle that contains two stars represents $\bar{x} \wedge z$. Thus the expression simplifies to

$$y \vee (\bar{x} \wedge z).$$

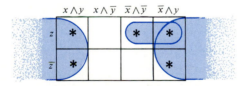

Figure 1.5.16

Example 1.5.8. Simplify the Boolean expression

$$(x \wedge y \wedge \bar{z}) \vee (x \wedge \bar{y} \wedge \bar{z}) \vee (\bar{x} \wedge \bar{y} \wedge z)$$
$$\vee (\bar{x} \wedge \bar{y} \wedge \bar{z}) \vee (\bar{x} \wedge y \wedge z).$$

The Karnaugh map is shown in Figure 1.5.17. The simplified expression is

$$(x \wedge \bar{z}) \vee (\bar{x} \wedge \bar{y}) \vee (\bar{x} \wedge z).$$

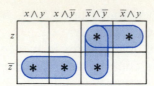

Figure 1.5.17

Example 1.5.9.   Use Karnaugh maps to design a combinatorial circuit whose logic table is

x	y	z	Output
1	1	1	1
1	1	0	0
1	0	1	0
1	0	0	1
0	1	1	0
0	1	0	0
0	0	1	1
0	0	0	1

The disjunctive normal form is

$$(x \wedge y \wedge z) \vee (x \wedge \bar{y} \wedge \bar{z}) \vee (\bar{x} \wedge \bar{y} \wedge z) \vee (\bar{x} \wedge \bar{y} \wedge \bar{z}).$$

The Karnaugh map of this expression is shown in Figure 1.5.18. The simplified expression is

$$(x \wedge y \wedge z) \vee (\bar{x} \wedge \bar{y}) \vee (\bar{y} \wedge \bar{z}).$$

The circuit corresponding to this expression is shown in Figure 1.5.19.

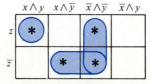

Figure 1.5.18

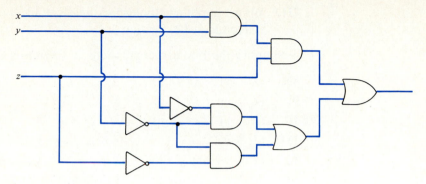

Figure 1.5.19

EXERCISES

Write the simplified Boolean expression given by the Karnaugh maps in Exercises 1–6.

1H.

2.

3.

4H.

5.

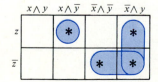

6.

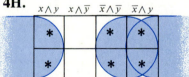

7H. Write simplified Boolean expressions for the logic tables of Exercises 1–10, Section 1.4.

Simplify the Boolean expressions in Exercises 8–16.

8H. $(\bar{x} \wedge y) \vee (\bar{x} \wedge \bar{y})$

9. $(x \wedge y) \vee (\bar{x} \wedge y)$

10. $(\bar{x} \wedge y) \vee (x \wedge \bar{y}) \vee (\bar{x} \wedge \bar{y})$

11H. $(\bar{x} \wedge \bar{y} \wedge z) \vee (\bar{x} \wedge y \wedge z) \vee (\bar{x} \wedge \bar{y} \wedge \bar{z}) \vee (\bar{x} \wedge y \wedge \bar{z})$

12. $(\bar{x} \wedge y \wedge z) \vee (\bar{x} \wedge \bar{y} \wedge \bar{z}) \vee (x \wedge \bar{y} \wedge \bar{z})$

13. $(x \wedge y \wedge z) \vee (x \wedge \bar{y} \wedge z) \vee (x \wedge \bar{y} \wedge \bar{z})$
$\vee (\bar{x} \wedge \bar{y} \wedge \bar{z}) \vee (\bar{x} \wedge y \wedge \bar{z})$

14H. $(x \wedge y \wedge z) \vee (x \wedge \bar{y} \wedge z) \vee (\bar{x} \wedge \bar{y} \wedge z)$
$\vee (\bar{x} \wedge y \wedge z) \vee (\bar{x} \wedge \bar{y} \wedge \bar{z})$

15. $(x \wedge y \wedge z) \vee (x \wedge \bar{y} \wedge z) \vee (x \wedge \bar{y} \wedge \bar{z})$
$\vee (\bar{x} \wedge \bar{y} \wedge \bar{z}) \vee (\bar{x} \wedge y \wedge \bar{z}) \vee (\bar{x} \wedge y \wedge z)$

16. $(x \wedge y \wedge z) \vee (\bar{x} \wedge y \wedge z) \vee (x \wedge \bar{y} \wedge z) \vee (x \wedge y \wedge \bar{z})$
$\vee (\bar{x} \wedge \bar{y} \wedge z) \vee (\bar{x} \wedge y \wedge \bar{z}) \vee (x \wedge \bar{y} \wedge \bar{z}) \vee (\bar{x} \wedge \bar{y} \wedge \bar{z})$

Simplify the combinatorial circuits of Exercises 17–19.

17H.

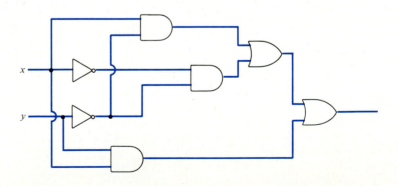

18.

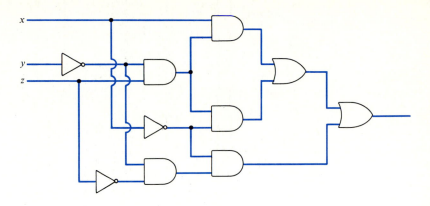

19.

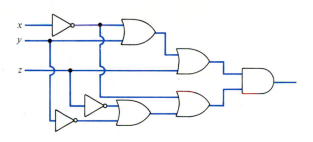

1.6 Number Systems and Circuits for Addition

We have seen that digital computers manipulate bits. For this reason, all programs and data must be encoded as bits. In this section we discuss the **binary number system**, a method for representing the nonnegative integers $\{0, 1, 2, 3, \ldots\}$ using bits. (We also look at some other number systems.) We also discuss a method for adding binary numbers and a combinatorial circuit that implements this method. Since these number systems are analogous to the familiar decimal system, we begin with a review of the **decimal number system**.

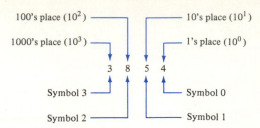

Figure 1.6.1

In the decimal number system, to represent nonnegative integers we use the ten symbols 0, 1, 2, 3, 4, 5, 6, 7, 8, and 9. In representing an arbitrary nonnegative integer, the position of the symbols is significant; reading from the right, the first symbol represents the number of 1's, the next symbol the number of 10's, the next symbol the number of 100's, and so on (see Figure 1.6.1). In general, the symbol in position n (with the rightmost symbol being in position 0) represents the number of 10^n's. Since $10^0 = 1$, the symbol in position 0 represents the number of 10^0's or 1's; since $10^1 = 10$, the symbol in position 1 represents the number of 10^1's or 10's; since $10^2 = 100$, the symbol in position 2 represents the number of 10^2's or 100's; and so on. We call the value on which the system is based (10 in the case of the decimal system) the **base** of the number system.

Example 1.6.1 Decimal Number System.   The decimal number 3854 represents the number consisting of four 1's, five 10's, eight 100's, and three 1000's (see Figure 1.6.1). This representation may be expressed

$$3854 = 3 \cdot 10^3 + 8 \cdot 10^2 + 5 \cdot 10^1 + 4 \cdot 10^0.$$

In the binary (base 2) number system, to represent nonnegative integers we need only two symbols, 0 and 1. In representing an arbitrary nonnegative integer, reading from the right, the first symbol represents the number of 1's, the next symbol the number of 2's, the next symbol the number of 4's, the next symbol the number of 8's, and so on (see Figure 1.6.2). In general, the symbol in position n (with the rightmost symbol being in position 0) represents the number of 2^n's. Since $2^0 = 1$, the

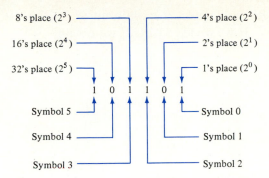

Figure 1.6.2

symbol in position 0 represents the number of 2^0's, or 1's; since $2^1 = 2$, the symbol in position 1 represents the number of 2^1's or 2's; since $2^2 = 4$, the symbol in position 2 represents the number of 2^2's or 4's; and so on.

Without knowing which number system is being used, a representation is ambiguous; for example, 101101 represents one number in decimal and quite a different number in binary. Often the context will make clear which number system is in effect, but when we want to be absolutely clear, we subscript the number to specify the base—the subscript 10 denotes the decimal system and the subscript 2 denotes the binary system. For example, the binary number 101101 can be written 101101_2.

Example 1.6.2 Binary to Decimal.

The binary number 101101_2 represents the number consisting of one 1, no 2's, one 4, one 8, no 16's, and one 32 (see Figure 1.6.2). This representation may be expressed

$$101101_2 = 1 \cdot 2^5 + 0 \cdot 2^4 + 1 \cdot 2^3 + 1 \cdot 2^2 + 0 \cdot 2^1 + 1 \cdot 2^0.$$

Computing the right-hand side in decimal, we find that

$$101101_2 = 1 \cdot 32 + 0 \cdot 16 + 1 \cdot 8 + 1 \cdot 4 + 0 \cdot 2 + 1 \cdot 1$$

$$= 32 + 8 + 4 + 1 = 45_{10}.$$

Example 1.6.2 shows how to convert a binary number to decimal. Consider the reverse problem—converting a decimal number to binary.

Suppose, for example, that we want to convert the decimal number 91 to binary. If we divide 91 by 2, we obtain

$$
\begin{array}{r}
45 \\
2\overline{)91} \\
8 \\
\overline{11} \\
10 \\
\overline{1}
\end{array}
$$

This computation shows that

$$91 = 2 \cdot 45 + 1. \tag{1.6.1}$$

We are beginning to express 91 in powers of 2. If we next divide 45 by 2, we find

$$45 = 2 \cdot 22 + 1. \tag{1.6.2}$$

Substituting this expression for 45 into (1.6.1), we obtain

$$
\begin{aligned}
91 &= 2 \cdot 45 + 1 \\
&= 2 \cdot (2 \cdot 22 + 1) + 1 \\
&= 2^2 \cdot 22 + 2 + 1.
\end{aligned} \tag{1.6.3}
$$

If we next divide 22 by 2, we find

$$22 = 2 \cdot 11.$$

Substituting this expression for 22 into (1.6.3), we obtain

$$
\begin{aligned}
91 &= 2^2 \cdot 22 + 2 + 1 \\
&= 2^2 \cdot (2 \cdot 11) + 2 + 1 \\
&= 2^3 \cdot 11 + 2 + 1.
\end{aligned} \tag{1.6.4}
$$

If we next divide 11 by 2, we find

$$11 = 2 \cdot 5 + 1.$$

Substituting this expression for 11 into (1.6.4), we obtain

$$91 = 2^4 \cdot 5 + 2^3 + 2 + 1. \tag{1.6.5}$$

If we next divide 5 by 2, we find

$$5 = 2 \cdot 2 + 1.$$

Substituting this expression for 5 into (1.6.5), we obtain

$$91 = 2^5 \cdot 2 + 2^4 + 2^3 + 2 + 1$$
$$= 2^6 + 2^4 + 2^3 + 2 + 1$$
$$= 1011011_2.$$

The preceding computation shows that the *remainders*, as N is successively divided by 2, give the bits in the binary representation of N. The first division by 2 in (1.6.1) gives the 1's bit; the second division by 2 in (1.6.2) gives the 2's bit; and so on. We illustrate with another example.

Example 1.6.3 Decimal to Binary.   Write the decimal number 130 in binary.

The computation shows the successive divisions by 2 with the remainders recorded at the right.

2) 130	remainder = 0	1's bit
2) 65	remainder = 1	2's bit
2) 32	remainder = 0	4's bit
2) 16	remainder = 0	8's bit
2) 8	remainder = 0	16's bit
2) 4	remainder = 0	32's bit
2) 2	remainder = 0	64's bit
2) 1	remainder = 1	128's bit
0		

We may stop when the dividend is 0. Remembering that the first remainder gives the number of 1's, the second remainder gives the number of 2's, and so on, we obtain

$$130_{10} = 10000010_2.$$

Next we turn our attention to addition of numbers in arbitrary bases. The same method that we use to add decimal numbers can be used to add binary numbers; however, we must replace the decimal addition table with the binary addition table

+	0	1
0	0	1
1	1	10

(In decimal, $1 + 1 = 2$, and $2_{10} = 10_2$; thus, in binary, $1 + 1 = 10$.)

Example 1.6.4 Binary Addition.   Add the binary numbers 10011011 and 1011011.

We write the problem as

$$
\begin{array}{r}
10011011 \\
+ \quad 1011011 \\
\hline
\end{array}
$$

As in decimal addition, we begin from the right adding 1 and 1. This sum is 10_2; thus we write 0 and carry 1. At this point the computation is

$$
\begin{array}{r}
1 \quad\quad\quad \\
10011011 \\
+ \quad 1011011 \\
\hline
0
\end{array}
$$

Next, we add 1 and 1 and 1, which is 11_2. We write 1 and carry 1. At this point, the computation is

$$
\begin{array}{r}
1 \quad\quad\quad \\
10011011 \\
+ \quad 1011011 \\
\hline
10
\end{array}
$$

You should verify that by continuing in this way, we obtain

$$
\begin{array}{r}
10011011 \\
+ \quad 1011011 \\
\hline
11110110
\end{array}
$$

Example 1.6.5.   The addition problem of Example 1.6.4, in decimal, is

$$
\begin{array}{r}
155 \\
+ \quad 91 \\
\hline
246
\end{array}
$$

Other important bases for number systems in computer science are base 8 or **octal** and base 16 or **hexadecimal** (sometimes shortened to **hex**). We will discuss the hexadecimal system and leave the octal system to the exercises (see Exercises 41–46).

In the hexadecimal number system, to represent nonnegative integers we use the symbols 0, 1, 2, 3, 4, 5, 6, 7, 8, 9, A, B, C, D, E, and F. The symbols A–F are interpreted as decimal 10–15. (In general, in the base N number system, N distinct symbols, representing 0, 1, 2, . . . , $N - 1$, will be required.) In representing an arbitrary nonnegative integer, reading from the right, the first symbol represents the number of 1's, the next symbol the number of 16's, the next symbol the number of 16^2's, and so on (see Figure 1.6.3). In general, the symbol in position n (with the rightmost symbol being in position 0) represents the number of 16^n's.

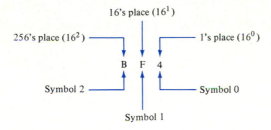

Figure 1.6.3

Example 1.6.6 Hexadecimal to Decimal.

Convert the hexadecimal number B4F to decimal.

We obtain

$$B4F_{16} = 11 \cdot 16^2 + 4 \cdot 16^1 + 15 \cdot 16^0$$

$$= 11 \cdot 256 + 4 \cdot 16 + 15 = 2816 + 64 + 15 = 2895_{10}.$$

To convert a decimal number to hexadecimal, we divide successively by 16. The remainders give the hexadecimal symbols.

Example 1.6.7 Decimal to Hexadecimal.   Convert the decimal number 20385 to hexadecimal.

The computation shows the successive divisions by 16 with the remainders recorded at the right.

$$
\begin{array}{lll}
16 \,)\, \underline{20385} & \text{remainder} = 1 & \text{1's place} \\
16 \,)\, \underline{1274} & \text{remainder} = 10 & \text{16's place} \\
16 \,)\, \underline{79} & \text{remainder} = 15 & \text{16}^2\text{'s place} \\
16 \,)\, \underline{4} & \text{remainder} = 4 & \text{16}^3\text{'s place} \\
 0 & &
\end{array}
$$

We may stop when the dividend is 0. The first remainder gives the number of 1's, the second remainder gives the number of 16's, and so on; thus we obtain

$$20385_{10} = 4FA1_{16}.$$

Our next example shows that we can add hexadecimal numbers in the same way that we add decimal or binary numbers.

Example 1.6.8 Hexadecimal Addition.   Add the hexadecimal numbers 84F and 42EA.

The problem may be written

$$
\begin{array}{r}
84F \\
+\ 42EA \\
\end{array}
$$

We begin in the rightmost column by adding F and A. Since F is 15_{10} and A is 10_{10}, $F + A = 15_{10} + 10_{10} = 25_{10} = 19_{16}$. We write 9 and carry 1:

$$
\begin{array}{r}
1 \\
84F \\
+\ 42EA \\
\hline
9 \\
\end{array}
$$

Next, we add 1, 4, and E, obtaining 13_{16}. We write 3 and carry 1:

$$
\begin{array}{r}
1 \\
84F \\
+\ 42EA \\
\hline
39 \\
\end{array}
$$

Continuing in this way, we obtain

$$
\begin{array}{r}
84F \\
+\ 42EA \\
\hline
4B39
\end{array}
$$

Example 1.6.9. The addition problem of Example 1.6.8, in decimal, is

$$
\begin{array}{r}
2127 \\
+\ 17130 \\
\hline
19257
\end{array}
$$

We conclude this section by discussing combinatorial circuits to do binary addition. We begin with a circuit known as the **half-adder** circuit.

Definition 1.6.10. A *half-adder circuit* is a circuit that accepts as input two bits x and y and produces as output two bits, c and s. The term cs is the sum of x and y expressed in binary. We call s the *sum bit* and c the *carry bit*.

Example 1.6.11. In the notation of Definition 1.6.10, if $x = 1$ and $y = 0$, then $x + y = 01 = cs$. Thus the carry bit is 0 and the sum bit is 1.

Example 1.6.12. In the notation of Definition 1.6.10, if $x = 1$ and $y = 1$, then $x + y = 10 = cs$. Thus the carry bit is 1 and the sum bit is 0.

Example 1.6.13 Half-Adder Circuit. Design a half-adder combinatorial circuit. The table for the half-adder circuit is

x	y	c	s
1	1	1	0
1	0	0	1
0	1	0	1
0	0	0	0

We observe that $c = x \wedge y$ and $s = x \oplus y$, the exclusive-OR of x and y. Thus we obtain the half-adder circuit of Figure 1.6.4. We used the circuit of Figure 1.4.1 to compute the exclusive-OR.

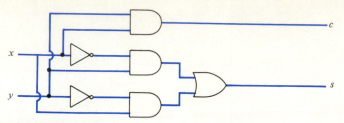

Figure 1.6.4

A **full-adder** sums three bits and is useful for adding two bits and a third carry bit from a previous addition.

Definition 1.6.14.   A *full-adder circuit* is a circuit that accepts as input three bits x, y, and z and produces as output two bits, c and s. The term cs is the sum of x, y, and z expressed in binary.

Example 1.6.15.   In the notation of Definition 1.6.14, if $x = 1$, $y = 0$, and $z = 1$, then $x + y + z = 10 = cs$. Thus the carry bit is 1 and the sum bit is 0.

Example 1.6.16.   In the notation of Definition 1.6.14, if $x = 1$, $y = 1$, and $z = 1$, then $x + y + z = 11 = cs$. Thus both the carry bit and the sum bit are 1.

We will show how to construct a full-adder circuit using half-adders and an OR gate.

Example 1.6.17 Full-Adder Circuit.   We are to sum the three bits x, y, and z. First, we can add x and y using one half-adder circuit (see Figure 1.6.5):

$$x + y = c's'.$$

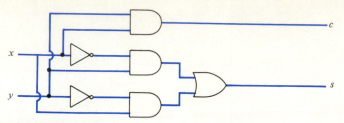

Figure 1.6.5

We have denoted the carry and sum bits by c' and s', respectively. To this sum, we must add z:

$$
\begin{array}{r}
c'' \\
c's' \\
+ \quad z \\
\hline
s''
\end{array}
\qquad (1.6.6)
$$

When we add z to $c's'$, we first add the bits s' and z to obtain the sum $c''s''$. This addition can be computed using a second half-adder circuit (see Figure 1.6.5). We write s'' and carry c''. Finally, we must add c'' and c'. Now $x + y + z$ cannot yield a three-bit sum since the largest possible value is $1 + 1 + 1 = 11_2$. Thus one of c' or c'' is 0 and the sum is simply $c' \lor c''$:

c'	c''	$c' + c'' = c' \lor c''$
0	0	0
0	1	1
1	0	1

If we let $x + y + z = cs$, the computation (1.6.6) becomes

$$
\begin{array}{r}
c'' \\
c's' \\
+ \quad z \\
\hline
c \ s'' = cs
\end{array}
$$

Thus we can construct a full-adder circuit that computes $x + y + z$ using two half-adder circuits and one OR gate. The circuit is shown in Figure 1.6.5.

Advances in solid-state technology have made it possible to manufacture very small components, called **integrated circuits**, which are themselves entire circuits. If we have a supply of integrated circuits, some of which are half-adders and others of which are full-adders, we will show how we may use these adders to construct circuits to add binary numbers. This modest example shows how circuit design consists of combining AND, OR, NOT, and other gates with integrated circuits to construct the desired circuits.

Example 1.6.18.   Using half-adder and full-adder circuits, design a combinatorial circuit that computes the sum of two 3-bit numbers.

We will let $M = x_2x_1x_0$ and $N = y_2y_1y_0$ denote the numbers, written in binary, to be added and let $z_3z_2z_1z_0$ denote the sum, also written in binary. The circuit that computes the sum of M and N is drawn in Figure 1.6.6. It is an implementation of the method of Example 1.6.4 for adding binary numbers, inasmuch as the "carry bit" is indeed *carried* into the next binary addition.

Overflow occurs in addition when the result is too large for the storage available. Suppose, for example, that we are using storage that can hold only three-bit numbers. If we add $101 + 001$, the result is 110, which can be represented in the available storage; no overflow occurs. But if we add $101 + 011$, the result is 1000, which is too large to be represented in the available storage; overflow occurs. Overflow occurs precisely when the bit in position 3 in the sum is 1.

If we were using storage that could hold only three-bit numbers, we could use the z_3 bit in Example 1.6.18 to test for overflow. If $z_3 = 1$, overflow occurs; if $z_3 = 0$, there is no overflow.

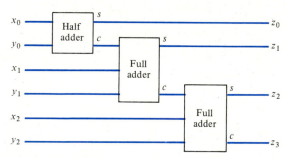

Figure 1.6.6

EXERCISES

In Exercises 1–6, express each decimal number in the form shown in Example 1.6.1.

1H. 9841	**2.** 6038	**3.** 95,042
4H. 48,702	**5.** 3,000,000	**6.** 100,030,024

In Exercises 7–12, express each binary number in decimal.

7H. 1001   **8.** 11011   **9.** 11011011

10H. 100000   **11.** 11111111   **12.** 110111011011

In Exercises 13–18, express each decimal number in binary.

13H. 34   **14.** 61   **15.** 223

16H. 400   **17.** 1024   **18.** 12,340

In Exercises 19–24, add the binary numbers.

19H. 1001 + 1111   **20.** 11011 + 1101

21.   110110 + 101101

22H. 101101 + 11011

23.   110110101 + 1101101

24.   1101 + 101100 + 11011011

In Exercises 25–30, express each hexadecimal number in decimal.

25H. 3A   **26.** 1E9   **27.** 3E7C

28H. A03   **29.** 209D   **30.** 4B07A

31H. Express each decimal number in Exercises 1–6 and 13–18 in hexadecimal.

32. Express each binary number in Exercises 7–12 in hexadecimal.

33. Express each hexadecimal number in Exercises 25, 26, and 28 in binary.

In Exercises 34–38, add the hexadecimal numbers.

34H. 4A + B4   **35.** 195 + 76E   **36.** 49F7 + C66

37H. 349CC + 922D **38.** 82054 + AEFA3

39H. Does 2010 represent a number in binary? in decimal? in hexadecimal?

40. Does 1101010 represent a number in binary? in decimal? in hexadecimal?

In the octal (base 8) number system, to represent nonnegative integers we use the symbols 0, 1, 2, 3, 4, 5, 6, and 7. In representing an arbitrary nonnegative integer, reading from the right, the first symbol represents the number of 1's, the next symbol the number of 8's, the next symbol the number of 8^2's, and so on. In general, the symbol in position n (with the rightmost symbol being in position 0) represents the number of 8^n's. In Exercises 41–46, express each octal number in decimal.

41H. 63   **42.** 7643   **43.** 7711

44H. 10732   **45.** 1007   **46.** 537261

47H. Express each decimal number in Exercises 13–18 in octal.

48. Express each binary number in Exercises 7–12 in octal.

49. Express each hexadecimal number in Exercises 25–30 in octal.

50H. Express each octal number in Exercises 41–46 in hexadecimal.

51. Does 1101010 represent a number in octal?

52. Does 30470 represent a number in binary? in octal? in decimal? in hexadecimal?

53H. Does 9450 represent a number in binary? in octal? in decimal? in hexadecimal?

In Exercises 54–58, find the sum and carry bits.

54H. $0 + 1$ **55.** $0 + 0$ **56.** $1 + 0 + 1$

57H. $0 + 1 + 0$ **58.** $0 + 0 + 0$

59H. Using half-adder and full-adder circuits, design a combinatorial circuit that computes the sum of two four-bit numbers.

60. Assuming that you can store at most four bits, how could you detect overflow in your circuit for Exercise 59?

61. Design a circuit with three inputs that outputs 1 precisely when two or three inputs have value 1.

62. Design a circuit that multiplies the binary numbers $x_1 x_0$ and $y_1 y_0$. The output will be of the form $z_3 z_2 z_1 z_0$.

63. By writing the table for each expression, show that

$$x \oplus y = (x \vee y) \wedge \overline{(x \wedge y)}.$$

64. Use the result of Exercise 63 to design a half-adder circuit with fewer gates than that of Figure 1.6.4.

The **two's complement** of a binary expression is obtained as follows. Starting from the right, copy the bits until the first 1 is copied. Thereafter, if the bit is 1, copy 0 and, if the bit is 0, copy 1.

EXAMPLE: The two's complement of 01100 is 10100. Working from the right, we copy the bits until the first 1 is copied:

$$0 \ 1 \ 1 \ 0 \ 0$$
$$\downarrow \ \downarrow \ \downarrow$$
$$1 \ 0 \ 0$$

Thereafter, we interchange 0's and 1's:

$$0\ 1\ 1\ 0\ 0$$
$$\downarrow\ \downarrow$$
$$1\ 0\ 1\ 0\ 0$$

Two's complement is used to represent negative integers in computers. Find the two's complement in each of Exercises 65–70.

65H. 0011   **66.** 1101   **67.** 00110111

68H. 11001011   **69.** 10000000   **70.** 1111110111011010

71H. Design a **two's module** combinatorial circuit that either copies an input bit b to the output y or outputs $\bar{b}$. There are two inputs, b and FLAGIN, and two outputs, y and FLAGOUT. FLAGIN tells whether to output b or $\bar{b}$. If FLAGIN $= 0$, $y = b$ and, if FLAGIN $= 1$, $y = \bar{b}$. FLAGOUT is 1 unless both b and FLAGIN are 0.

72. Using two's modules (see Exercise 71), design a combinatorial circuit that computes the two's complement $y_2 y_1 y_0$ of the three-bit binary number $x_2 x_1 x_0$. There will be three inputs x_0, x_1, and x_2, and three outputs y_0, y_1, and y_2.

1.7 Notes

General references on discrete mathematics are [Johnsonbaugh; Lipschutz, 1976, 1982; Liu, 1985; Roberts; Ross; and Tucker]. [Knuth, 1973 Vols. 1 and 3, 1981] is the classic reference on much of this material.

[Barker; Copi; Gustason; Jeffrey; and Resnik] are introductory logic texts. The first chapter of the geometry text by [Jacobs] is devoted to basic logic. [Solow] addresses the problem of how to construct proofs. For a history of logic, see [Kline]. For a brief history of the computer and its relationship to logic, see Part One of [Goldstine]. This book also includes a very interesting and highly personal account of recent computer history.

General references on circuit design include [D'Angelo; Hill; and Kohavi]. [D'Angelo] is a very nice introduction to circuit hardware. The introduction to Chapter 3 contains a brief but enlightening discussion of logic design. The book by [Reid, 1984] is primarily a history of the invention of the integrated circuit, but it also contains an introduction to the physics behind transistors and integrated circuits and some discussion of binary numbers and Boolean logic.

The use of Karnaugh maps to simplify Boolean expressions is restricted to expressions in a limited number of variables. Other techniques, such as the Quine-McCluskey method (see [Mendelson]), can be used to simplify arbitrary Boolean expressions and can be executed by a computer.

Technical books on Boolean algebra are [Halmos, 1967; Hohn; and Mendelson].

Computer Exercises

1. Write a program that inputs a logical expression in p and q and prints the truth table of the expression. Use a higher-level language that has the capability of evaluating a logical expression.

2. Write a program that inputs a logical expression in p, q, and r and prints the truth table of the expression.

3. Write a program that tests whether two logical expressions in p and q are logically equivalent.

4. Write a program that tests whether two logical expressions in p, q, and r are logically equivalent.

5. Write a program that outputs the disjunctive normal form of a logical expression in x, y, and z.

6. Write a program that inputs a number in binary, octal, decimal, or hexadecimal and writes the number in binary, octal, decimal, and hexadecimal. (One of the inputs must indicate the base used.)

7. Write a program that inputs two numbers written in one of binary, octal, decimal, or hexadecimal and outputs the sum of the numbers in the same base in which they were input. (One of the inputs must indicate the base used.)

8. Write a program that computes the two's complement of an n-bit binary expression.

Chapter Review

SECTION 1.1
Logic
Proposition
Conjunction: p and q, $p \wedge q$
Disjunction: p or q, $p \vee q$
Negation: not p, $\bar{p}$
Compound proposition
Truth table
Exclusive-OR of propositions p, q: p or q, but not both

SECTION 1.2
Conditional proposition: if p, then q; $p \rightarrow q$
Necessary condition
Sufficient condition
Converse of $p \rightarrow q$: $q \rightarrow p$
Biconditional proposition: p if and only if q, $p \leftrightarrow q$
Logical equivalence: $P \equiv Q$
De Morgan's laws for logic: $\overline{p \vee q} \equiv \bar{p} \wedge \bar{q}$, $\overline{p \wedge q} \equiv \bar{p} \vee \bar{q}$
Contrapositive of $p \rightarrow q$: $\bar{q} \rightarrow \bar{p}$

SECTION 1.3
Bit: 0 or 1
Combinatorial circuit
Gate
AND gate
OR gate
NOT gate (inverter)
Logic table
Boolean expression

SECTION 1.4
Exclusive-OR of bits
Circuit to compute the exclusive-OR
Disjunctive normal form
Minterm

SECTION 1.5
Equal Boolean expressions
Equivalent combinatorial circuits
Karnaugh map
Simplifying a Boolean expression

SECTION 1.6
Decimal number system
Binary number system
Hexadecimal number system
Octal number system
Base of a number system
Convert binary to decimal
Convert decimal to binary
Convert hexadecimal to decimal
Convert decimal to hexadecimal
Add binary numbers
Add hexadecimal numbers
Half-adder circuit
Full-adder circuit
Integrated circuit
Overflow

Chapter Self-Test

SECTION 1.1

1H. If p, q, and r are true, find the truth value of the proposition $(p \vee q) \wedge ((\overline{p} \wedge r) \vee q)$.

2H. Write the truth table of the proposition $\overline{(p \wedge q)} \vee (p \vee \overline{r})$.

3H. Formulate the proposition $p \wedge (\overline{q} \vee r)$ in words using

p: I take calculus. q: I take philosophy. r: I take art.

4H. Assume that a, b, and c are real numbers. Represent the statement

$$a < b \text{ or } (b < c \text{ and } a \geq c)$$

symbolically, letting

$$p: a < b, \qquad q: b < c, \qquad r: a < c.$$

SECTION 1.2

5H. Restate the proposition "A necessary condition for Leah to get an A in discrete mathematics is to study hard" in the form of a conditional proposition.

6H. Write the converse and contrapositive of the proposition of Exercise 5.

7H. If p is true and q and r are false, find the truth value of the proposition $(p \lor q) \to \bar{r}$.

8H. Represent the statement

$$\text{If } (a \geq c \text{ or } b < c), \text{ then } b \geq c$$

symbolically using the definitions of Exercise 4.

SECTION 1.3

9H. Write a Boolean expression that represents the combinatorial circuit and write the logic table.

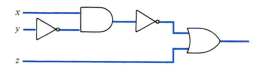

10H. Find the value of the Boolean expression

$$(x_1 \land x_2) \lor (\bar{x_2} \land x_3)$$

if $x_1 = x_2 = 0$ and $x_3 = 1$.

11H. Find a combinatorial circuit corresponding to the Boolean expression of Exercise 10.

12H. Show that the circuit is not a combinatorial circuit.

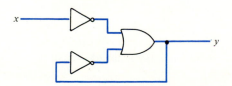

SECTION 1.4

In Exercises 13–16, find the disjunctive normal form of a Boolean expression having a logic table the same as the given table and draw the combinatorial circuit corresponding to the disjunctive normal form.

13H.

x_1	x_2	x_3	y
1	1	1	0
1	1	0	0
1	0	1	0
1	0	0	1
0	1	1	0
0	1	0	0
0	0	1	0
0	0	0	0

14H.

x_1	x_2	x_3	y
1	1	1	0
1	1	0	1
1	0	1	0
1	0	0	1
0	1	1	0
0	1	0	0
0	0	1	0
0	0	0	0

15H.

x_1	x_2	x_3	y
1	1	1	1
1	1	0	0
1	0	1	0
1	0	0	1
0	1	1	0
0	1	0	0
0	0	1	0
0	0	0	1

16H.

x_1	x_2	x_3	y
1	1	1	0
1	1	0	1
1	0	1	0
1	0	0	1
0	1	1	1
0	1	0	0
0	0	1	1
0	0	0	0

SECTION 1.5

17H. Which term does the circle in the Karnaugh map represent?

18H. Write the simplified Boolean expression given by the Karnaugh map.

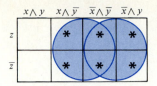

Use Karnaugh maps to simplify the Boolean expressions in Exercises 19 and 20.

19H. $(x \wedge y \wedge z) \vee (x \wedge \bar{y} \wedge z) \vee (\bar{x} \wedge \bar{y} \wedge z)$
$\vee (\bar{x} \wedge y \wedge z) \vee (\bar{x} \wedge y \wedge \bar{z})$

20H. $(x \wedge y \wedge z) \vee (\bar{x} \wedge y \wedge z) \vee (x \wedge y \wedge \bar{z}) \vee (x \wedge \bar{y} \wedge \bar{z})$
$\vee (\bar{x} \wedge y \wedge \bar{z}) \vee (\bar{x} \wedge \bar{y} \wedge \bar{z})$

SECTION 1.6

21H. Write the binary number 10010110 in decimal.

22H. Write the decimal number 430 in binary and in hexadecimal.

23H. Add the binary numbers 11001 and 101001.

24H. Write the hexadecimal number C39 in decimal.

25H. Design a full-adder circuit directly; that is, without using half-adders.

2

When *I* use a word, it means just what I choose it to mean—neither more nor less.

—from *Alice in Wonderland*

The Language of Mathematics

The present chapter deals with the language of mathematics. The topics, some of which may be familiar, are sets, sequences, mathematical induction, functions, and matrices. These concepts will be used throughout the remainder of this book. In fact, these ideas are also used throughout various branches of mathematics and in subjects that rely on mathematics such as computer science, engineering, and others.

2.1 Sets

The concept of set is basic to all of mathematics and mathematical applications. Indeed, one can take "set" and "membership" as undefined terms and develop the rest of mathematics from these fundamental terms. A **set** is simply any collection of objects. A notation for sets is useful for thinking about sets and for communicating about sets to other persons or to machines. If a set is finite and not too large, we can describe it by listing the elements in it. For example, the equation

$$A = \{1, 2, 3, 4\} \tag{2.1.1}$$

describes a set A made up of the four elements 1, 2, 3, and 4. A set is determined by its elements and not any particular order in which the elements might be listed. Thus A might just as well be specified as

$$A = \{1, 3, 4, 2\}.$$

The elements making up a set are assumed to be distinct, and although for some reason we may have duplicates in our list, only one occurrence of each element is in the set. For this reason we may also describe the set A defined in (2.1.1) as

$$A = \{1, 2, 2, 3, 4\}.$$

If a set is a large finite set or an infinite set, we can describe it by listing a property necessary for membership. For example, the equation

$$B = \{x \mid x \text{ is a positive, even integer}\} \tag{2.1.2}$$

describes the set B made up of all positive, even integers; that is, B consists of the integers 2, 4, 6, and so on. The vertical bar "$|$" is read "such that." Equation (2.1.2) would be read "B equals the set of all x such that x is a positive, even integer." Here the property necessary for membership is "is a positive, even integer." Note that the property appears after the vertical bar. The method (2.1.2) of specifying a set is called **set-builder notation**.

Example 2.1.1. The set of squares 1, 4, 9, 16, . . . can be written in set-builder notation as

$$\{x^2 \mid x \text{ is a positive integer}\}.$$

If X is a finite set, we let

$$|X| = \text{number of elements in } X.$$

Notice that in this context, the vertical bars have nothing to do with absolute value.

Example 2.1.2. Let

$$A = \{x \mid x^2 + 2x - 3 = 0\}.$$

Solving $x^2 + 2x - 3 = 0$, we see that $x^2 + 2x - 3 = 0$ if and only if $x = 1$ or $x = -3$. Thus

$$A = \{1, -3\}$$

and

$$|A| = 2.$$

Given a description of a set X such as (2.1.1) or (2.1.2) and an element x, we can determine whether or not x belongs to X. If the members of X are listed as in (2.1.1), we simply look to see whether or not x appears in the listing. In a description such as (2.1.2), we check to see whether the element x has the property listed. If x is in the set X, we write $x \in X$ and if x is not in X, we write $x \notin X$.

Example 2.1.3. Let A and B be defined by equations (2.1.1) and (2.1.2). If $x = 1$, then $x \in A$, but $x \notin B$.

The set with no elements is called the **empty** (or **null** or **void**) **set** and is denoted $\varnothing$. Thus $\varnothing = \{\ \ \}$.

Example 2.1.4.   If

$$A = \{x \mid x^2 + 2 = 0 \text{ and } x \text{ is a real number}\},$$

then

$$A = \varnothing$$

since no real number x satisfies $x^2 + 2 = 0$.

Two sets X and Y are **equal** and we write $X = Y$ if X and Y have the same elements. To put it another way, $X = Y$ if whenever $x \in X$, then $x \in Y$ and whenever $x \in Y$, then $x \in X$.

Example 2.1.5.   If

$$A = \{x \mid x^2 + x - 6 = 0\}, \qquad B = \{2, -3\},$$

then $A = B$.

Notice that $\varnothing \neq \{\varnothing\}$. The empty set $\varnothing$ has no members whereas $\{\varnothing\}$ has one member, namely $\varnothing$.

Suppose that X and Y are sets. If every element of X is an element of Y, we say that X is a **subset** of Y and write $X \subseteq Y$.

Example 2.1.6.   If

$$C = \{1, 3\} \qquad \text{and} \qquad A = \{1, 2, 3, 4\},$$

then C is a subset of A.

In order for a set A *not* to be a subset of B,

there must be some element in A that is not in B.   (2.1.3)

(Otherwise, every element in A would also be in B and we would have $A \subseteq B$.) For example, $A = \{1, 3, 5\}$ is not a subset of $B = \{2, 3, 5\}$ because the element 1 is in A but not in B. If A is the empty set, condition

(2.1.3) will not be true for any set B. Thus the empty set is a subset of any set B.

Any set X is a subset of itself, since any element in X is in X. If X is a subset of Y and X does not equal Y, we say that X is a **proper subset** of Y. The set of all subsets (proper or not) of a set X, denoted $\mathcal{P}(X)$, is called the **power set** of X. In Section 2.4 (Theorem 2.4.5) we will show that if $|X| = n$, then $|\mathcal{P}(X)| = 2^n$.

Example 2.1.7. If $A = \{a, b, c\}$, the members of $\mathcal{P}(A)$ are

$$\emptyset, \{a\}, \{b\}, \{c\}, \{a, b\}, \{a, c\}, \{b, c\}, \{a, b, c\}.$$

All but $\{a, b, c\}$ are proper subsets of A. For this example,

$$|A| = 3, \qquad |\mathcal{P}(A)| = 2^3 = 8.$$

Given two sets X and Y, there are various ways to combine X and Y to form a new set. The set

$$X \cup Y = \{x \mid x \in X \text{ or } x \in Y\}$$

is called the **union** of X and Y. The union consists of all elements belonging to either X or Y (or both).

Example 2.1.8. If $A = \{1, 3, 5\}$ and $B = \{4, 5, 6\}$, then

$$A \cup B = \{1, 3, 4, 5, 6\}.$$

The set

$$X \cap Y = \{x \mid x \in X \text{ and } x \in Y\}$$

is called the **intersection** of X and Y. The intersection consists of all elements belonging to both X and Y.

Example 2.1.9. If $A = \{1, 3, 5\}$ and $B = \{4, 5, 6\}$, then

$$A \cap B = \{5\}.$$

Sets X and Y are **disjoint** if $X \cap Y = \emptyset$.

Example 2.1.10.   If $A = \{1, 3, 5\}$ and $B = \{2, 4, 9\}$, then A and B are disjoint.

The set

$$X - Y = \{x \mid x \in X \text{ and } x \notin Y\}$$

is called the **difference** (or **relative complement**). The difference $X - Y$ consists of all elements in X which are not in Y.

Example 2.1.11.   If $A = \{1, 3, 5\}$ and $B = \{4, 5, 6\}$, then

$$A - B = \{1, 3\},$$

$$B - A = \{4, 6\}.$$

Sometimes we are dealing with sets all of which are subsets of a set U. This set U is called a **universal set** or a **universe**. The set U must be explicitly given or inferred from the context. Given a universal set U and a subset X of U, the set $U - X$ is called the **complement** of X and is written $\overline{X}$.

Example 2.1.12.   Let $A = \{1, 3, \underline{5}\}$. If U, a universal set, is specified as $U = \{1, 2, 3, 4, 5\}$, then $\overline{A} = \{2, 4\}$. If, on the other hand, a universal set is specified as $U = \{1, 3, 5, 7, 9\}$, then $\overline{A} = \{7, 9\}$. The complement obviously depends on the universe in which we are working.

†COMPUTER NOTES FOR SECTION 2.1

A **data type** in a higher-level programming language is a set. For example, the data type *integer* refers to the set

$$\mathbf{Z} = \{\ldots, -3, -2, -1, 0, 1, 2, 3, \ldots\}.$$

In implementing a particular higher-level language on a specific computer, *integer* will refer to some finite subset of $\mathbf{Z}$. For example, in Turbo Pascal on the IBM Personal Computer, the data type *integer* refers to the set

$$\{x \mid x \text{ is an integer and } -32,768 \leq x \leq 32,767\}.$$

†This section can be omitted without loss of continuity.

In VAX-11 PL/I, the data type *FIXED* refers to the set of integers

$$\{x \mid x \text{ is an integer and } -2{,}147{,}483{,}648 \le x \le 2{,}147{,}483{,}647\}.$$

EXERCISES

In Exercises 1–16, let the universe be the set $U = \{1, 2, 3, \ldots, 10\}$.
Let $A = \{1, 4, 7, 10\}$, $B = \{1, 2, 3, 4, 5\}$, and $C = \{2, 4, 6, 8\}$.
List the elements of each set.

1H. $A \cup B$ **2.** $B \cap C$

3. $A - B$ **4H.** $B - A$

5. $\overline{A}$ **6.** $U - C$

7H. $\overline{U}$ **8.** $A \cup \varnothing$

9. $B \cap \varnothing$ **10H.** $A \cup U$

11. $B \cap U$ **12.** $A \cap (B \cup C)$

13H. $\overline{B} \cap (C - A)$ **14.** $(A \cap B) - C$

15. $\overline{(A \cap B)} \cup C$ **16H.** $(A \cup B) - (C - B)$

In Exercises 17–20, answer true or false.

17H. $\{x\} \subseteq \{x\}$ **18.** $\{x\} \in \{x\}$

19. $\{x\} \in \{x, \{x\}\}$ **20H.** $\{x\} \subseteq \{x, \{x\}\}$

In Exercises 21–25, determine whether each pair of sets is equal.

21H. $\{1, 2, 3\}$, $\{1, 3, 2\}$

22. $\{1, 2, 2, 3\}$, $\{1, 2, 3\}$

23. $\{1, 1, 3\}$, $\{3, 3, 1\}$

24H. $\{x \mid x^2 + x = 2\}$, $\{1, -2\}$

25. $\{x \mid x \text{ is a real number and } 0 < x \le 2\}$, $\{1, 2\}$

26. True or false? For any sets A and B, either A is a subset of B or B is a subset of A. Give a reason for your conclusion.

27H. List the members of $\mathscr{P}(\{a, b\})$. Which are proper subsets of $\{a, b\}$?

28. List the members of $\mathscr{P}(\{a, b, c, d\})$. Which are proper subsets of $\{a, b, c, d\}$?

29. If X has 10 members, how many members does $\mathscr{P}(X)$ have? How many proper subsets does X have?

30H. If X has n members, how many proper subsets does X have?

John Venn (1834–1923)

John Venn was an English mathematician and cleric. He graduated from Cambridge in mathematics in 1857 and took holy orders in 1858. Besides his work in mathematics, especially in logic and probability, he was interested in botany and mountaineering. Having something of a mechanical bent, he once constructed a successful bowling machine. [*Photo used by permission of The Royal Society, London*]

For each condition in Exercises 31–34, what relation must hold between the sets A and B?

31H. $A \cap B = A$ **32.** $A \cup B = A$

33. $\overline{A} \cap U = \varnothing$ **34H.** $\overline{A} \cap \overline{B} = \overline{B}$

The **symmetric difference** of two sets A and B is the set

$$A \vartriangle B = (A \cup B) - (A \cap B).$$

35H. If $A = \{1, 2, 3\}$ and $B = \{2, 3, 4, 5\}$, find $A \vartriangle B$.

36. Describe the symmetric difference of sets A and B in words.

37. Given a universe U, describe $A \vartriangle A$, $A \vartriangle \overline{A}$, $U \vartriangle A$, and $\varnothing \vartriangle A$.

2.2   More on Sets

In this section we continue the discussion of sets begun in Section 2.1. We discuss Venn diagrams, Cartesian products of sets, n-tuples, and partitions. Venn diagrams provide visual representations of sets and relationships among sets. Unlike sets, Cartesian products and n-tuples take into account the order in which elements appear. A partition of a set X subdivides X into nonoverlapping subsets.

Venn diagrams provide pictorial views of sets. In a Venn diagram, a rectangle depicts a universal set (see Figure 2.2.1). Subsets of the universal set are drawn as circles. The inside of a circle represents the members of that set. In Figure 2.2.1 we see two sets A and B within the universal set U. The elements in neither A nor B are in region 1. Elements in region 2 are in A but not in B. Region 3 represents $A \cap B$, the elements common to both A and B. The elements in B but not in A comprise region 4.

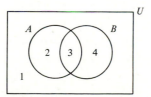

Figure 2.2.1

Example 2.2.1.   Particular regions in Venn diagrams are depicted by shading. $A \cup B$ is shown in Figure 2.2.2, and Figure 2.2.3 represents $A - B$.

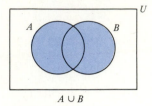

$A \cup B$

Figure 2.2.2

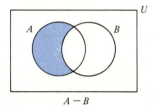

$A - B$

Figure 2.2.3

To represent three sets, we use three overlapping circles (see Figure 2.2.4).

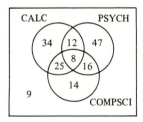

Figure 2.2.4

Example 2.2.2.   Among a group of 165 students, eight are taking calculus, psychology, and computer science; 33 are taking calculus and computer science; 20 are taking calculus and psychology; 24 are taking psychology and computer science; 79 are taking calculus; 83 are taking psychology; and 63 are taking computer science. How many are taking none of the three subjects?

Let CALC, PSYCH, and COMPSCI denote the sets of students taking calculus, psychology, and computer science, respectively. Let U denote the set of all 165 students (see Figure 2.2.4). Since 8 students are taking calculus and psychology and computer science, we write 8 in the region representing CALC $\cap$ PSYCH $\cap$ COMPSCI. Of the 33 students taking calculus and computer science, 8 are also taking psychology; thus 25 are taking calculus and computer science but not psychology. We write 25 in the region representing CALC $\cap$ $\overline{\text{PSYCH}}$ $\cap$ COMPSCI. Similarly, we write 12 in the region representing CALC $\cap$ PSYCH $\cap$ $\overline{\text{COMPSCI}}$ and 16 in the region representing $\overline{\text{CALC}}$ $\cap$ PSYCH $\cap$ COMPSCI. Of the 79 students taking calculus, 45 have now been accounted for. This leaves 34 students taking only calculus. We write 34 in the region representing CALC $\cap$ $\overline{\text{PSYCH}}$ $\cap$ $\overline{\text{COMPSCI}}$. Similarly, we write 47 in the region representing $\overline{\text{CALC}}$ $\cap$ PSYCH $\cap$ $\overline{\text{COMPSCI}}$ and 14 in the region representing $\overline{\text{CALC}}$ $\cap$ $\overline{\text{PSYCH}}$ $\cap$ COMPSCI. At this point, 156 students have been accounted for. This leaves 9 students taking none of the three subjects.

Venn diagrams can also be used to establish certain properties of sets.

Theorem 2.2.3. *Let U be a universal set and let A, B, and C be subsets of U. The following properties hold.*

(a) Associative laws:

$$(A \cup B) \cup C = A \cup (B \cup C), \qquad (A \cap B) \cap C = A \cap (B \cap C)$$

(b) Commutative laws:

$$A \cup B = B \cup A, \qquad A \cap B = B \cap A$$

(c) Distributive laws:

$$A \cap (B \cup C) = (A \cap B) \cup (A \cap C)$$
$$A \cup (B \cap C) = (A \cup B) \cap (A \cup C)$$

(d) Identity laws:

$$A \cup \varnothing = A, \qquad A \cap U = A$$

(e) Complement laws:

$$A \cup \overline{A} = U, \qquad A \cap \overline{A} = \varnothing$$

René Descartes (1596–1650)

The Cartesian product is named after the French mathematician and philosopher René Descartes. He is one of the founders of analytic geometry. Analytic geometry represents points in the plane as ordered pairs of numbers and thus links algebra and geometry. Descartes advocated spending all morning in bed as the optimal way to do mathematics and maintain good health. [*Photo courtesy of Brown Brothers*]

(f) Idempotent laws:

$$A \cup A = A, \qquad A \cap A = A$$

(g) Bound laws:

$$A \cup U = U, \qquad A \cap \varnothing = \varnothing$$

(h) Absorption laws:

$$A \cup (A \cap B) = A, \qquad A \cap (A \cup B) = A$$

(i) Involution law:

$$\overline{\overline{A}} = A$$

(j) 0/1 laws:

$$\overline{\varnothing} = U, \qquad \overline{U} = \varnothing$$

(k) De Morgan's laws for sets:

$$\overline{(A \cup B)} = \overline{A} \cap \overline{B}, \qquad \overline{(A \cap B)} = \overline{A} \cup \overline{B}$$

PROOF. We prove only the first of De Morgan's laws [part (k)].

By sketching both $\overline{(A \cup B)}$ and $\overline{A} \cap \overline{B}$ (see Figure 2.2.5), we see that these sets are identical.

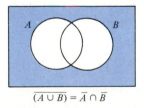

$\overline{(A \cup B)} = \overline{A} \cap \overline{B}$

Figure 2.2.5

Notice the similarity of De Morgan's laws for sets (Theorem 2.2.3k) and De Morgan's laws for logic (Example 1.2.11).

In Section 2.1 we pointed out that a set is an unordered collection of elements; that is, a set is determined by its elements and not by any particular order in which the elements are listed. Sometimes, however, we do want to take order into account. An **ordered pair** of elements, written (a, b), is considered distinct from the ordered pair (b, a), unless, of course, $a = b$. To put it another way, $(a, b) = (c, d)$ if and only if $a = c$ and $b = d$. If X and Y are sets we let $X \times Y$ denote the set of all ordered pairs (x, y) where $x \in X$ and $y \in Y$. We call $X \times Y$ the **Cartesian product** of X and Y.

Example 2.2.4.   If $X = \{1, 2, 3\}$ and $Y = \{a, b\}$, then

$$X \times Y = \{(1, a), (1, b), (2, a), (2, b), (3, a), (3, b)\},$$

$$Y \times X = \{(a, 1), (b, 1), (a, 2), (b, 2), (a, 3), (b, 3)\},$$

$$X \times X = \{(1, 1), (1, 2), (1, 3), (2, 1), (2, 2), (2, 3),$$
$$(3, 1), (3, 2), (3, 3)\},$$

$$Y \times Y = \{(a, a), (a, b), (b, a), (b, b)\}.$$

Example 2.2.4 shows that, in general, $X \times Y \neq Y \times X$.

Example 2.2.5.   A restaurant serves four appetizers

$$r = \text{ribs}, \qquad n = \text{nachos}, \qquad s = \text{shrimp}, \qquad f = \text{fried cheese}$$

and three main courses

$$c = \text{chicken}, \qquad b = \text{beef}, \qquad t = \text{trout}.$$

If we let $A = \{r, n, s, f\}$ and $M = \{c, b, t\}$, the Cartesian product $A \times M$ lists the 12 possible dinners consisting of one appetizer and one main course.

Ordered lists need not be restricted to two elements. An **n-tuple**, written $(a_1, a_2, \ldots, a_n)$, takes order into account:

$$(a_1, a_2, \ldots, a_n) = (b_1, b_2, \ldots, b_n)$$

if and only if

$$a_1 = b_1, a_2 = b_2, \ldots, a_n = b_n.$$

The Cartesian product $X_1 \times X_2 \times \cdots \times X_n$ of sets $X_1, X_2, \ldots, X_n$ is defined to be the set of all n-tuples $(x_1, x_2, \ldots, x_n)$, where $x_i \in X_i$ for $i = 1, \ldots, n$.

Example 2.2.6.   If

$$X = \{1, 2\}, \qquad Y = \{a, b\}, \qquad Z = \{\alpha, \beta\},$$

then

$$X \times Y \times Z = \{(1, a, \alpha), (2, a, \alpha), (1, b, \alpha), (2, b, \alpha),$$
$$(1, a, \beta), (2, a, \beta), (1, b, \beta), (2, b, \beta)\}.$$

Example 2.2.7. If A is a set of appetizers, M is a set of main courses, and D is a set of desserts, the Cartesian product $A \times M \times D$ lists all possible dinners consisting of one appetizer, one main course, and one dessert.

We conclude our discussion of sets by discussing partitions. A partition of a set X divides X into nonoverlapping subsets. More formally, a collection $\mathscr{C}$ of sets is said to be **pairwise disjoint** if whenever X and Y are distinct sets in $\mathscr{C}$, X and Y are disjoint (that is, $X \cap Y = \varnothing$). A **partition** of X is a pairwise disjoint collection $\mathscr{C}$ of nonempty subsets of X having the property that each element of X is in some member of $\mathscr{C}$.

Example 2.2.8. Consider the collection of sets

$$\mathscr{C} = \{\{1, 4, 5\}, \{2, 6\}, \{3\}, \{7, 8\}\}.$$

The distinct pairs in $\mathscr{C}$ are

$$\{1, 4, 5\}, \{2, 6\}$$
$$\{1, 4, 5\}, \{3\}$$
$$\{1, 4, 5\}, \{7, 8\}$$
$$\{2, 6\}, \{3\}$$
$$\{2, 6\}, \{7, 8\}$$
$$\{3\}, \{7, 8\}.$$

As we can see by inspection, each distinct pair is disjoint. Thus $\mathscr{C}$ is pairwise disjoint. Since each element of

$$X = \{1, 2, 3, 4, 5, 6, 7, 8\}$$

is in some member of $\mathscr{C}$, $\mathscr{C}$ is a partition of X.

Example 2.2.9. Let

$$X = \{0000, 0001, 0010, 0011, 0100, 0101, 0110, 0111, 1000,$$
$$1001, 1010, 1011, 1100, 1101, 1110, 1111\}$$

be the set of all four-bit strings. If a and b are bits, we let X_{ab} be the set of all four-bit strings that begin ab. For example,

$$X_{01} = \{0100, 0101, 0110, 0111\}.$$

Since the first two bits of a string are unique, if $ab \neq cd$, X_{ab} and X_{cd} are disjoint. Thus

$$\mathcal{C} = \{X_{00}, X_{01}, X_{10}, X_{11}\}$$

is pairwise disjoint. Since each string in X belongs to one of X_{00}, X_{01}, X_{10}, X_{11}, $\mathcal{C}$ is a partition of X.

EXERCISES

In Exercises 1–8, draw a Venn diagram and shade the given set.

1H. $A \cap \overline{B}$ **2.** $\overline{A} - B$

3. $B \cup (B - A)$ **4H.** $(A \cup B) - B$

5. $B \cap \overline{(C \cup A)}$ **6.** $(\overline{A} \cup B) \cap (\overline{C} - A)$

7H. $((C \cap A) - \overline{(B - A)}) \cap C$ **8.** $(B - \overline{C}) \cup ((B - \overline{A}) \cap (C \cup B))$

Exercises 9–13 refer to a group of 191 students, of which 10 are taking French, business, and music; 36 are taking French and business; 20 are taking French and music; 18 are taking business and music; 65 are taking French; 76 are taking business; and 63 are taking music.

9H. How many are taking French and music but not business?

10. How many are taking business and neither French nor music?

11. How many are taking French or business (or both)?

12H. How many are taking music or French (or both) but not business?

13. How many are taking none of the three subjects?

14. A television poll of 151 persons found that 68 watched "M∗E∗S∗S"; 61 watched "Leave It to Seaver" (a baseball show); 52 watched "The Yuppie Hour"; 16 watched both "M∗E∗S∗S" and "Leave It to Seaver"; 25 watched both "M∗E∗S∗S" and "The Yuppie Hour"; 19 watched both "Leave It to Seaver" and "The Yuppie Hour"; and 26 watched none of these shows. How many persons watched all three shows?

In Exercises 15–20, using Venn diagrams or otherwise, answer true or false. The sets X, Y, and Z are subsets of a universal set U.

15H. $X \cap (Y - Z) = (X \cap Y) - (X \cap Z)$ for all sets X, Y, and Z

16. $(X - Y) \cap (Y - X) = \varnothing$ for all sets X and Y

17. $X - (Y \cup Z) = (X - Y) \cup Z$ for all sets X, Y, and Z

18H. $\overline{X - Y} = \overline{Y - X}$ for all sets X and Y

19. $\overline{X \cap Y} \subseteq \overline{X}$ for all sets X and Y

20. $(X \cap Y) \cup (Y - X) = Y$ for all sets X and Y

21. Draw a Venn diagram that represents $X \Delta Y$, the symmetric difference of X and Y.

In Exercises 22–27, let $X = \{1, 2\}$ and $Y = \{a, b, c\}$. List the elements in each set.

22H. $X \times Y$ **23.** $Y \times X$ **24.** $X \times X$

25H. $Y \times Y$ **26.** $X \times (Y - \{b\})$ **27.** $Y \times \varnothing$

In Exercises 28–33, let $X = \{1, 2\}$, $Y = \{a\}$, and $Z = \{\alpha, \beta\}$. List the elements of each set.

28H. $X \times Y \times Z$ **29.** $X \times Y \times Y$ **30.** $X \times X \times X$

31H. $Y \times X \times Y \times Z$ **32.** $(X \cup Y) \times Y \times Z$ **33.** $X \times Y \times \varnothing$

34. Suppose that C is a set of computers, M is a set of modems, and P is a set of printers. What interpretation can be given to the set $C \times M \times P$?

35H. Partition the set $\{1, 2, 3, 4, 5, 6, 7\}$ into exactly four subsets.

36. List the elements of each set X_{ab} in Example 2.2.9.

37. Let X be the set of all four-bit strings. Let X_{ad} be the set of all four-bit strings that start with a and end with d. Explain why

$$\mathscr{C} = \{X_{00}, X_{10}, X_{01}, X_{11}\}$$

is a partition of X. List the members of each set in $\mathscr{C}$.

In Exercises 38–41, list all partitions of the set.

38H. $\{1\}$ **39.** $\{\alpha, \beta\}$

40. $\{a, b, c\}$ **41H.** $\{a, b, c, d\}$

42H. Use a Venn diagram to show that

$$|A \cup B| = |A| + |B| - |A \cap B|.$$

43. Find a formula for $|A \cup B \cup C|$ similar to the formula of Exercise 42. Use a Venn diagram to show that your formula holds for all sets A, B, and C.

2.3 Sequences

Blue Taxi, Inc. charges $1 for the first mile and 50 cents for each additional mile. The following table shows the cost of traveling from 1 to 10 miles.

Mileage	Cost
1	$1.00
2	1.50
3	2.00
4	2.50
5	3.00
6	3.50
7	4.00
8	4.50
9	5.00
10	5.50

In general, the cost C_n of traveling n miles is 1.00 (the cost of traveling the first mile) plus 0.50 times the number $(n - 1)$ of additional miles. That is,

$$C_n = 1 + 0.5(n - 1).$$

As examples,

$$C_1 = 1 + 0.5(1 - 1) = 1 + 0.5 \cdot 0 = 1,$$

$$C_5 = 1 + 0.5(5 - 1) = 1 + 0.5 \cdot 4 = 1 + 2 = 3.$$

A **sequence** is a list in which order is taken into account. In the previous example, the list of fares

$$1.00, \ 1.50, \ 2.00, \ 2.50, \ 3.00, \ . \ . \ .$$

is a sequence. Notice that order is indeed important. For example, if the first and fifth numbers were interchanged, a 1-mile fare would then cost $3.00—quite different from a $1.00 1-mile fare.

We next formally define a sequence and give several additional examples.

Definition 2.3.1.   A *sequence* is an ordered list of elements. If s is a sequence, we let s_1 denote the first element of the sequence, s_2 denote the second element of the sequence, and so on. In general, s_n denotes the *n*th element of the sequence. We call n the *index* of the sequence.

Example 2.3.2.   The ordered list

$$2, 4, 6, \ldots, 2n, \ldots$$

is a sequence. The first element of the sequence is 2, the second element of the sequence is 4, and so on. The *n*th element of the sequence is $2n$. If we let s denote this sequence, we have

$$s_1 = 2, \quad s_2 = 4, \quad s_3 = 6, \quad \ldots, \quad s_n = 2n, \quad \ldots.$$

Example 2.3.3.   The ordered list

$$a, a, b, a, b$$

is a sequence. The first element of the sequence is a, the second element of the sequence is a, and so on. If we let t denote this sequence, we have

$$t_1 = a, \quad t_2 = a, \quad t_3 = b, \quad t_4 = a, \quad t_5 = b.$$

Example 2.3.3 shows that a sequence (unlike a set) can have repetitions. A sequence may have an infinite number of elements (like the sequence of Example 2.3.2) or a finite number of elements (like the sequence of Example 2.3.3).

An alternative notation for the sequence s is $\{s_n\}$. Here s or $\{s_n\}$ denotes the entire sequence

$$s_1, s_2, s_3, \ldots.$$

We use the notation s_n to denote the single, *n*th element of the sequence s.

Example 2.3.4.   Define a sequence $\{t_n\}$ by the rule

$$t_n = n^2 - 1, \quad n \geq 1.$$

The first five terms of this sequence are

$$0, 3, 8, 15, 24.$$

The 55th term is

$$t_{55} = 55^2 - 1 = 3024.$$

Example 2.3.5.   Define a sequence u by the rule u_n is the nth letter in the word *digital*. Then $u_1 = d$, $u_2 = u_4 = i$, and $u_7 = l$. This sequence is a finite sequence.

Although in this book we often denote the first element of a sequence s as s_1, in general the first element may be indexed by any integer. For example, if v is a sequence whose first element is v_0, the elements of v would be

$$v_0, v_1, v_2, \ldots .$$

When we want to explicitly mention the initial index of an infinite sequence s, we write $\{s_n\}_{n=1}^{\infty}$. An infinite sequence v whose initial index is 0 is denoted $\{v_n\}_{n=0}^{\infty}$. A finite sequence x indexed from -1 to 4 is denoted $\{x_n\}_{n=-1}^{4}$.

Example 2.3.6.   If x is the sequence defined by

$$x_n = \frac{1}{2^n}, \qquad -1 \leq n \leq 4,$$

the elements of x are

$$2, 1, \tfrac{1}{2}, \tfrac{1}{4}, \tfrac{1}{8}, \tfrac{1}{16}.$$

Example 2.3.7 Factorial.   If n is a positive integer, we define **n factorial** to be

$$n! = n(n - 1)(n - 2) \cdots 1.$$

We also define $0! = 1$. For example,

$$1! = 1, \qquad 2! = 2 \cdot 1 = 2, \qquad 3! = 3 \cdot 2 \cdot 1 = 6.$$

If we define a sequence $\{c_n\}$ by the rule

$$c_n = n!, \qquad n \geq 0,$$

the first six terms of this sequence are

$$1, 1, 2, 6, 24, 120.$$

Two important ways to create new sequences from numerical sequences are by adding and multiplying the terms together. More formally, if $\{a_n\}$ is a sequence, we define

$$\sum_{i=1}^{n} a_i = a_1 + a_2 + \cdots + a_n,$$

$$\prod_{i=1}^{n} a_i = a_1 \cdot a_2 \cdot \cdots \cdot a_n.$$

(2.3.1)

Example 2.3.8. Let a be the sequence defined by $a_n = 2n$, $n \geq 1$. Then

$$\sum_{i=1}^{3} a_i = a_1 + a_2 + a_3 = 2 + 4 + 6 = 12,$$

$$\prod_{i=1}^{3} a_i = a_1 \cdot a_2 \cdot a_3 = 2 \cdot 4 \cdot 6 = 48.$$

The name of the subscript in (2.3.1) is irrelevant. For example,

$$\sum_{i=1}^{n} a_i = \sum_{j=1}^{n} a_j \quad \text{and} \quad \prod_{i=1}^{n} a_i = \prod_{x=1}^{n} a_x.$$

Example 2.3.9. Let a be the sequence defined by the rule $a_n = 2(-1)^n$, $n \geq 0$. Find a formula for the sequence s defined by

$$s_n = \sum_{i=0}^{n} a_i.$$

We find that

$$s_n = 2(-1)^0 + 2(-1)^1 + 2(-1)^2 + \cdots + 2(-1)^n$$

$$= 2 - 2 + 2 - \cdots \pm 2$$

$$= \begin{cases} 2 & \text{if } n \text{ is even} \\ 0 & \text{if } n \text{ is odd.} \end{cases}$$

Leonardo Fibonacci *(ca. 1170–1250)*

Leonardo Fibonacci was an Italian merchant and mathematician. After returning from the Orient in 1202, he wrote his most famous work *Liber Abaci*, which, in addition to containing what we now call the Fibonacci sequence, advocated the use of Hindu–Arabic numerals. This book was one of the main influences in bringing our modern number system to western Europe. Fibonacci signed much of his work Leonardo Bigollo. *Bigollo* translates as "traveler" or "blockhead." There is some evidence that Fibonacci enjoyed having his contemporaries consider him a blockhead for advocating the new number system. [*Photo courtesy of the Italian Cultural Center, New York*]

Sometimes sequences are defined recursively; that is, the nth element is defined in terms of certain of its predecessors.

Example 2.3.10 Fibonacci Sequence.   The Fibonacci sequence $\{f_n\}_{n=0}^{\infty}$ is defined by the rules

$$f_0 = f_1 = 1, \qquad f_n = f_{n-1} + f_{n-2}, \qquad n \geq 2. \quad (2.3.2)$$

The sequence originally arose in a puzzle about rabbits (see Exercises 24 and 25). We find

$$f_2 = f_1 + f_0 = 1 + 1 = 2,$$

$$f_3 = f_2 + f_1 = 2 + 1 = 3,$$

$$f_4 = f_3 + f_2 = 3 + 2 = 5.$$

The first 10 terms of the Fibonacci sequence are

$$1, 1, 2, 3, 5, 8, 13, 21, 34, 55.$$

In words, for $n \geq 2$, f_n is the sum of the two previous terms.

Fibonacci numbers occur frequently in nature. For example, the number of clockwise spirals and the number of counterclockwise spirals formed by the seeds of certain varieties of sunflowers are Fibonacci numbers. Figure 2.3.1 shows a hypothetical sunflower with 13 clockwise spirals and 8 counterclockwise spirals.

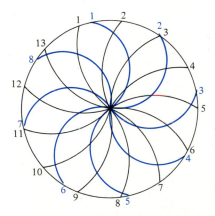

Figure 2.3.1

Definitions like (2.3.2) in which the nth element is defined in terms of its predecessors are known as **recurrence relations** and will be studied in detail in Chapter 7.

Example 2.3.11.   For the sequence h defined by

$$h_n = 3 + h_{\lfloor n/2 \rfloor}, \qquad n \geq 2; \qquad h_1 = 1,$$

find h_2, h_3, h_4, and h_5. We let $\lfloor x \rfloor$ denote the greatest integer less than or equal to x. For example, $\lfloor 2.67 \rfloor = 2$, $\lfloor 5/2 \rfloor = 2$, $\lfloor 3 \rfloor = 3$, $\lfloor -3.2 \rfloor = -4$. $\lfloor x \rfloor$ is called the **floor** of x.

We have

$$h_2 = 3 + h_{\lfloor 2/2 \rfloor} = 3 + h_1 = 3 + 1 = 4,$$

$$h_3 = 3 + h_{\lfloor 3/2 \rfloor} = 3 + h_1 = 3 + 1 = 4,$$

$$h_4 = 3 + h_{\lfloor 4/2 \rfloor} = 3 + h_2 = 3 + 4 = 7,$$

$$h_5 = 3 + h_{\lfloor 5/2 \rfloor} = 3 + h_2 = 3 + 4 = 7.$$

Example 2.3.12.   The sequence 2, 3, 5, 8, 12, . . . can be defined by the equations

$$a_n = a_{n-1} + n - 1, \qquad n \geq 2; \qquad a_1 = 2.$$

The first equation says, given the $(n - 1)$st term, to get the nth term, add $n - 1$. The last equation says that the first term of the sequence is 2.

Example 2.3.13.   Consider the sequence s defined by

$$s_n = 4 \cdot 2^n + 2 \cdot 3^n, \qquad n \geq 0. \qquad (2.3.3)$$

Find s_0, s_1, s_m, s_{n-1}, and s_{n-2} and show that $\{s_n\}$ satisfies the recurrence relation

$$s_n = 5s_{n-1} - 6s_{n-2}, \qquad n \geq 2. \qquad (2.3.4)$$

We find s_0 by replacing n by 0 in (2.3.3):

$$s_0 = 4 \cdot 2^0 + 2 \cdot 3^0 = 4 \cdot 1 + 2 \cdot 1 = 6.$$

We find s_1 by replacing n by 1 in (2.3.3):

$$s_1 = 4 \cdot 2^1 + 2 \cdot 3^1 = 4 \cdot 2 + 2 \cdot 3 = 14.$$

We find s_m by replacing n by m in (2.3.3):

$$s_m = 4 \cdot 2^m + 2 \cdot 3^m.$$

We find s_{n-1} by replacing n by $n-1$ in (2.3.3):

$$s_{n-1} = 4 \cdot 2^{n-1} + 2 \cdot 3^{n-1}. \qquad (2.3.5)$$

We find s_{n-2} by replacing n by $n-2$ in (2.3.3):

$$s_{n-2} = 4 \cdot 2^{n-2} + 2 \cdot 3^{n-2}. \qquad (2.3.6)$$

To verify (2.3.4), we will compute the right-hand side of equation (2.3.4) and show that we obtain s_n. Using the fact that

$$2^{n-2} = \frac{2^{n-1}}{2}, \qquad 3^{n-2} = \frac{3^{n-1}}{3}, \qquad 2^{n-1} = \frac{2^n}{2}, \qquad 3^{n-1} = \frac{3^n}{3},$$

and equations (2.3.5) and (2.3.6), we find that

$$
\begin{aligned}
5s_{n-1} - 6s_{n-2} &= 5(4 \cdot 2^{n-1} + 2 \cdot 3^{n-1}) \\
&\quad - 6(4 \cdot 2^{n-2} + 2 \cdot 3^{n-2}) \\
&= 20 \cdot 2^{n-1} + 10 \cdot 3^{n-1} \\
&\quad - 24 \cdot 2^{n-2} - 12 \cdot 3^{n-2} \\
&= 20 \cdot 2^{n-1} - 12 \cdot 2^{n-1} \\
&\quad + 10 \cdot 3^{n-1} - 4 \cdot 3^{n-1} \\
&= 8 \cdot 2^{n-1} + 6 \cdot 3^{n-1} \\
&= 4 \cdot 2^n + 2 \cdot 3^n = s_n.
\end{aligned}
$$

In certain contexts, a finite sequence is called a **string**.

Definition 2.3.14. A *string over X* is a finite sequence of elements from X.

Example 2.3.15. Let $X = \{a, b, c\}$. If we let

$$\beta_1 = b, \quad \beta_2 = a, \quad \beta_3 = a, \quad \beta_4 = c,$$

we obtain a string over X. This string is written *baac*.

Since a string is a sequence, order is taken into account. For example, the string *baac* is different from the string *acab*.

Repetitions in a string can be specified by superscripts. For example, the string *bbaaac* may be written b^2a^3c.

The string with no elements is called the **null string** and is denoted λ. We let X^* denote the set of all strings over X, including the null string.

Example 2.3.16.   Let $X = \{a, b\}$. Some elements in X^* are

$$\lambda, \; a, \; b, \; abab, \; a^{20}b^5ab.$$

The **length** of a string α is the number of elements in α. The length of α is denoted $|\alpha|$.

Example 2.3.17.   If $\alpha = aabab$ and $\beta = a^3b^4a^{32}$, then

$$|\alpha| = 5 \qquad \text{and} \qquad |\beta| = 39.$$

If α and β are two strings, the string consisting of α followed by β, written $\alpha\beta$, is called the **concatenation** of α and β.

Example 2.3.18.   If $\gamma = aab$ and $\theta = cabd$, then

$$\gamma\theta = aabcabd,$$
$$\theta\gamma = cabdaab,$$
$$\gamma\lambda = \gamma = aab,$$
$$\lambda\gamma = \gamma = aab.$$

---

†COMPUTER NOTES FOR SECTION 2.3

In computer science, an **array** can be used to represent a finite sequence. For example, the sequence t of Example 2.3.3 can be represented as an array A indexed from 1 to n and stored in memory in consecutive cells as shown in Figure 2.3.2. The ith element of the sequence is stored in the ith cell. In Pascal, element i of the array is referenced by writing A[i]. [Notations such as A(i) and A⟨i⟩ are used by some languages.] For example, $t_1 = a$ is stored in the first cell and, in Pascal, is referenced

---

†This section can be omitted without loss of continuity.

A

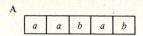

Figure 2.3.2

as A[1]. Many higher-level languages allow the index of an array to begin with any integer.

EXERCISES

1H. Answer (a)–(c) for the sequence s defined by

$$c, d, d, c, d, c.$$

(a) Find s_1.   (b) Find s_4.
(c) Write s as a string.

2. Answer (a)–(i) for the sequence t defined by

$$t_n = 2n - 1, \qquad n \geq 1.$$

(a) Find t_3.   (b) Find t_7.
(c) Find t_{100}.   (d) Find t_{2077}.

(e) Find $\displaystyle\sum_{i=1}^{3} t_i$.   (f) Find $\displaystyle\sum_{i=3}^{7} t_i$.

(g) Find $\displaystyle\prod_{i=1}^{3} t_i$.   (h) Find $\displaystyle\prod_{i=3}^{6} t_i$.

(i) Find a formula that represents this sequence as a sequence whose first index is 0.

3. Answer (a)–(d) for the sequence v defined by

$$v_n = n! + 2, \qquad n \geq 1.$$

(a) Find v_3.   (b) Find v_4.

(c) Find $\displaystyle\sum_{i=1}^{4} v_i$.   (d) Find $\displaystyle\sum_{i=3}^{3} v_i$.

4H. Compute the given quantity using the sequence a defined by

$$a_n = n^2 - 3n + 3.$$

(a) $\displaystyle\sum_{i=1}^{4} a_i$ (b) $\displaystyle\sum_{j=3}^{5} a_j$

(c) $\displaystyle\sum_{i=4}^{4} a_i$ (d) $\displaystyle\sum_{k=1}^{6} a_k$

(e) $\displaystyle\prod_{i=1}^{2} a_i$
(f) $\displaystyle\prod_{i=1}^{3} a_i$

(g) $\displaystyle\prod_{n=2}^{3} a_n$
(h) $\displaystyle\prod_{x=3}^{4} a_x$

5. Answer (a)–(d) for the sequence b defined by $b_n = n(-1)^n$.

(a) Find $\displaystyle\sum_{i=1}^{4} b_i$.
(b) Find $\displaystyle\sum_{i=1}^{10} b_i$.

(c) Find a formula for the sequence c defined by

$$c_n = \sum_{i=1}^{n} b_i.$$

(d) Find a formula for the sequence d defined by

$$d_n = \prod_{i=1}^{n} b_i.$$

6. Answer (a)–(d) for the sequence Ω defined by $\Omega_n = 3$ for all n.

(a) Find $\displaystyle\sum_{i=1}^{3} \Omega_i$.
(b) Find $\displaystyle\sum_{i=1}^{10} \Omega_i$.

(c) Find a formula for the sequence c defined by

$$c_n = \sum_{i=1}^{n} \Omega_i.$$

(d) Find a formula for the sequence d defined by

$$d_n = \prod_{i=1}^{n} \Omega_i.$$

7H. Answer (a)–(c) for the sequence x defined by

$$x_1 = 2 \qquad x_n = 3 + x_{n-1}, \qquad n \geq 2.$$

(a) Find $\displaystyle\sum_{i=1}^{3} x_i$.
(b) Find $\displaystyle\sum_{i=1}^{10} x_i$.

(c) Find a formula for the sequence c defined by

$$c_n = \sum_{i=1}^{n} x_i.$$

8. Find the term using the definitions

$$a_n = 1 + a_{\lfloor n/2 \rfloor} + a_{\lfloor (n+1)/2 \rfloor}, \qquad n \geq 2; \qquad a_1 = 2.$$

(a) a_2 (b) a_3
(c) a_4 (d) a_5
(e) a_6 (f) a_7

9. Answer (a)–(d) for the sequence w defined by

$$w_n = \frac{1}{n} - \frac{1}{n+1}, \qquad n \geq 1.$$

(a) Find $\displaystyle\sum_{i=1}^{3} w_i$.   (b) Find $\displaystyle\sum_{i=1}^{10} w_i$.

(c) Find a formula for the sequence c defined by

$$c_n = \sum_{i=1}^{n} w_i.$$

(d) Find a formula for the sequence d defined by

$$d_n = \prod_{i=1}^{n} w_i.$$

10H. Let u be the sequence defined by

$$u_1 = 3 \qquad u_n = 3 + u_{n-1}, \qquad n \geq 2.$$

Find a formula for the sequence d defined by

$$d_n = \prod_{i=1}^{n} u_i.$$

11. Answer (a)–(d) using the sequences y and z defined by

$$y_n = 2^n - 1, \qquad z_n = n(n-1).$$

(a) Find $\displaystyle\left(\sum_{i=1}^{3} y_i \right)\left(\sum_{i=1}^{3} z_i \right)$.   (b) Find $\displaystyle\left(\sum_{i=1}^{5} y_i \right)\left(\sum_{i=1}^{4} z_i \right)$.

(c) Find $\displaystyle\sum_{i=1}^{3} y_i z_i$.   (d) Find $\displaystyle\left(\sum_{i=3}^{4} y_i \right)\left(\prod_{i=2}^{4} z_i \right)$.

12. Find a formula for a sequence that begins with the terms given.
(a) $\frac{3}{1}, \frac{3}{2}, \frac{3}{3}, \frac{3}{4}, \ldots$

(b) $-2, -3, -4, -5, \ldots$

(c) $\frac{1}{2}, \frac{2}{3}, \frac{3}{4}, \frac{4}{5}, \ldots$

(d) $-\frac{1}{2}, 1, -\frac{3}{2}, 2, -\frac{5}{2}, 3, \ldots$

*(e)† $3, \frac{7}{2}, \frac{11}{3}, \frac{15}{4}, \ldots$

13H. Find a formula similar to the formula of Example 2.3.12 for a sequence that begins with the terms given.

(a) $2, 3, 5, 9, 17, \ldots$

(b) $21, 20, 18, 15, 11, \ldots$

14.   Answer (a)–(h) for the sequence r defined by

$$r_n = 3 \cdot 2^n - 4 \cdot 5^n, \qquad n \geq 0.$$

(a) Find r_0.   (b) Find r_1.

(c) Find r_2.   (d) Find r_3.

(e) Find a formula for r_p.

(f) Find a formula for r_{n-1}.

(g) Find a formula for r_{n-2}.

(h) Show that $\{r_n\}$ satisfies the recurrence relation

$$r_n = 7r_{n-1} - 10r_{n-2}.$$

15.   Answer (a)–(h) for the sequence z defined by

$$z_n = (2 + n)3^n, \qquad n \geq 0.$$

(a) Find z_0.   (b) Find z_1.

(c) Find z_2.   (d) Find z_3.

(e) Find a formula for z_i.

(f) Find a formula for z_{n-1}.

(g) Find a formula for z_{n-2}.

(h) Show that $\{z_n\}$ satisfies the recurrence relation

$$z_n = 6z_{n-1} - 9z_{n-2}.$$

16H. Find b_i, $i = 1, \ldots, 6$, where

$$b_n = 2[1 + (n-1)(n-2)(n-3)(n-4)(n-5)]$$
$$+ \frac{(n-1)n}{2}.$$

17.   Compute the given quantity using the strings

$$\alpha = baab, \qquad \beta = caaba, \qquad \gamma = bbab.$$

†A starred exercise indicates a problem of above-average difficulty.

(a) $\alpha\beta$ (b) $\beta\alpha$ (c) $\alpha\alpha$ (d) $\beta\beta$
(e) $|\alpha\beta|$ (f) $|\beta\alpha|$ (g) $|\alpha\alpha|$ (h) $|\beta\beta|$
(i) $\alpha\lambda$ (j) $\lambda\beta$ (k) $\alpha\beta\gamma$ (l) $\beta\beta\gamma\alpha$

18H. List all strings over $X = \{0, 1\}$ of length 2.

19. List all strings over $X = \{0, 1\}$ of length 2 or less.

20. List all strings over $X = \{0, 1\}$ of length 3.

21H. List all strings over $X = \{0, 1\}$ of length 3 or less.

22. A string s is a **substring** of t if there are strings u and v with $t = usv$. Find all substrings of the string $babc$.

23. Find all substrings of the string $aabaabb$.

24H. Suppose that at the beginning of the year, there is one pair of rabbits and every month each pair produces a new pair that becomes productive after one month. Suppose further that no deaths occur. Let a_n denote the number of pairs of rabbits at the end of the nth month. Show that $a_0 = a_1 = 1$ and $a_n - a_{n-1} = a_{n-2}$. Explain why $a_n = f_n$, $n = 0, 1, \ldots$, where f denotes the Fibonacci sequence.

25. Fibonacci's original question was: Under the conditions of Exercise 24, how many pairs of rabbits are there after one year? Answer Fibonacci's question.

2.4 Mathematical Induction

Mathematical induction is a proof technique that is frequently used in both mathematics and computer science. To show how mathematical induction can be used to prove a statement about a sequence, let S_n denote the sum of the first n positive integers

$$S_n = 1 + 2 + 3 + \cdots + n. \qquad (2.4.1)$$

Suppose that someone claims that

$$S_n = \frac{n(n + 1)}{2}, \qquad \text{for } n = 1, 2, \ldots. \qquad (2.4.2)$$

A sequence of statements is really being made, namely

$$S_1 = \frac{1(2)}{2} = 1$$

$$S_2 = \frac{2(3)}{2} = 3$$

$$S_3 = \frac{3(4)}{2} = 6$$

$$\cdot$$
$$\cdot \qquad\qquad\qquad\qquad\qquad\qquad (2.4.3)$$
$$\cdot$$

$$S_n = \frac{n(n+1)}{2}$$

$$S_{n+1} = \frac{(n+1)(n+2)}{2}$$

$$\cdot$$
$$\cdot$$
$$\cdot$$

Our goal is to prove that each equation (2.4.3) is true. We can immediately verify that the first equation is true. Now suppose we can show that

for any equation E, if equation E is true,
then the equation immediately following E is also true.

It will then follow that all of the equations are true: for since the first equation is true, the equation following it, namely the second equation, is also true; now, since the second equation is true, the equation following it, namely the third equation, is also true; but since the third equation is true, the equation following it, namely the fourth equation, is also true; we can continue the argument indefinitely and thus conclude that all of the equations are true.

We have already observed that the first equation is true. The remaining part of the proof is to show that if any equation is true, say the nth, the following equation, namely the $(n + 1)$st, is also true. If the nth equation is true, then

$$S_n = \frac{n(n+1)}{2}. \qquad\qquad (2.4.4)$$

We must show that the $(n + 1)$st equation

$$S_{n+1} = \frac{(n + 1)(n + 2)}{2}$$

is true. According to the definition (2.4.1) of the sequence S,

$$S_{n+1} = 1 + 2 + \cdots + n + (n + 1).$$

We note that S_n is contained within S_{n+1} in the sense that

$$S_{n+1} = 1 + 2 + \cdots + n + (n + 1) \qquad (2.4.5)$$
$$= S_n + (n + 1).$$

Because of (2.4.4) and (2.4.5), we have

$$S_{n+1} = S_n + n + 1 = \frac{n(n + 1)}{2} + n + 1 = \frac{(n + 1)(n + 2)}{2}.$$

We have proved that the $(n + 1)$st equation is true. Since we have shown that the first equation is true and, if any equation E is true, the equation immediately following E is also true, it follows that all of the equations are true. We have proved formula (2.4.2).

Our proof using mathematical induction consisted of two steps. First, we verified that the statement corresponding to $n = 1$ was true. Second, we *assumed* that the nth statement was true, and then *proved* that the $(n + 1)$st statement was also true. In proving the $(n + 1)$st statement, we were permitted to make use of the nth statement; indeed, the trick in constructing a proof using mathematical induction is to relate the nth statement to the $(n + 1)$st statement.

We next formally state the Principle of Mathematical Induction.

Principle of Mathematical Induction.   *Suppose that for each positive integer n we have a statement S(n) which is either true or false. Suppose that*

$$S(1) \text{ is true}; \qquad (2.4.6)$$

$$\text{if } S(n) \text{ is true, then } S(n + 1) \text{ is true}. \qquad (2.4.7)$$

Then S(n) is true for every positive integer n.

Condition (2.4.6) is sometimes called the **basis step** and condition (2.4.7) is sometimes called the **inductive step**. Hereafter, "induction" will mean "mathematical induction."

At this point, we illustrate the Principle of Mathematical Induction with several more examples.

Example 2.4.1.   Use induction to show that

$$\frac{1}{1 \cdot 2} + \frac{1}{2 \cdot 3} + \frac{1}{3 \cdot 4} + \cdots + \frac{1}{n(n+1)} = \frac{n}{n+1} \quad (2.4.8)$$

for $n = 1, 2, \ldots$.

BASIS STEP [Condition (2.4.6)]. We must show that (2.4.8) is true if $n = 1$. Since

$$\frac{1}{1 \cdot 2} = \frac{1}{1 + 1},$$

(2.4.8) is true if $n = 1$.

INDUCTIVE STEP [Condition (2.4.7)]. We must show that if (2.4.8) is true for n, then (2.4.8) with n everywhere replaced by $n + 1$ is also true. That is, we must show that

$$\frac{1}{1 \cdot 2} + \frac{1}{2 \cdot 3} + \frac{1}{3 \cdot 4} + \cdots + \frac{1}{n(n+1)} \quad (2.4.9)$$
$$+ \frac{1}{(n+1)(n+2)} = \frac{n+1}{n+2}.$$

But,

$$\frac{1}{1 \cdot 2} + \frac{1}{2 \cdot 3} + \frac{1}{3 \cdot 4} + \cdots + \frac{1}{n(n+1)}$$
$$+ \frac{1}{(n+1)(n+2)} = \frac{n}{n+1} + \frac{1}{(n+1)(n+2)}$$

since we assume that (2.4.8) is true for n. A little algebra shows that

$$\frac{n}{n+1} + \frac{1}{(n+1)(n+2)} = \frac{n+1}{n+2},$$

and so we have verified (2.4.9) and completed the inductive step.

Since the basis step and the inductive step have been verified, the Principle of Mathematical Induction tells us that (2.4.8) is true for every positive integer n.

Example 2.4.2. Use induction to show that

$$n! \geq 2^{n-1} \qquad \text{for } n = 1, 2, \ldots . \qquad (2.4.10)$$

BASIS STEP. We must show that (2.4.10) is true if $n = 1$. This is easily accomplished, since

$$1! = 1 \geq 1 = 2^{1-1}.$$

INDUCTIVE STEP. We must show that if $n! \geq 2^{n-1}$, then

$$(n + 1)! \geq 2^n. \qquad (2.4.11)$$

We can relate (2.4.10) and (2.4.11) by observing that

$$(n + 1)! = (n + 1)(n!).$$

Now

$$\begin{aligned}
(n + 1)! &= (n + 1)(n!) \\
&\geq (n + 1)2^{n-1} \qquad \text{by (2.4.10)} \\
&\geq 2 \cdot 2^{n-1} \qquad \text{since } n + 1 \geq 2 \\
&= 2^n.
\end{aligned}$$

Therefore, (2.4.11) is true. We have completed the inductive step.

Since the basis step and the inductive step have been verified, the Principle of Mathematical Induction tells us that (2.4.10) is true for every positive integer n.

Mathematical induction can be used to prove that a sequence S of statements is true even if the first statement is indexed with an integer different from 1. If the index of the first statement is k, to verify the basis step we must prove that $S(k)$ is true. The inductive step is unchanged.

Example 2.4.3 Geometric Sum. Use induction to show that if $r \neq 1$,

$$a + ar^1 + ar^2 + \cdots + ar^n = \frac{a(r^{n+1} - 1)}{r - 1} \qquad (2.4.12)$$

for $n = 0, 1, \ldots .$

The sum on the left is called the **geometric sum**.

BASIS STEP. The basis step, which in this case is obtained by setting $n = 0$, is

$$a = \frac{a(r^1 - 1)}{r - 1},$$

which is true.

INDUCTIVE STEP. Assume that statement (2.4.12) is true for n. Now

$$a + ar^1 + ar^2 + \cdots + ar^n + ar^{n+1}$$

$$= \frac{a(r^{n+1} - 1)}{r - 1} + ar^{n+1}$$

$$= \frac{a(r^{n+1} - 1)}{r - 1} + \frac{ar^{n+1}(r - 1)}{r - 1}$$

$$= \frac{a(r^{n+2} - 1)}{r - 1}.$$

Since the modified basis step and the inductive step have been verified, the Principle of Mathematical Induction tells us that (2.4.12) is true for $n = 0, 1, \ldots$.

As an example of the use of the geometric sum, if we take $a = 1$ and $r = 2$ in (2.4.12), we obtain the formula

$$1 + 2 + 2^2 + 2^3 + \cdots + 2^n = \frac{2^{n+1} - 1}{2 - 1} = 2^{n+1} - 1.$$

Example 2.4.4.   Show that

$$\sum_{k=0}^{n} f_k = f_{n+2} - 1, \qquad n \geq 0,$$

where f denotes the Fibonacci sequence.

BASIS STEP. $(n = 0)$

$$\sum_{k=0}^{0} f_k = f_0 = 1 = 2 - 1 = f_2 - 1$$

INDUCTIVE STEP. Assume that the formula holds for n. Then

$$\sum_{k=0}^{n+1} f_k = \sum_{k=0}^{n} f_k + f_{n+1}$$

$$= (f_{n+2} - 1) + f_{n+1} \qquad \text{by the inductive assumption}$$

$$= (f_{n+1} + f_{n+2}) - 1$$

$$= f_{n+3} - 1 \qquad \text{since } f_{n+1} + f_{n+2} = f_{n+3}.$$

Since the basis step and the inductive step have been verified, by the Principle of Mathematical Induction, the formula holds for $n \geq 0$.

We close by using induction to prove that the power set of a set with n elements has 2^n elements.

Theorem 2.4.5. *If $|X| = n$, then*

$$|\mathcal{P}(X)| = 2^n. \qquad (2.4.13)$$

PROOF. The proof is by induction on n.

BASIS STEP. If $n = 0$, X is the empty set. The only subset of the empty set is the empty set itself; thus

$$|\mathcal{P}(X)| = 1 = 2^0 = 2^n.$$

Thus (2.4.13) is true for $n = 0$.

INDUCTIVE STEP. Assume that (2.4.13) holds for n. Let X be a set with $|X| = n + 1$. Choose $x \in X$. We partition $\mathcal{P}(X)$ into two classes. The first class consists of those subsets of X that include x, and the second class consists of those subsets of X that do not include x. If we list the subsets of X that do not include x,

$$X_1, X_2, \ldots, X_k,$$

then

$$X_1 \cup \{x\}, X_2 \cup \{x\}, \ldots, X_k \cup \{x\}$$

is a list of the subsets of X that do include x. Thus the number of subsets of X that include x is equal to the number of subsets of X that do not include x. If $Y = X - \{x\}$, $\mathcal{P}(Y)$ consists precisely of those subsets of X that do not include x. Thus $|\mathcal{P}(X)| = 2|\mathcal{P}(Y)|$. Since $|Y| = n$, by the inductive assumption, $|\mathcal{P}(Y)| = 2^n$. Therefore,

$$|\mathcal{P}(X)| = 2|\mathcal{P}(Y)| = 2 \cdot 2^n = 2^{n+1}.$$

Thus (2.4.13) holds for $n + 1$ and the inductive step is complete. By the Principle of Mathematical Induction, (2.4.13) holds for all $n \geq 0$. ■

In Section 6.1 (see Example 6.1.5), we give another proof of Theorem 2.4.5.

EXERCISES

In Exercises 1–12, using induction, verify that the given equation is true for every positive integer n.

1H. $1 + 3 + 5 + \cdots + 2n - 1 = n^2$

2. $2 + 5 + 8 + \cdots + (3n - 1) = \dfrac{n(3n + 1)}{2}$

3. $1 \cdot 2 + 2 \cdot 3 + 3 \cdot 4 + \cdots + n(n + 1) = \dfrac{n(n + 1)(n + 2)}{3}$

4H. $1(1!) + 2(2!) + \cdots + n(n!) = (n + 1)! - 1$

5. $1^2 + 2^2 + 3^2 + \cdots + n^2 = \dfrac{n(n + 1)(2n + 1)}{6}$

6. $1^2 - 2^2 + 3^2 - \cdots + (-1)^{n+1}n^2 = \dfrac{(-1)^{n+1}n(n + 1)}{2}$

7H. $1^3 + 2^3 + 3^3 + \cdots + n^3 = \left[\dfrac{n(n + 1)}{2}\right]^2$

8. $\dfrac{1}{1 \cdot 3} + \dfrac{1}{3 \cdot 5} + \dfrac{1}{5 \cdot 7} + \cdots + \dfrac{1}{(2n - 1)(2n + 1)} = \dfrac{n}{2n + 1}$

9. $\dfrac{1}{2 \cdot 4} + \dfrac{1 \cdot 3}{2 \cdot 4 \cdot 6} + \dfrac{1 \cdot 3 \cdot 5}{2 \cdot 4 \cdot 6 \cdot 8} + \cdots$

$\qquad + \dfrac{1 \cdot 3 \cdot 5 \cdot \cdots \cdot (2n - 1)}{2 \cdot 4 \cdot 6 \cdot \cdots \cdot (2n + 2)}$

$\qquad = \dfrac{1}{2} - \dfrac{1 \cdot 3 \cdot 5 \cdot \cdots \cdot (2n + 1)}{2 \cdot 4 \cdot 6 \cdot \cdots \cdot (2n + 2)}$

10H. $\dfrac{1}{2^2 - 1} + \dfrac{1}{3^2 - 1} + \cdots + \dfrac{1}{(n + 1)^2 - 1}$

$$= \frac{3}{4} - \frac{1}{2(n + 1)} - \frac{1}{2(n + 2)}$$

11. $\quad 3 + 3^2 + 3^3 + \cdots + 3^n = \dfrac{3^{n+1} - 3}{2}$

12. $\quad 1 + r^1 + r^2 + \cdots + r^n = \dfrac{r^{n+1} - 1}{r - 1}, \quad r \neq 1$

In Exercises 13–16, using induction, verify the inequality.

13H. $\dfrac{1}{2n} \leq \dfrac{1 \cdot 3 \cdot 5 \cdot \cdots \cdot (2n - 1)}{2 \cdot 4 \cdot 6 \cdot \cdots \cdot (2n)}, \quad n = 1, 2, \ldots$

14. $\quad 2n + 1 \leq 2^n, \quad n = 3, 4, \ldots$

*15. $\quad 2^n \geq n^2, \quad n = 4, 5, \ldots$

16H. $(1 + x)^n \geq 1 + nx \quad$ for $x \geq -1$ and $n = 1, 2, \ldots$

17. Use induction to show that if $X_1, \ldots, X_n$ and X are sets, then

$$X \cap (X_1 \cup X_2 \cup \cdots \cup X_n)$$
$$= (X \cap X_1) \cup (X \cap X_2) \cup \cdots \cup (X \cap X_n).$$

18. Use induction to show that if $X_1, \ldots, X_n$ are sets, then

$$\overline{X_1 \cap X_2 \cap \cdots \cap X_n} = \overline{X}_1 \cup \overline{X}_2$$
$$\cup \cdots \cup \overline{X}_n.$$

19H. Use induction to show that if $X_1, \ldots, X_n$ are sets, then

$$\overline{X_1 \cup X_2 \cup \cdots \cup X_n} = \overline{X}_1 \cap \overline{X}_2 \cap \cdots \cap \overline{X}_n.$$

Exercises 18 and 19 generalize De Morgan's laws for sets (Theorem 2.2.3k).

20. Show that $f_n^2 = f_{n-1}f_{n+1} + (-1)^n, n \geq 1$, where f denotes the Fibonacci sequence.

21. Show that $\Sigma_{k=0}^{n} f_k^2 = f_n f_{n+1}, k \geq 0$, where f denotes the Fibonacci sequence.

22H. Suppose that $S_n = (n + 2)(n - 1)$ is (incorrectly) proposed as a formula for

$$2 + 4 + \cdots + 2n.$$

Show that the inductive step is satisfied but that the basis step fails.

23. What is wrong with the following argument, which allegedly shows that any two positive integers are equal?

We use induction on n to "prove" that if a and b are positive integers and $n = \max\{a, b\}$, then $a = b$.

BASIS STEP ($n = 1$). If a and b are positive integers and $1 = \max\{a, b\}$, we must have $a = b$. (In this case, $a = b = 1$.)

INDUCTIVE STEP. Assume that if a' and b' are positive integers and $n = \max\{a', b'\}$, then $a' = b'$. Suppose that a and b are positive integers and that $n + 1 = \max\{a, b\}$. Now $n = \max\{a - 1, b - 1\}$. By the inductive hypothesis, $a - 1 = b - 1$. But this implies that $a = b$.

Since we have verified the basis step and the inductive step, by the Principle of Mathematical Induction, any two positive integers are equal!

2.5 Functions

When we travel, the distance we cover depends on the time. For example, if we travel for t hours at a constant speed of 55 miles per hour, from the distance = rate $\times$ time formula, we find that we will travel $55t$ miles. Formally,

$$D = 55t, \tag{2.5.1}$$

where t is the time and D is the distance traveled.

Equation (2.5.1) defines a **function**. A function assigns each member of some set a member of some other set. The function defined by (2.5.1) assigns every nonnegative real number t the value $55t$. For example, the number $t = 1$ is assigned the value 55; the number $t = 3.45$ is assigned the value 189.75; and so on. We can represent these assignments as ordered pairs: (1, 55), (3.45, 189.75). Formally, we *define* a function to be a particular kind of set of ordered pairs.

Definition 2.5.1.   Let X and Y be sets. A *function* from X to Y is a subset of the Cartesian product $X \times Y$ having the property that for each $x \in X$, there is exactly one $y \in Y$ with $(x, y) \in f$. We sometimes denote a function f from X to Y as $f: X \rightarrow Y$.

The set X is called the *domain* of f. The set

$$\{y \mid (x, y) \in f\}$$

is called the *range* of f.

If f is a function from X to Y, f consists of ordered pairs of the form (x, y), $x \in X$, $y \in Y$. If (x, y) is in f, the interpretation is that the element x is assigned the value y. Definition 2.5.1 requires *every* element x in X to be assigned a value y in Y and requires the value y assigned to x to be *unique*.

Example 2.5.2. The domain of the function defined by (2.5.1) is the set of all non-negative real numbers. The range is also equal to the set of all non-negative real numbers.

The next example shows that the domain and range are not always equal.

Example 2.5.3. The set

$$f = \{(1, a), (2, b), (3, a)\}$$

is a function from $X = \{1, 2, 3\}$ to $Y = \{a, b, c\}$. Each element of X is assigned a unique value in Y: 1 is assigned the unique value a; 2 is assigned the unique value b; and 3 is assigned the unique value a. We can depict the situation as shown in Figure 2.5.1, where an arrow from j to x means that we assign the integer j the letter x. We call a picture like Figure 2.5.1 an **arrow diagram**. Notice that Defi-

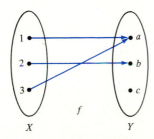

Figure 2.5.1

nition 2.5.1 allows us to reuse elements in Y. For this function, the element a in Y is used twice. Further, Definition 2.5.1 does *not* require us to use all the elements in Y. For this function, the element c in Y is not assigned to any element in X. The domain of f is X and the range of f is the set $\{a, b\}$.

A sequence is a special type of function. A sequence whose smallest index is 1 is a function whose domain is either the set of all positive integers or a set of the form $\{1, 2, \ldots, n\}$. The sequence of Example 2.3.3 has domain $\{1, 2, 3, 4, 5\}$. The sequence of Example 2.3.4 has domain equal to the set of all positive integers.

Example 2.5.4.   The set

$$\{(3, c), (5, a), (6, b)\}$$

is not a function from $X = \{1, 2, 3, 4, 5, 6\}$ to $Y = \{a, b, c\}$ because some elements in X (namely 1, 2, and 4) have not been assigned elements in Y.

Example 2.5.5.   The set

$$\{(\alpha, 5), (\beta, 2), (\Gamma, 2), (\beta, 1)\}$$

is not a function from $X = \{\alpha, \beta, \Gamma\}$ to $Y = \{1, 2, 3, 4, 5\}$ because β is not assigned a *unique* element in Y. (β is assigned the values 2 *and* 1.)

Given a function f from X to Y, according to Definition 2.5.1, for each element x in X, there is exactly one y in Y with (x, y) in f. This unique value y is denoted $f(x)$. In other words, $y = f(x)$ is another way to write $(x, y) \in f$.

Example 2.5.6.   For the function f of Example 2.5.3, we may write

$$f(1) = a, \qquad f(2) = b, \qquad f(3) = a.$$

Example 2.5.7.   For the function D defined by (2.5.1), we have

$$D(1) = 55, \qquad D(3.45) = 189.75.$$

The next example shows how we sometimes use the $f(x)$ notation to define a function.

Example 2.5.8. Let f be the function defined by the rule

$$f(x) = x^2.$$

For example,

$$f(2) = 4, \qquad f(-3.5) = 12.25, \qquad f(0) = 0.$$

Although we frequently find functions defined in this way, the definition is incomplete since the domain is not specified. If we are told that the domain is the set of all real numbers, in ordered pair notation, we would have

$$f = \{(x, x^2) \mid x \text{ is a real number}\}.$$

The range of f is the set of all nonnegative real numbers.

Example 2.5.9. Many hand-held electronic calculators have a $1/x$ key. If you enter a number and hit the $1/x$ key, (an approximation to) the reciprocal of the number entered is displayed. This function can be defined by the rule

$$R(x) = \frac{1}{x}.$$

The domain is the set of all numbers that can be entered into the calculator and whose reciprocals can be computed and displayed by the calculator. The range is the set of all the reciprocals that can be computed and displayed. Notice that by the nature of the calculator, the domain and range are finite sets.

Another way to visualize a function is to draw its graph. The **graph of a function** f whose domain and range are subsets of the real numbers is obtained by plotting points in the plane that correspond to the elements in f. The domain is contained in the horizontal axis and the range is contained in the vertical axis.

Example 2.5.10. The graph of the function of Example 2.5.8 is shown in Figure 2.5.2 and the graph of the function (2.5.1) is shown in Figure 2.5.3.

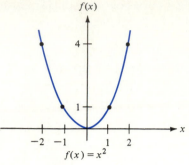

Figure 2.5.2   **Figure 2.5.3**

We note that a set S of points in the plane defines a function precisely when each vertical line intersects at most one point of S. If some vertical line contains two or more points of some set, the domain point does not assign a *unique* range point and the set does not define a function (see Figure 2.5.4).

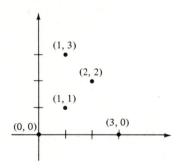

Figure 2.5.4

We next define the **floor** and **ceiling** of a real number.

Definition 2.5.11.   The *floor* of x, denoted $\lfloor x \rfloor$, is the greatest integer less than or equal to x. The *ceiling* of x, denoted $\lceil x \rceil$, is the least integer greater than or equal to x.

Example 2.5.12

$$\lfloor 8.3 \rfloor = 8, \quad \lceil 9.1 \rceil = 10, \quad \lfloor -8.7 \rfloor = -9,$$
$$\lceil -11.3 \rceil = -11, \quad \lfloor 6 \rfloor = 6, \quad \lceil -8 \rceil = -8$$

The floor of x "rounds x down" while the ceiling of x "rounds x up." We will use the floor and ceiling functions throughout the book.

Example 2.5.13. Figure 2.5.5 shows the graphs of the floor and ceiling functions. A bracket, [or], indicates that the point is to be included in the graph and a parenthesis, (or), indicates that the point is to be excluded from the graph.

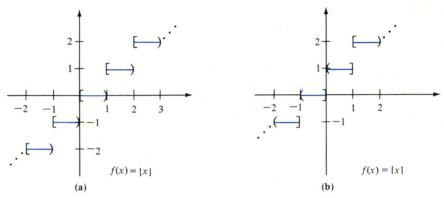

$f(x) = \lfloor x \rfloor$ $f(x) = \lceil x \rceil$

(a)   (b)

Figure 2.5.5

Example 2.5.14. In 1986, the U.S. first-class postage rate for up to 12 ounces was 22 cents for the first ounce or fraction thereof and 17 cents for each additional ounce or fraction thereof. The postage $P(w)$ as a function of weight w is given by the equation

$$P(w) = 22 + 17\lceil w - 1 \rceil, \qquad 12 \geq w > 0.$$

The expression $\lceil w - 1 \rceil$ counts the number of additional ounces beyond 1 with a fraction counting as one additional ounce. As examples,

$$P(3.7) = 22 + 17\lceil 3.7 - 1 \rceil = 22 + 17\lceil 2.7 \rceil$$
$$= 22 + 17 \cdot 3 = 73,$$

$$P(2) = 22 + 17\lceil 2 - 1 \rceil = 22 + 17\lceil 1 \rceil$$
$$= 22 + 17 \cdot 1 = 39.$$

The graph of the function P is shown in Figure 2.5.6.

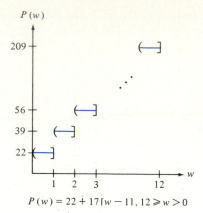

$$P(w) = 22 + 17\lceil w - 1 \rceil, 12 \geqslant w > 0$$

Figure 2.5.6

Functions involving the **modulus operator** play an important role in mathematics and computer science.

Definition 2.5.15.   If x is a nonnegative integer and y is a positive integer, we define $x\ mod\ y$ to be the remainder when x is divided by y.

Example 2.5.16

$$6\ mod\ 2 = 0, \quad 5\ mod\ 1 = 0, \quad 8\ mod\ 12 = 8,$$
$$199{,}673\ mod\ 2 = 1$$

Example 2.5.17.   What day of the week will it be 365 days from Wednesday?
Seven days after Wednesday it is Wednesday again; 14 days after Wednesday it is Wednesday again; and in general, if n is a positive integer, after $7n$ days it is Wednesday again. Thus we need to remove as many 7's as possible from 365 and see how many days are left. But this is simply $365\ mod\ 7 = 1$. Thus 365 days from Wednesday, it will be one day later, namely Thursday. This explains why, except for leap year when an extra day is added to February, the identical month and date in consecutive years moves forward one day of the week.

†COMPUTER NOTES FOR SECTION 2.5

In a higher-level programming language, certain functions are provided to the programmer. For example, in Pascal, the function $sqr(x)$ computes the square of x as in Example 2.5.8. The programmer can also write new functions. For example, a Pascal function to compute the distance given the rate and time might begin

$$\text{\textbf{function} } Distance \text{ } (Rate: real; Time: real): real; \qquad (2.5.2)$$

Here the domain consists of pairs of real numbers called *Rate* and *Time*. The domain can be thought of as allowable inputs (see Figure 2.5.7). The term *real* at the end of (2.5.2) states that the range is a subset of real numbers. The range can be thought of as the set of all possible outputs. The output can be calculated as *Rate * Time*, * denoting multiplication.

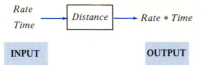

INPUT   **OUTPUT**

Figure 2.5.7

PL/I furnishes floor and ceiling functions. The floor function is called *FLOOR* and the ceiling function is called *CEIL*. *FLOOR* and *CEIL* are functions from the set of real numbers to the set of integers.

Suppose that we have cells in a computer memory indexed from 0 to 10 (see Figure 2.5.8). We wish to store and retrieve arbitrary nonnegative integers in these cells. One approach is to use a **hash function**. A hash function takes a data item to be stored or retrieved and computes the first choice for a location for the item. For example, for our problem, to store

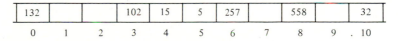

132			102	15	5	257		558		32
0	1	2	3	4	5	6	7	8	9	10

Figure 2.5.8

†This section can be omitted without loss of continuity.

or retrieve the number n, we might take as the first choice for a location, n *mod* 11. Our hash function becomes

$$h(n) = n \bmod 11.$$

Figure 2.5.8 shows the result of storing 15, 558, 32, 132, 102, and 5, in this order, in initially empty cells.

Now suppose that we want to store 257. Since $h(257) = 4$, item 257 should be stored at location 4; however, this position is already occupied. In this case, we say that a **collision** has occurred. More precisely, a collision occurs for a hash function H if $H(x) = H(y)$, but $x \neq y$. To handle collisions, a **collision resolution policy** is required. One simple collision resolution policy is to find the next highest (with 0 assumed to follow 10) unoccupied cell. If we use this collision resolution policy, we would store 257 at location 6 (see Figure 2.5.8).

If we want to locate a value n, we compute $m = h(n)$ and begin looking at location m. If n is not at this position, we look in the next highest position (again, 0 is assumed to follow 10); if n is not in this position, we proceed to the next highest position; and so on. If we reach an empty cell or we return to our original position, we conclude that n is not present; otherwise, we obtain the position of n.

Because the amount of potential data is usually so much larger than the available memory, hash functions are usually not one-to-one. In other words, most hash functions produce collisions.

EXERCISES

Determine whether each set in Exercises 1–6 is a function from $X = \{1, 2, 3, 4\}$ to $Y = \{a, b, c, d\}$. If it is a function, find its domain and range, find $f(1)$ and $f(2)$, and draw its arrow diagram. If it is not a function from X to Y, explain why.

1H. $\{(1, a), (2, a), (3, c), (4, b)\}$
2. $\{(1, c), (2, a), (3, b), (4, c), (2, d)\}$
3. $\{(1, c), (2, d), (3, a), (4, b)\}$
4H. $\{(1, d), (2, d), (4, a)\}$
5. $\{(1, b), (2, b), (3, b), (4, b)\}$
6. $\{(1, d), (2, a), (3, c), (4, b)\}$

In Exercises 7–18, compute the indicated quantity.

7H. $\lfloor 13.5 \rfloor$ **8.** $\lfloor 13.5 \rfloor$ **9.** $\lfloor -13.5 \rfloor$

10H. $\lceil -13.5 \rceil$ **11.** $\lceil 16 \rceil$ **12.** $\lfloor 16 \rfloor$

13H. 8 *mod* 3   **14.** 12 *mod* 4   **15.** 21 *mod* 1

16H. 7735 *mod* 2   **17.** 7735 *mod* 3   **18.** 283393 *mod* 394

Disprove each of the equations in Exercises 19–27 by exhibiting values for x and y for which the statement does not hold.

19H. $\lceil x + y \rceil = \lceil x \rceil + \lceil y \rceil$

20. $\lfloor x + y \rfloor = \lfloor x \rfloor + \lfloor y \rfloor$

21. $\lceil x + y \rceil = \lfloor x \rfloor + \lceil y \rceil$

22H. $\lfloor xy \rfloor = \lfloor x \rfloor \lfloor y \rfloor$

23. $\lceil xy \rceil = \lceil x \rceil \lceil y \rceil$

24. $\lfloor xy \rfloor = \lceil x \rceil \lfloor y \rfloor$

25H. $\lfloor x^2 \rfloor = \lfloor x \rfloor^2$

26. $\lceil x^2 \rceil = \lceil x \rceil^2$

27. $\lceil \sqrt{x} \rceil = \sqrt{\lceil x \rceil}$

28H. A particular hand-held electronic calculator has a % key. If you enter a number and hit the % key, the number you entered is considered a percentage and is converted to decimal. Write a formula for this function.

29. A particular hand-held electronic calculator has a factorial key. If you enter a number n and hit the factorial key, the number you entered is replaced by n factorial. Write a formula for this function. Speculate on the domain of this function.

*30H. Show that if n is an odd integer,

$$\left\lfloor \frac{n^2}{4} \right\rfloor = \left(\frac{n-1}{2} \right)\left(\frac{n+1}{2} \right).$$

*31. Show that if n is an odd integer,

$$\left\lceil \frac{n^2}{4} \right\rceil = \frac{n^2 + 3}{4}.$$

Draw the graphs of the functions in Exercises 32–35.

32H. $f(x) = \lceil x \rceil - \lfloor x \rfloor$ **33.** $f(x) = x - \lfloor x \rfloor$

34. $f(x) = \lceil x^2 \rceil$ **35H.** $f(x) = \lfloor x^2 - x \rfloor$

36H. What day of the week will it be 629 days from Saturday?

37.   What day of the week will it be 220 days from Tuesday?

38.   What day of the week will it be 1445 days from Friday?

January 1 in year x occurs on the day of the week shown in the second column of row

$$y = \left(x + \left\lfloor \frac{x-1}{4} \right\rfloor - \left\lfloor \frac{x-1}{100} \right\rfloor + \left\lfloor \frac{x-1}{400} \right\rfloor \right) \bmod 7$$

in the following table (see [Ritter]).

y	January 1	Non-Leap Year	Leap Year
0	Sunday	January, October	January, April, July
1	Monday	April, July	September, December
2	Tuesday	September, December	June
3	Wednesday	June	March, November
4	Thursday	February, March, November	February, August
5	Friday	August	May
6	Saturday	May	October

The months with Friday the 13th in year x are found in row y in the appropriate column.

39H.   Find the months with Friday the 13th in 1987.

40.   Find the months with Friday the 13th in the present year.

41.   Find the months with Friday the 13th in 2000.

42H.   Verify the entries in the last two columns of the first row.

43.   Verify the entries in the last two columns of the third row.

44.   Verify the entries in the last two columns of the last row.

For each hash function in Exercises 45–48, show how the data would be inserted in the order given in initially empty cells. Use the collision resolution policy described in the Computer Notes section.

45H. $h(x) = x \bmod 11$; cells indexed 0 to 10; data: 53, 13, 281, 743, 377, 20, 10, 796

46. $h(x) = x \bmod 17$; cells indexed 0 to 16; data: 714, 631, 26, 373, 775, 906, 509, 2032, 42, 4, 136, 1028

47. $h(x) = x^2 \bmod 11$; cells and data as in Exercise 45

48H. $h(x) = (x^2 + x) \bmod 17$; cells and data as in Exercise 46

2.6 More on Functions

In Section 2.5 we defined the term "function" and gave several examples of functions. In the present section we consider several types of functions, including one-to-one functions, onto functions, and inverse functions.

Definition 2.6.1.   A function f from X to Y is said to be *one-to-one* if for each $y \in Y$, there is at most one $x \in X$ with $f(x) = y$.

Example 2.6.2.   The function

$$f = \{(1, b), (3, a), (2, c)\}$$

from $X = \{1, 2, 3\}$ to $Y = \{a, b, c\}$ is one-to-one.

Example 2.6.3.   The reciprocal function $R(x) = 1/x$ of Example 2.5.9 and the distance function $D(t) = 55t$ (2.5.1) are one-to-one.

Example 2.6.4.   The function

$$f = \{(1, a), (2, b), (3, a)\} \qquad (2.6.1)$$

from $X = \{1, 2, 3\}$ to $Y = \{a, b, c\}$ is *not* one-to-one since we have $f(1) = a$ and $f(3) = a$. The function $f(x) = x^2$ (with domain equal to the set of all real numbers) is not one-to-one since, for example, $f(2) = 4$ and $f(-2) = 4$.

Example 2.6.5.   If X is the set of persons who have social security numbers and we assign each person $x \in X$ his or her social security number $SS(x)$, we obtain a one-to-one function. It is because this correspondence is one-to-one that we can use social security numbers as identifiers.

Example 2.6.6.   If a function from X to Y is one-to-one, each element in Y in its arrow diagram will have at most one arrow pointing to it (see Figure 2.6.1,

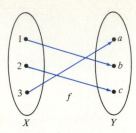

Figure 2.6.1

which is the arrow diagram for the function of Example 2.6.2). If a function is not one-to-one, some element in Y in its arrow diagram will have two or more arrows pointing to it [see Figure 2.6.2, which is the arrow diagram of the function (2.6.1)].

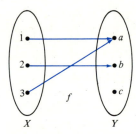

Figure 2.6.2

If the range of a function f from X to Y "fills up" Y, the function is said to be **onto** Y.

Definition 2.6.7.   If f is a function from X to Y and the range of f is equal to Y, f is said to be *onto* Y (or an *onto function*).

Example 2.6.8.   The function

$$f = \{(1, a), (2, c), (3, b), (4, a)\}$$

from $X = \{1, 2, 3, 4\}$ to $Y = \{a, b, c\}$ is onto Y.

Example 2.6.9.   The function of Example 2.6.2 is onto Y.

Example 2.6.10.   We define a function *reverse* from the set X^* of all strings over $\{a, b\}$ to X^* as follows. If $\alpha \in X^*$, *reverse*(α) is the same as α except that the characters are written in reverse order. As examples,

$$reverse(aabab) = babaa, reverse(baab) = baab, reverse(\lambda) = \lambda.$$

The function *reverse* is both one-to-one and onto.

Example 2.6.11.   The function f defined in (2.6.1) is *not* onto $Y = \{a, b, c\}$ since there is no element $x \in X = \{1, 2, 3\}$ with $f(x) = c$.

Example 2.6.12.   If a function from X to Y is onto, each element in Y in its arrow diagram will have at least one arrow pointing to it. (See Figure 2.6.1, which is the arrow diagram for the function of Example 2.6.2.) If a function from X to Y is not onto, some element in Y in its arrow diagram will fail to have an arrow pointing to it (see Figure 2.6.2).

Suppose that f is a one-to-one, onto function from X to Y. It can be shown (see Exercise 41) that

$$\{(y, x) \mid y = f(x)\}$$

is a function from Y to X. This new function, denoted f^{-1}, is called f **inverse**.

Example 2.6.13.   For the function f of Example 2.6.2, we have

$$f^{-1} = \{(b, 1), (a, 3), (c, 2)\}.$$

Example 2.6.14.   Given the arrow diagram for a one-to-one, onto function f from X to Y, we can obtain the arrow diagram for f^{-1} simply by reversing

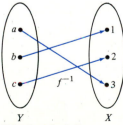

Figure 2.6.3

the direction of each arrow (see Figure 2.6.3, which is the arrow diagram for f^{-1}, where f is the function of Figure 2.6.1).

Example 2.6.15.   The function

$$f(x) = 2^x$$

is a one-to-one function from the set **R** of all real numbers onto the set $\mathbf{R}^+$ of all positive real numbers. We will derive a formula for $f^{-1}(y)$.

Suppose that (y, x) is in f^{-1}, that is,

$$f^{-1}(y) = x. \tag{2.6.2}$$

Then $(x, y) \in f$. Thus

$$y = 2^x.$$

By the definition of logarithm,

$$\log_2 y = x. \tag{2.6.3}$$

Combining (2.6.2) and (2.6.3), we have

$$f^{-1}(y) = x = \log_2 y.$$

That is, for each $y \in \mathbf{R}^+$, $f^{-1}(y)$ is the logarithm to the base 2 of y. We can summarize the situation by saying that the inverse of the exponential function is the logarithm function.

Let g be a function from X to Y and let f be a function from Y to Z. Given $x \in X$, we may apply g to determine a unique element $y = g(x)$ in Y. We may then apply f to determine a unique element $z = f(y) = f(g(x))$ in Z. This compound action is called **composition**.

Definition 2.6.16.   Let g be a function from X to Y and let f be a function from Y to Z. The *composition of f with g*, denoted $f \circ g$, is the function

$$(f \circ g)(x) = f(g(x))$$

from X to Z.

Example 2.6.17.   Given

$$g = \{(1, a), (2, a), (3, c)\},$$

a function from $X = \{1, 2, 3\}$ to $Y = \{a, b, c\}$, and

$$f = \{(a, y), (b, x), (c, z)\},$$

a function from Y to $Z = \{x, y, z\}$, the composition function from X to Z is the function

$$f \circ g = \{(1, y), (2, y), (3, z)\}.$$

Example 2.6.18. Given the arrow diagram for a function g from X to Y and the arrow diagram for a function f from Y to Z, we can obtain the arrow diagram for the composition $f \circ g$ simply by "following the arrows" (see Figure 2.6.4, which gives the arrow diagrams for the functions of Example 2.6.17).

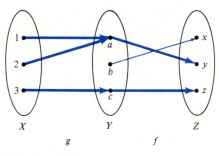

Figure 2.6.4

Example 2.6.19. If $f(x) = \log_3 x$ and $g(x) = x^4$,

$$f(g(x)) = \log_3(x^4) \qquad g(f(x)) = (\log_3 x)^4.$$

Example 2.6.20. Composition sometimes allows us to decompose complicated functions into simpler functions. For example, the function

$$f(x) = \sqrt{\sin 2x}$$

can be decomposed into the functions

$$g(x) = \sqrt{x}, \qquad h(x) = \sin x, \qquad w(x) = 2x.$$

We can then write

$$f(x) = g(h(w(x))).$$

This decomposition technique is important in differential calculus since there are rules for differentiating simple functions such as g, h, and w, and also rules about how to differentiate the composition of functions. Combining these rules, we can differentiate more complicated functions.

EXERCISES

1H. Determine whether each function in Exercises 1–6, Section 2.5 (some are not functions!) is one-to-one or onto or neither. If it is both one-to-one and onto, give the description of the inverse function as a set of ordered pairs, give the domain and range of the inverse function, and draw the arrow diagram of the inverse function.

Determine whether each function in Exercises 2–7 is one-to-one. The domain of each function is the set of all real numbers. If the function is not one-to-one, exhibit distinct numbers a and b with $f(a) = f(b)$. Also, determine whether f is onto the set of all real numbers. If f is not onto, exhibit a number y for which $f(x) \neq y$ for all real x.

2H. $f(x) = 6x - 9$ **3.** $f(x) = 3x^2 - 2x + 1$

4. $f(x) = \sin x$ **5H.** $f(x) = 2x^3 - 4$

6. $f(x) = 3^x - 2$ **7.** $f(x) = \dfrac{x}{1 + x^2}$

8H. Give an example of a function that is one-to-one but not onto.

9. Give an example of a function that is onto but not one-to-one.

10. Give an example of a function that is neither one-to-one nor onto.

11H. For which strings α do we have *reverse*$(\alpha) = \alpha$? (The function *reverse* is defined in Example 2.6.10.)

12. Let X^* denote the set of all strings over $\{a, b\}$. Is the function f from X^* to X^* defined by

$$f(\alpha) = \alpha aab$$

one-to-one? onto?

13. Let X^* denote the set of all strings over $\{a, b\}$. Is the function f from X^* to the set of nonnegative integers defined by

$$f(\alpha) = |\alpha|$$

one-to-one? onto?

14H. Let f be the function from $X = \{0, 1, 2, 3, 4\}$ to X defined by

$$f(x) = 4x \text{ mod } 5.$$

Write f as a set of ordered pairs. Is f one-to-one or onto?

15. Let f be the function from $X = \{0, 1, 2, 3\}$ to X defined by

$$f(x) = 3x \text{ mod } 4.$$

Write f as a set of ordered pairs. Is f one-to-one or onto?

Each function in Exercises 16–21 is one-to-one. Find each inverse function.

16H. $f(x) = 4x + 2$ **17.** $f(x) = 3^x$

18. $f(x) = 3 \log_2 x$ **19H.** $f(x) = 3 + \dfrac{1}{x}$

20. $f(x) = 4x^3 - 5$ **21.** $f(x) = 6 + 2^{7x-1}$

22H. Given

$$g = \{(1, b), (2, c), (3, a)\},$$

a function from $X = \{1, 2, 3\}$ to $Y = \{a, b, c, d\}$, and

$$f = \{(a, x), (b, x), (c, z), (d, w)\},$$

a function from Y to $Z = \{w, x, y, z\}$; write $f \circ g$ as a set of ordered pairs and draw the arrow diagrams of f, g, and $f \circ g$.

23. Given

$$f = \{(a, b), (b, a), (c, b)\},$$

a function from $X = \{a, b, c\}$ to X.
(a) Write $f \circ f$ and $f \circ f \circ f$ as sets of ordered pairs.
(b) Define

$$f^n = f \circ f \circ \cdots \circ f$$

to be the n-fold composition of f with itself. Find f^9 and f^{623}.

24. Draw arrow diagrams of functions f, g, and $f \circ g$ in which f is one-to-one but $f \circ g$ is not one-to-one.

25H. Draw arrow diagrams of functions f, g, and $f \circ g$ in which $f \circ g$ is one-to-one but f is not one-to-one.

26. Draw arrow diagrams of functions f, g, and $f \circ g$ in which g is onto but $f \circ g$ is not onto.

27. Draw arrow diagrams of functions f, g, and $f \circ g$ in which $f \circ g$ is onto but g is not onto.

28H. How many functions are there from $\{1, 2\}$ into $\{a, b\}$? Which are one-to-one? Which are onto?

In Exercises 29–34, find $f \circ g$ and $g \circ f$.

29H. $f(x) = x^2$, $g(x) = 2x + 3$

30. $f(x) = g(x) = 3x^2 + 2x - 4$

31. $f(x) = \sin x$, $g(x) = 3x$

32H. $f(x) = \sqrt{2x}$, $g(x) = 4x - 5$

33. $f(x) = g(x) = \dfrac{1}{x}$

34. $f(x) = \dfrac{1}{2x}$, $g(x) = 3x^2$

In Exercises 35–40, decompose the function into simpler functions as in Example 2.6.20.

35H. $f(x) = \log_2(x^2 + 2)$ **36.** $f(x) = \dfrac{1}{2x^2}$

37. $f(x) = \sin 2x$ **38H.** $f(x) = 2 \sin x$

39. $f(x) = (3 + \sin x)^4$ **40.** $f(x) = \dfrac{1}{(\cos 6x)^3}$

41. Show that if f is a one-to-one, onto function from X to Y, then

$$\{(y, x) \mid (x, y) \in f\}$$

is a one-to-one, onto function from Y to X.

2.7 Matrices

Table 2.7.1 gives the value of some Series I Four Color comic books (in dollars). In mathematics, such an organization of data into rows and columns is called a **matrix**. Because matrices are so important in mathematics, entire books have been devoted to the subject. Many mathematics courses (e.g., linear algebra) deal with matrices.

Table 2.7.1

	Condition		
Title	Good	Fine	Mint
Dick Tracy (No. 1)	75	200	400
Don Winslow (No. 2)	35	100	215
Myra North (No. 3)	8	24	48
Donald Duck (No. 4)	220	650	1400
Smilin' Jack (No. 5)	17	50	100
Dick Tracy (No. 6)	40	120	240

Definition 2.7.1.   Let S be a set. A *matrix over S*

$$A = \begin{pmatrix} a_{11} & a_{12} & \cdots & a_{1n} \\ a_{21} & a_{22} & \cdots & a_{2n} \\ \cdot & \cdot & & \cdot \\ \cdot & \cdot & & \cdot \\ \cdot & \cdot & & \cdot \\ a_{m1} & a_{m2} & \cdots & a_{mn} \end{pmatrix} \tag{2.7.1}$$

is a rectangular array where each a_{ij} is an element of S.

If A has m rows and n columns, we say that the *size of A is m by n* (written $m \times n$).

We will often abbreviate equation (2.7.1) to $A = (a_{ij})$. In this equation, a_{ij} denotes the element of A appearing in the ith row and jth column.

Example 2.7.2.   In matrix notation Table 2.7.1 becomes

$$\begin{pmatrix} 75 & 200 & 400 \\ 35 & 100 & 215 \\ 8 & 24 & 48 \\ 220 & 650 & 1400 \\ 17 & 50 & 100 \\ 40 & 120 & 240 \end{pmatrix}.$$

This matrix has six rows and three columns so its size is 6×3. If we denote this matrix as A and write $A = (a_{ij})$, we would have, for example,

$$a_{11} = 75 \quad a_{21} = 35, \quad a_{43} = 1400.$$

Definition 2.7.3.   Two matrices A and B are *equal*, written $A = B$, if they are the same size and their corresponding entries are equal.

Example 2.7.4.   Determine w, x, y, and z so that

$$\begin{pmatrix} x + y & y \\ w + z & w - z \end{pmatrix} = \begin{pmatrix} 5 & 2 \\ 4 & 6 \end{pmatrix}.$$

According to Definition 2.7.3, since the matrices are the same size, they will be equal provided that

$$x + y = 5, \qquad y = 2$$
$$w + z = 4, \qquad w - z = 6.$$

Solving these equations, we obtain

$$w = 5, \qquad x = 3, \qquad y = 2, \qquad z = -1.$$

We describe next some operations that can be performed on matrices whose entries are numbers. The **sum** of two matrices is obtained by adding the corresponding entries. The **scalar product** is obtained by multiplying each entry in the matrix by a fixed number.

Definition 2.7.5.   Let $A = (a_{ij})$ and $B = (b_{ij})$ be two $m \times n$ matrices. The *sum* of A and B is defined as

$$A + B = (a_{ij} + b_{ij}).$$

The *scalar product* of a number c and a matrix $A = (a_{ij})$ is defined as

$$cA = (ca_{ij}).$$

If A and B are matrices, we define $-A = (-1)A$ and $A - B = A + (-B)$.

Example 2.7.6.   If

$$A = \begin{pmatrix} 4 & 2 \\ -1 & 0 \\ 6 & -2 \end{pmatrix}, \qquad B = \begin{pmatrix} 1 & -3 \\ 4 & 4 \\ -1 & -3 \end{pmatrix},$$

then

$$A + B = \begin{pmatrix} 5 & -1 \\ 3 & 4 \\ 5 & -5 \end{pmatrix}, \qquad 2A = \begin{pmatrix} 8 & 4 \\ -2 & 0 \\ 12 & -4 \end{pmatrix},$$

$$-B = \begin{pmatrix} -1 & 3 \\ -4 & -4 \\ 1 & 3 \end{pmatrix}.$$

A **zero matrix** is a matrix in which every entry is 0.

Example 2.7.7.   Examples of zero matrices are

$$\begin{pmatrix} 0 & 0 & 0 \\ 0 & 0 & 0 \end{pmatrix}, \qquad \begin{pmatrix} 0 & 0 \\ 0 & 0 \\ 0 & 0 \end{pmatrix}, \qquad \begin{pmatrix} 0 & 0 \\ 0 & 0 \end{pmatrix}.$$

If A is any matrix and $\mathbf{0}$ denotes the zero matrix that is the same size as A, $A + \mathbf{0} = A = \mathbf{0} + A$. We state this result and others about matrix addition; the proof is given as an exercise (Exercise 28).

Theorem 2.7.8.   *Let A, B, C, and $\mathbf{0}$ be matrices all of the same size. ($\mathbf{0}$ denotes the zero matrix.) Then*

$$A + B = B + A, \qquad (A + B) + C = A + (B + C),$$
$$A + \mathbf{0} = A = \mathbf{0} + A.$$

The definition of the sum of matrices and the definition of scalar multiplication are reasonably straightforward in contrast to the definition of the product of matrices. For this reason, before giving the formal definition of matrix multiplication, we motivate the definition with an example.

Example 2.7.9.   Ginger, Garth, Dirk, and Kristin paid a visit to Hearty Hanna's Health Hutch. Ginger bought four bottles of carob cream soda (CCS), two bags of unhulled barley (UB), and two bottles of salmon oil capsules (SOC). Garth bought three bottles of carob cream soda, six bags of unhulled barley, and one bottle of salmon oil capsules. Dirk bought eight bottles of carob cream soda and three bags of unhulled barley.

(Dirk did not need any salmon oil capsules.) Kristin bought two bottles of carob cream soda, five bags of unhulled barley, and three bottles of salmon oil capsules. Carob cream soda costs 85 cents, unhulled barley costs 65 cents, and salmon oil capsules cost $9.89. We may compute the amount spent by Ginger as

$$4 \cdot 0.85 + 2 \cdot 0.65 + 2 \cdot 9.89 = 24.48;$$

we may compute the amount spent by Garth as

$$3 \cdot 0.85 + 6 \cdot 0.65 + 1 \cdot 9.89 = 16.34;$$

we may compute the amount spent by Dirk as

$$8 \cdot 0.85 + 3 \cdot 0.65 + 0 \cdot 9.89 = 8.75;$$

and we may compute the amount spent by Kristin as

$$2 \cdot 0.85 + 5 \cdot 0.65 + 3 \cdot 9.89 = 34.62.$$

We show how with an appropriate definition of matrix multiplication the amounts spent by each person can be computed using matrices.

The purchases are summarized by the following matrix:

$$P = \begin{array}{c} \\ \\ \\ \\ \end{array} \begin{array}{ccc} \text{CCS} & \text{UB} & \text{SOC} \\ \end{array} \\ P = \left(\begin{array}{ccc} 4 & 2 & 2 \\ 3 & 6 & 1 \\ 8 & 3 & 0 \\ 2 & 5 & 3 \end{array} \right) \begin{array}{l} \text{Ginger} \\ \text{Garth} \\ \text{Dirk} \\ \text{Kristin.} \end{array}$$

The costs of the items are summarized by the following matrix:

$$C = \left(\begin{array}{c} 0.85 \\ 0.65 \\ 9.89 \end{array} \right) \begin{array}{l} \text{CCS} \\ \text{UB} \\ \text{SOC.} \end{array}$$

Using matrix multiplication, we find that the amounts spent by the persons are computed as

$$PC = \left(\begin{array}{ccc} 4 & 2 & 2 \\ 3 & 6 & 1 \\ 8 & 3 & 0 \\ 2 & 5 & 3 \end{array} \right) \left(\begin{array}{c} 0.85 \\ 0.65 \\ 9.89 \end{array} \right)$$

$$= (4 \cdot 0.85 + 2 \cdot 0.65 + 2 \cdot 9.89 \quad 3 \cdot 0.85 + 6 \cdot 0.65$$
$$+ 1 \cdot 9.89 \quad 8 \cdot 0.85 + 3 \cdot 0.65 + 0 \cdot 9.89 \quad 2 \cdot 0.85$$
$$+ 5 \cdot 0.65 + 3 \cdot 9.89)$$

$$= (24.48 \quad 16.34 \quad 8.75 \quad 34.62).$$

We note that the product PC gives the amounts spent by the four persons. In general, the ikth entry in the product AB of the matrices A and B is obtained by multiplying, consecutively, each element in the ith row of A by each element in the kth column of B and then summing.

Definition 2.7.10.   Let $A = (a_{ij})$ be an $m \times n$ matrix and let $B = (b_{jk})$ be an $n \times p$ matrix. The *matrix product* is defined as the $m \times p$ matrix

$$AB = (c_{ik}),$$

where

$$c_{ik} = \sum_{j=1}^{n} a_{ij}b_{jk}.$$

To multiply the matrix A by the matrix B, Definition 2.7.10 requires that the number of columns of A be equal to the number of rows of B.

Example 2.7.11.   Let

$$A = \begin{pmatrix} 1 & 6 \\ 4 & 2 \\ 3 & 1 \end{pmatrix}, \qquad B = \begin{pmatrix} 1 & 2 & -1 & -2 \\ 4 & 7 & 0 & -3 \end{pmatrix}.$$

The matrix product AB is defined since the number of columns of A is the same as the number of rows of B; both are equal to 2. Entry c_{ik} in the product AB is obtained by using the ith row of A and the kth column of B. For example, the entry c_{31} will be computed using the third row

$$(3 \quad 1)$$

of A and the first column

$$\begin{pmatrix} 1 \\ 4 \end{pmatrix}$$

of B. We then multiply, consecutively, each element in the third row of A by each element in the first column of B and then sum to obtain

$$3 \cdot 1 + 1 \cdot 4 = 7.$$

Since the number of columns of A is the same as the number of rows of B, the elements pair up correctly. Proceeding in this way, we obtain the product

$$AB = \begin{pmatrix} 25 & 44 & -1 & -20 \\ 12 & 22 & -4 & -14 \\ 7 & 13 & -3 & -9 \end{pmatrix}.$$

Example 2.7.12.   In general, even if both AB and BA are defined, they may not be equal. For example, if

$$A = \begin{pmatrix} 3 & 2 \\ 2 & 1 \end{pmatrix}, \qquad B = \begin{pmatrix} 0 & 1 \\ 1 & 0 \end{pmatrix},$$

then

$$AB = \begin{pmatrix} 2 & 3 \\ 1 & 2 \end{pmatrix} \neq \begin{pmatrix} 2 & 1 \\ 3 & 2 \end{pmatrix} = BA.$$

Example 2.7.13.   The matrix product

$$\begin{pmatrix} a & b \\ c & d \end{pmatrix} \begin{pmatrix} x \\ y \end{pmatrix}$$

is

$$\begin{pmatrix} ax + by \\ cx + dy \end{pmatrix}.$$

We let I_n denote the $n \times n$ matrix $I_n = (a_{ij})$, where

$$a_{ij} = \begin{cases} 1 & \text{if } i = j \\ 0 & \text{if } i \neq j. \end{cases}$$

By direct calculation, we see that if A is an $n \times n$ matrix,

$$AI_n = A = I_nA.$$

For this reason, we call I_n the $n \times n$ **identity matrix**.

Example 2.7.14. If

$$A = \begin{pmatrix} 3 & -2 \\ 4 & 1 \end{pmatrix}, \qquad B = \begin{pmatrix} 5 & 0 \\ 3 & -7 \end{pmatrix},$$

then

$$AI_2 = \begin{pmatrix} 3 & -2 \\ 4 & 1 \end{pmatrix}\begin{pmatrix} 1 & 0 \\ 0 & 1 \end{pmatrix} = \begin{pmatrix} 3 & -2 \\ 4 & 1 \end{pmatrix} = A,$$

$$I_2B = \begin{pmatrix} 1 & 0 \\ 0 & 1 \end{pmatrix}\begin{pmatrix} 5 & 0 \\ 3 & -7 \end{pmatrix} = \begin{pmatrix} 5 & 0 \\ 3 & -7 \end{pmatrix} = B.$$

The following theorem states some properties of scalar multiplication and matrix multiplication; the proof is given as an exercise (Exercise 29).

Theorem 2.7.15. *Let A, B, and C be n × n matrices and let u and t be numbers. Then*

$$A(BC) = (AB)C, \qquad\qquad A(B + C) = AB + AC$$
$$(A + B)C = AC + BC, \qquad\qquad u(tA) = (ut)A$$
$$(u + t)A = uA + tA, \qquad\qquad u(A + B) = uA + uB$$
$$A(uB) = u(AB) = (uA)B.$$

†COMPUTER NOTES FOR SECTION 2.7

In computer science, a **two-dimensional array** may be used to represent a matrix. For example, if we represent the 2 × 3 matrix

$$\begin{pmatrix} 2 & 1 & 0 \\ -1 & 6 & 14 \end{pmatrix}$$

as an array Q, in Pascal we would reference element ij of the array by writing $Q[i, j]$. The array Q might be stored by rows in a (one-dimensional) memory as shown in Figure 2.7.1. In this case, array element $Q[i, j]$ is stored at position $3(i - 1) + j$. For example, array element $Q[2, 1]$ is stored at position $3(2 - 1) + 1 = 4$.

†This section can be omitted without loss of continuity.

2	1	0	−1	6	14
$Q[1,1]$	$Q[1,2]$	$Q[1,3]$	$Q[2,1]$	$Q[2,2]$	$Q[2,3]$

Position: 1   2   3   4   5   6

Figure 2.7.1

EXERCISES

Exercises 1–6 refer to the matrix

$$A = (a_{ij}) = \begin{pmatrix} 11 & -6 & 2 \\ 3 & 0 & 1 \\ -4 & 6 & 9 \end{pmatrix}.$$

1H. Find a_{11}.   **2.** Find a_{21}.   **3.** Find a_{12}.

4H. Find a_{33}.   **5.** Find a_{31}.   **6.** Find a_{23}.

In Exercises 7–12, let

$$A = \begin{pmatrix} 1 & 6 & 9 \\ 0 & 4 & -2 \end{pmatrix}, \qquad B = \begin{pmatrix} 4 & 1 & -2 \\ -7 & 6 & 1 \end{pmatrix}$$

and compute the given expression.

7H. $A + B$ **8.** $B + A$ **9.** $-A$

10H. $3A$ **11.** $-2B$ **12.** $2B + A$

In Exercises 13–18, compute the products.

13H. $\begin{pmatrix} 1 & 2 & 3 \\ -1 & 2 & 3 \\ 0 & 1 & 4 \end{pmatrix} \begin{pmatrix} 2 & 8 \\ -1 & 1 \\ 6 & 0 \end{pmatrix}$

14. AI_3, where A is the matrix for Exercises 7–12.

15. $\begin{pmatrix} 1 & 6 \\ -8 & 2 \\ 4 & 1 \end{pmatrix} \begin{pmatrix} 4 & 1 \\ 7 & -6 \end{pmatrix}$

16H. $A^2 (= AA)$, where

$$A = \begin{pmatrix} 1 & -2 \\ 6 & 2 \end{pmatrix}.$$

17. $(2 \quad -4 \quad 6 \quad 1 \quad 3) \begin{pmatrix} 1 \\ 3 \\ -2 \\ 6 \\ 4 \end{pmatrix}$

18. $\begin{pmatrix} 2 & 4 & 1 \\ 6 & 9 & 3 \\ 1 & -1 & 6 \end{pmatrix} \begin{pmatrix} a & b \\ c & d \\ e & f \end{pmatrix}$

19H. (a) Give the size of each matrix.

$$A = \begin{pmatrix} 1 & 4 & 6 \\ 0 & 1 & 7 \end{pmatrix}, \quad B = \begin{pmatrix} 1 & 4 & 7 \\ 8 & 2 & 1 \\ 0 & 1 & 6 \end{pmatrix}, \quad C = \begin{pmatrix} 4 & 2 \\ 0 & 0 \\ 2 & 9 \end{pmatrix}.$$

(b) Using the matrices of part (a), decide which of the products

$$A^2, AB, BA, AC, CA, AB^2, BC, CB, C^2$$

are defined and then compute the products that are defined.

20. Determine x, y, and z so that the equation

$$\begin{pmatrix} x + y & 3x + y \\ x + z & x + y - 2z \end{pmatrix} = \begin{pmatrix} -1 & 1 \\ 9 & -17 \end{pmatrix}$$

holds.

21. Determine w, x, y, and z so that the equation

$$\begin{pmatrix} 2 & 1 & -1 & 7 \\ 6 & 8 & 0 & 3 \end{pmatrix} \begin{pmatrix} x & 2x \\ y & -y + z \\ x + w & w - 2y + x \\ z & z \end{pmatrix} = -\begin{pmatrix} 45 & 46 \\ 3 & 87 \end{pmatrix}$$

holds.

22H. Let

$$A = \begin{pmatrix} 1 & 1 \\ 1 & 0 \end{pmatrix}.$$

Use mathematical induction to show that

$$A^n = \begin{pmatrix} f_n & f_{n-1} \\ f_{n-1} & f_{n-2} \end{pmatrix}, \quad n \geq 2,$$

where f denotes the Fibonacci sequence.

An $n \times n$ matrix A is said to be **invertible** if there exists an $n \times n$ matrix B satisfying

$$AB = I_n = BA.$$

23. Show that the matrix

$$\begin{pmatrix} 2 & 1 \\ 1 & 1 \end{pmatrix}$$

is invertible by solving the equation

$$\begin{pmatrix} 2 & 1 \\ 1 & 1 \end{pmatrix}\begin{pmatrix} a & b \\ c & d \end{pmatrix} = \begin{pmatrix} 1 & 0 \\ 0 & 1 \end{pmatrix}$$

for a, b, c, and d.

*24. Show that the matrix

$$\begin{pmatrix} a & b \\ c & d \end{pmatrix}$$

is invertible if and only if $ad - bc \neq 0$.

25H. Suppose that we want to solve the system

$$AX = C,$$

where

$$A = \begin{pmatrix} a_{11} & a_{12} \\ a_{21} & a_{22} \end{pmatrix}, \qquad X = \begin{pmatrix} x \\ y \end{pmatrix}, \qquad C = \begin{pmatrix} c_1 \\ c_2 \end{pmatrix}$$

for x and y.

Show that if A is invertible, the system has a solution.

The **transpose** of a matrix $A = (a_{ij})$ is the matrix $A^T = (a'_{ji})$, where $a'_{ji} = a_{ij}$.

EXAMPLE:

$$\begin{pmatrix} 1 & 3 \\ 4 & 6 \end{pmatrix}^T = \begin{pmatrix} 1 & 4 \\ 3 & 6 \end{pmatrix}.$$

26. Find the transpose of each matrix of Exercise 19(a).

27. If A and B are $m \times k$ and $k \times n$ matrices, respectively, show that

$$(AB)^T = B^T A^T.$$

28H. Prove Theorem 2.7.8.

29. Prove Theorem 2.7.15.

30. Show how a two-dimensional array that represents the matrix A of Exercise 19 can be stored in a one-dimensional computer memory by rows. Find a formula for the position of the ijth array element.

31H. Show how a two-dimensional array that represents the matrix A of Exercise 19 can be stored in a one-dimensional computer memory by columns. Find a formula for the position of the ijth array element.

2.8 Notes

Most general references on discrete mathematics address the topics of this chapter. [Halmos, 1974; Lipschutz, 1964; and Stoll] are recommended to the reader wanting to study set theory and functions in more detail. [Gardner, 1979] contains a chapter about the Fibonacci numbers. *The Fibonacci Quarterly* is a journal devoted to the Fibonacci numbers and similar sequences. Hash functions are treated in virtually any data structures book; especially recommended are [Horowitz, 1976; Kruse; and Standish].

Computer Exercises

In Exercises 1–4, assume that a set of n elements is represented as an array of size at least $n + 1$. The elements are listed consecutively in the array starting in the first position and terminating with 0. Assume further that no set contains 0.

1. Write a program to represent the sets $X \cup Y$, $X \cap Y$, $X - Y$, and $X \Delta Y$, given the arrays A and B representing X and Y. (The definition of Δ is given before Exercise 35, Section 2.1.)

   2. Assuming a universe represented as an array, write a program to represent the set $\overline{X}$, given the array A representing X.

   3. Given an element E and the array A, which represents X, write a program that determines whether $E \in X$.

   4. Given the array A representing X, write a program that lists all subsets of X.

   5. Write a program that calculates the nth Fibonacci number.

   In Exercises 6 and 7, assume that a sequence from $\{1, \ldots, N\}$ to the real numbers is represented as an array A, indexed from 1 to N.

   6. Write a program that tests whether A is one-to-one.

   7. Write a program that tests whether A is onto a given set.

   8. Implement as a program the hashing system described in the Computer Notes for Section 2.5. Your program should be able to both store and retrieve data.

   9. Write a program that computes the sum of two matrices.

   10. Write a program that computes the scalar product of a number and a matrix.

   11. Write a program that computes the product of two matrices.

Chapter Review

SECTION 2.1

Set

Set-builder notation: $\{x \mid x \text{ has property } P\}$

$|X|$: the number of elements in the set X

$x \in X$: x is an element of the set X

$x \notin X$: x is not an element of the set X

Empty set: $\varnothing$ or $\{\ \ \}$

$X = Y$, where X and Y are sets: X and Y have the same elements

$X \subseteq Y$, X is a subset of Y: every element in X is also in Y

X is a proper subset of Y: $X \subseteq Y$ and $X \neq Y$

$\mathscr{P}(X)$, the power set of X: Set of all subsets of X

$X \cup Y$, X union Y: Set of elements in X *or* Y

$X \cap Y$, X intersect Y: Set of elements in X *and* Y

Disjoint sets X and Y: $X \cap Y = \varnothing$

$X - Y$, difference of X and Y, relative complement: Set of elements in X but not in Y

Universal set

Complement, $\overline{X}$: $U - X$, where U is a universal set

SECTION 2.2

Venn diagram

Laws of sets (see Theorem 2.2.3)

De Morgan's laws for sets: $\overline{(A \cup B)} = \overline{A} \cap \overline{B}, \overline{(A \cap B)} = \overline{A} \cup \overline{B}$

Ordered pair: (x, y)

Cartesian product of X and Y: $X \times Y = \{(x, y) \mid x \in X, y \in Y\}$

n-tuple: $(a_1, a_2, \ldots, a_n)$

Cartesian product of $X_1, X_2, \ldots, X_n$: $X_1 \times X_2 \times \cdots \times X_n$ $= \{(a_1, a_2, \ldots, a_n) \mid a_i \in X_i\}$

Family $\mathscr{C}$ of pairwise disjoint sets: If $X, Y \in \mathscr{C}$ and $X \neq Y$, then $X \cap Y = \varnothing$

Partition of X: Pairwise disjoint collection of nonempty subsets of X whose union is X

SECTION 2.3

Sequence

Index of a sequence

Recursive sequence

n factorial: $n! = n(n - 1) \cdot \cdots \cdot 2 \cdot 1, n \geq 1; 0! = 1$

$$\sum_{i=m}^{n} a_i: a_m + a_{m+1} + \cdots + a_n$$

$$\prod_{i=m}^{n} a_i: a_m \cdot a_{m+1} \cdot \cdots \cdot a_n$$

Fibonacci sequence

Recurrence relation

Floor

String over X

Null string

$|\alpha|$, length of the string α

Concatenation of strings

SECTION 2.4

Principle of Mathematical Induction

Basis step: Prove true for the first index

Inductive step: Assume true for n, then prove true for $n + 1$

Formula for the sum of the first n positive integers:

$$1 + 2 + \cdots + n = n(n + 1)/2$$

Geometric sum: $a + ar^1 + ar^2 + \cdots + ar^n = a(r^{n+1} - 1)/(r - 1)$

If X is a finite set and $|X| = n$, then $|\mathcal{P}(X)| = 2^n$

SECTION 2.5

Function from X to Y

Domain of a function f: $\{x \mid (x, y) \in f\}$

Range of a function f: $\{y \mid (x, y) \in f\}$

Arrow diagram

Graph of a function

Floor of x: $\lfloor x \rfloor$ = greatest integer less than or equal to x

Ceiling of x: $\lceil x \rceil$ = least integer greater than or equal to x

$x \bmod y$: remainder when x is divided by y

SECTION 2.6

One-to-one function f from X to Y: For each $y \in Y$, there is at most one
    $x \in X$ with $f(x) = y$

Onto function f from X to Y: Range of $f = Y$

f inverse, f^{-1}: $f^{-1} = \{(y, x) \mid (x, y) \in f\}$

Composition of functions, $f \circ g$: $(f \circ g)(x) = f(g(x))$

SECTION 2.7

Matrix

Size of a matrix

Equal matrices

Sum of matrices

Scalar product

Product of two matrices

Zero matrix

Identity matrix

Chapter Self-Test

SECTION 2.1

1H. If $A = \{1, 3, 4, 5, 6, 7\}$, $B = \{x \mid x$ is an even integer$\}$,
    $C = \{2, 3, 4, 5, 6\}$, find $(A \cap B) - C$.

2H. If X is a set and $|X| = 8$, how many members does $\mathscr{P}(X)$ have? How many proper subsets does X have?

3H. If $A \cup B = B$, what relation must hold between A and B?

4H. Are the sets

$$\{3,\ 2,\ 2\}, \qquad \{x \mid x \text{ is an integer and } 1 < x \le 3\}$$

equal? Explain.

SECTION 2.2

5H. Draw a Venn diagram and shade the set $A \cap (B - C)$.

6H. List the elements of the set $\{a,\ b\} \times \{a,\ c\}$.

7H. Find a partition of $\{a,\ b,\ c,\ d\}$ containing exactly two sets.

8H. A survey of 76 college students found that 41 read the *National Enquirer*, 36 read *Rolling Stone*, 49 read *Byte*, 9 read both the *National Enquirer* and *Rolling Stone*, 18 read both the *National Enquirer* and *Byte*, and 29 read both *Rolling Stone* and *Byte*. Each of the 76 students reads at least one of these publications. Represent these data as a Venn diagram and determine how many students read all three publications.

SECTION 2.3

9H. For the sequence $a_n = 2n + 2$, find

   (a) a_6 (b) $\displaystyle\sum_{i=1}^{3} a_i$ (c) $\displaystyle\prod_{i=1}^{3} a_i$

10H. Find f_{10}, where f is the Fibonacci sequence.

11H. Let

$$b_n = \sum_{i=1}^{n} (i + 1)^2 - i^2.$$

   (a) Find b_5 and b_{10}.
   (b) Find a formula for b_n.

12H. Let $\alpha = ccddc$ and $\beta = c^3 d^2$. Find

   (a) $\alpha\beta$ (b) $\beta\alpha$ (c) $|\alpha|$ (d) $|\alpha\alpha\beta\alpha|$

SECTION 2.4

Use mathematical induction to prove the statements in Exercises 13–16 are true for every positive integer n.

13H. $2 + 4 + \cdots + 2n = n(n + 1)$

14H. $2^2 + 4^2 + \cdots + (2n)^2 = \dfrac{2n(n + 1)(2n + 1)}{3}$

15H. $\dfrac{1}{2!} + \dfrac{2}{3!} + \cdots + \dfrac{n}{(n + 1)!} = 1 - \dfrac{1}{(n + 1)!}$

16H. $2^{n+1} < 1 + (n + 1)2^n$

SECTION **2.5**

17H. Explain why

$$f = \{(1, \$), (2, \&), (3, \$)\}$$

is a function from $X = \{1, 2, 3\}$ into $Y = \{@, \#, \$, \&\}$. Find the domain and range of f and draw the arrow diagram of f.

18H. Find 200 *mod* 3.

19H. Find real numbers x and y satisfying $\lfloor x \rfloor \lfloor y \rfloor = \lfloor xy \rfloor - 1$.

20H. Let X be the set of strings over $\{a, b\}$ of length 4 and let Y be the set of strings over $\{a, b\}$ of length 3. Define a function f from X to Y by the rule

$f(\alpha)$ = string consisting of the first three characters of α.

EXAMPLE: $f(abaa) = aba$.
  (a) Describe the domain of f.
  (b) Describe the range of f.

SECTION **2.6**

21H. Find the domain and range of the function of Exercise 17. Is f onto Y?

22H. List all one-to-one functions from $\{s, t\}$ into $\{0, 1, 2\}$.

23H. For the function f of Exercise 20,
  (a) Is f one-to-one? Explain.
  (b) Let g be the function from Y to Y that reverses a string.

EXAMPLE: $g(aab) = baa$. Describe, in words, the function $g \circ f$.

24H. If $f(x) = 3x^2$ and $g(x) = 6 + \tan x$, find $f \circ g$ and $g \circ f$.

SECTION **2.7**

25H. For the matrix

$$A = \begin{pmatrix} 2 & 4 \\ 9 & -3 \end{pmatrix},$$

find a_{22}, where $A = (a_{ij})$. If AB is a 2×8 matrix, what size is B?

26H. Let

$$A = \begin{pmatrix} 3 & 0 \\ -2 & 1 \end{pmatrix}, \qquad B = \begin{pmatrix} 2 & 4 \\ 3 & -3 \end{pmatrix}.$$

Compute

(a) $-3B$ (b) $2A + 3B$ (c) AB (d) BA

27H. Give examples of matrices A and B where AB and BA are both defined but AB and BA are not the same size.

28H. Determine x, y, and z so that the equation

$$\begin{pmatrix} x + y & x + z \\ y - 2z & -x + 3y + z \end{pmatrix} = \begin{pmatrix} -3 & -1 \\ 1 & -20 \end{pmatrix}$$

holds.

3

Graphs

Well, I got on the road, and I went north to Providence. Met the Mayor.

The Mayor of Providence!

He was sitting in the hotel lobby.

What'd he say?

He said, "Morning!" And I said, "You got a fine city here, Mayor." And then he had coffee with me. And then I went to Waterbury. Waterbury is a fine city. Big clock city, the famous Waterbury clock. Sold a nice bill there. And then Boston—Boston is the cradle of the Revolution. A fine city. And a couple of other towns in Mass., and on to Portland and Bangor and straight home!

—from *Death of a Salesman*

Although the first paper in graph† theory goes back to 1736 (see Section 3.2) and several important results in graph theory were obtained in the nineteenth century, it is only since the 1920s that there has been a sustained, widespread, intense interest in graph theory. Indeed, the first text on graph theory ([König]) appeared in 1936. Undoubtedly, one of the reasons for the recent interest in graph theory is its applicability in many diverse fields, including computer science, chemistry, operations research, electrical engineering, linguistics, and economics.

This chapter begins with some basic graph terminology and examples. In Section 3.2 we discuss certain important concepts in graph theory, including paths and cycles. Two classical graph problems, the existence of Hamiltonian cycles and the traveling salesperson problem, are the topics of optional Section 3.3. In Section 3.4 we discuss one possible way to represent a graph in a computer. Such representations of graphs are also useful in the theoretical analysis of graphs. We study the question of when two graphs are essentially the same (i.e., when two graphs are isomorphic) in Section 3.5. In optional Section 3.6, we present a solution based on a graph model to the Instant Insanity puzzle.

3.1 Introduction

Figure 3.1.1 shows the highway system in Wyoming that a particular person is responsible for inspecting. Specifically, this road inspector must travel all of these roads and file reports on road conditions, visibility of lines on the roads, status of traffic signs, and so on. Since the road inspector lives in Greybull, the most economical way to inspect all of the roads would be to start in Greybull, travel each of the roads exactly once, and return to Greybull. Is this possible? See if you can decide before reading on.

This problem can be modeled as a **graph**. In fact, since graphs are

---

†"Graph" in this chapter has a meaning quite distinct from "graph" (of a function) that we encountered in Section 2.5.

Figure 3.1.1

drawn with dots and lines, they look like road maps. In Figure 3.1.2, we have drawn a graph G that models the map of Figure 3.1.1. The dots in Figure 3.1.2 are called **vertices** and the lines that connect the vertices are called **edges**. (Later in this section we will define all of these terms carefully.) We have labeled each vertex with the first three letters of the city to which it corresponds. We have labeled the edges $e_1, \ldots, e_{13}$. In drawing a graph, the only information of importance is which vertices are connected by which edges. For this reason, the graph of Figure 3.1.2 could just as well be drawn as in Figure 3.1.3.

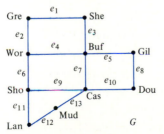

Figure 3.1.2

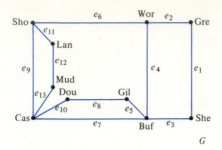

Figure 3.1.3

If we start at a vertex v_0, travel along an edge to vertex v_1, travel along another edge to vertex v_2, and so on, and eventually arrive at vertex v_n, we call the complete tour a **path** from v_0 to v_n. The path that starts at She, then goes to Buf, and ends at Gil corresponds to a trip on the map of Figure 3.1.1 that begins in Sheridan, goes to Buffalo, and ends at Gillette. The road inspector's problem can be rephrased for the graph model G in the following way: Is there a path from vertex Gre to vertex Gre that traverses every edge exactly once?

We can show that the road inspector cannot start in Greybull, travel each of the roads exactly once, and return to Greybull. To put the answer in graph terms, there is no path from vertex Gre to vertex Gre in Figure 3.1.2 that traverses every edge exactly once. To see this, suppose that there is such a path and consider vertex Wor. Each time we arrive at Wor on some edge, we must leave Wor on a different edge. Furthermore, every edge that touches Wor must be used. Thus the edges at Wor occur in pairs. It follows that an even number of edges must touch Wor. Since three edges touch Wor, we have a contradiction. Therefore, there is no path from vertex Gre to vertex Gre in Figure 3.1.2 that traverses every edge exactly once. The argument applies to an arbitrary graph G. If G has a path from vertex v to v that traverses every edge exactly once, an even number of edges must touch each vertex. We discuss this problem in greater detail in Section 3.2.

At this point we give some formal definitions.

Definition 3.1.1.   A *graph* (or *undirected graph*) G consists of a set V of *vertices* (or *nodes*) and a set E of *edges* (or *arcs*) such that each edge $e \in E$ is associated with an unordered pair of vertices. If there is a unique edge e associated with the vertices v and w, we write $e = (v, w)$ or $e = (w, v)$. In this context, (v, w) denotes an edge in an undirected graph and *not* an ordered pair.

A *directed graph* (or *digraph*) G consists of a set V of *vertices* (or *nodes*) and a set E of *edges* (or *arcs*) such that each edge $e \in E$ is associated with an ordered pair of vertices. If there is a unique edge e associated with the ordered pair (v, w) of vertices, we write $e = (v, w)$.

An edge e in a graph (undirected or directed) that is associated with the pair of vertices v and w is said to be *incident on v* and w and v and w are said to be *incident on e* and to be *adjacent vertices*.

If G is a graph (undirected or directed) with vertices V and edges E, we write $G = (V, E)$.

Unless specified otherwise, the sets E and V are assumed finite.

Example 3.1.2. In Figure 3.1.2 the (undirected) graph G consists of the set

$$V = \{Gre, She, Wor, Buf, Gil, Sho, Cas, Dou, Lan, Mud\}$$

of vertices and the set

$$E = \{e_1, e_2, \ldots, e_{13}\}$$

of edges. Edge e_1 is associated with the unordered pair {Gre, She} of vertices and edge e_{10} is associated with the unordered pair {Cas, Dou} of vertices. Edge e_1 is denoted (Gre, She) or (She, Gre) and edge e_{10} is denoted (Cas, Dou) or (Dou, Cas). Edge e_4 is incident on Wor and Buf and the vertices Wor and Buf are adjacent.

Example 3.1.3. A directed graph is shown in Figure 3.1.4. The directed edges are indicated by arrows. Edge e_1 is associated with the ordered pair (v_2, v_1) of vertices and edge e_7 is associated with the ordered pair

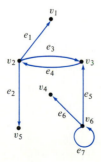

Figure 3.1.4

(v_6, v_6) of vertices. Edge e_1 is denoted (v_2, v_1) and edge e_7 is denoted (v_6, v_6).

Definition 3.1.1 allows distinct edges to be associated with the same pair of vertices. For example, in Figure 3.1.5 edges e_1 and e_2 are both associated with the vertex pair $\{v_1, v_2\}$. Such edges are called **parallel edges**. An edge incident on a single vertex is called a **loop**. For example, in Figure 3.1.5 edge $e_3 = (v_2, v_2)$ is a loop. A vertex, like vertex v_4 in Figure 3.1.5, that is not incident on any edge is called an **isolated vertex**. A graph with neither loops nor parallel edges is called a **simple graph**.

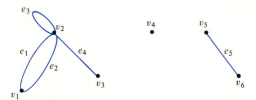

Figure 3.1.5

Example 3.1.4.   Since the graph of Figure 3.1.2 has neither parallel edges nor loops, it is a simple graph.

Some authors do not permit loops and parallel edges when they define graphs. One would expect that if agreement has not been reached on the definition of graph, most other terms in graph theory would also not have standard definitions. This is indeed the case. In reading articles and books about graphs, it is necessary to check on the definitions being used.

We turn next to an example that shows how a graph model can be used to analyze a manufacturing problem.

Example 3.1.5.   Frequently in manufacturing, it is necessary to bore many holes in sheets of metal (see Figure 3.1.6). Components can then be bolted to these sheets of metal. The holes can be drilled using a drill press under the control of a computer. To save time and money, the drill press should be moved as quickly as possible. We model the situation as a graph.

The vertices of the graph correspond to the holes (see Figure 3.1.7). Every pair of vertices is connected by an edge. We write on each edge the time to move the drill press between the corresponding holes. A

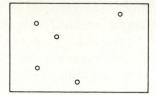

Figure 3.1.6

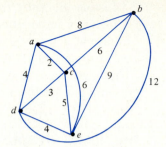

Figure 3.1.7

graph with numbers on the edges (like the graph of Figure 3.1.7) is called a **weighted graph**. If edge e is labeled k, we say that the **weight of edge** e is k. For example, in Figure 3.1.7 the weight of edge (c, e) is 5. In a weighted graph, the **length of a path** is the sum of the weights of the edges in the path. For example, in Figure 3.1.7 the length of the path that starts at a, visits c, and terminates at b, is 8. In this problem, the length of a path that starts at vertex v_1 and then visits $v_2, v_3, \ldots$, in this order, and terminates at v_n represents the time it takes the drill press to start at hole h_1 and then move to $h_2, h_3, \ldots$, in this order, and terminate at h_n, where hole h_i corresponds to vertex v_i. A path of minimum length that visits every vertex exactly one time represents the optimal path for the drill press to follow.

Suppose that in this problem the path is required to begin at vertex a and end at vertex e. We can find the minimum length path by listing all possible paths from a to e that pass through every vertex exactly one time and choose the shortest one (see Table 3.1.1). We see that the path that visits the vertices a, b, c, d, e, in this order, has minimum length. Of course, a different pair of starting and ending vertices might produce an even shorter path.

Table 3.1.1

Path	Length
a, b, c, d, e	21
a, b, d, c, e	28
a, c, b, d, e	24
a, c, d, b, e	26
a, d, b, c, e	27
a, d, c, b, e	22

Listing all paths from vertex v to vertex w, as we did in Example 3.1.5, is a rather time-consuming way to find a minimum-length path from v to w that visits every vertex exactly one time. Unfortunately, no one knows a method that is much more practical for arbitrary graphs. This problem is a version of the **traveling salesperson problem**. We discuss that problem further in Section 3.3.

Example 3.1.6 Similarity Graphs.

This example deals with the problem of grouping ''like'' objects into classes based on properties of the objects. For example, suppose that five persons write essays on the same theme and we want to group ''like'' essays into classes based on certain properties of the essays (see Table 3.1.2). Suppose that we select as properties

1. The average number of words per paragraph in the essay.
2. The number of paragraphs in the essay.
3. The number of grammatical errors in the essay.

Table 3.1.2

Essay	Average Number of Words per Paragraph	Number of Paragraphs	Number of Errors
1	66	20	1
2	41	10	2
3	68	5	8
4	90	34	5
5	75	12	14

A **similarity graph** G is constructed as follows. The vertices correspond to essays. A vertex is denoted (p_1, p_2, p_3), where p_i is the value of property i. We define a **dissimilarity function** s as follows. For each pair of vertices $v = (p_1, p_2, p_3)$ and $w = (q_1, q_2, q_3)$, we set

$$s(v, w) = |p_1 - q_1| + |p_2 - q_2| + |p_3 - q_3|.$$

If we let v_i be the vertex corresponding to essay i, we obtain

$$s(v_1, v_2) = 36, \ s(v_1, v_3) = 24, \ s(v_1, v_4) = 42, \ s(v_1, v_5) = 30,$$
$$s(v_2, v_3) = 38, \ s(v_2, v_4) = 76, \ s(v_2, v_5) = 48, \ s(v_3, v_4) = 54,$$
$$s(v_3, v_5) = 20, \ s(v_4, v_5) = 46.$$

If v and w are vertices corresponding to two essays, $s(v, w)$ is a measure of how dissimilar the essays are. A large value of $s(v, w)$ indicates dissimilarity, while a small value indicates similarity.

For a fixed number S, we insert an edge between vertices v and w if $s(v, w) < S$. (In general, there will be different similarity graphs for different values of S.) We say that v and w are **in the same class** if $v = w$ or there is a path from v to w. In Figure 3.1.8 we show the graph corresponding to the essays of Table 3.1.2 with $S = 25$. In this graph, the essays are grouped into three classes $\{1, 3, 5\}$, $\{2\}$, and $\{4\}$. In a real problem, an appropriate value for S might be selected by trial and error or the value of S might be selected automatically according to some predetermined criteria.

Figure 3.1.8

Example 3.1.6 belongs to the subject called **pattern recognition**. Pattern recognition is concerned with grouping data into classes based on properties of the data. Pattern recognition by computer has much practical significance. For example, computers have been programmed to detect cancer from X-rays, to select tax returns to be audited, to analyze satellite pictures, to recognize text, and to forecast weather.

EXERCISES

Explain why none of the graphs in Exercises 1–3 has a path from a to a that passes through each edge exactly one time.

1H.

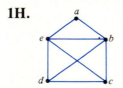

2.

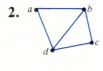

3.

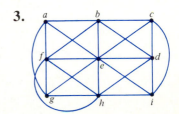

Show that each graph in Exercises 4–6 has a path from a to a that passes through each edge exactly one time by finding such a path by inspection.

4H.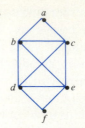

5.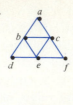

6.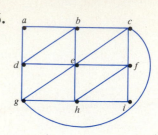

For each graph $G = (V, E)$ in Exercises 7–12, find V, E, all parallel edges, all loops, all isolated vertices, and tell whether G is a simple graph. Also, tell on which vertices edge e_1 is incident.

7H.

8.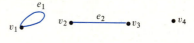

9.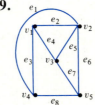

10H.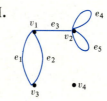

11.

12.

13H. The graph K_n, called the **complete graph on n vertices**, has n vertices and every vertex is joined to every other vertex by an edge. (There are no loops or parallel edges.) Draw K_3, K_4, and K_5.

A graph in which the vertices can be partitioned into disjoint sets V_1 and V_2 with every edge incident on one vertex in V_1 and one vertex in V_2 is called a **bipartite graph**. The following graph is a bipartite graph; we may take the disjoint vertex sets to be $V_1 = \{v_1, v_2, v_3\}$ and $V_2 = \{v_4, v_5\}$.

14. Give another example of a bipartite graph. Specify the disjoint vertex sets.

State which graphs in Exercises 15–21 are bipartite graphs. If the graph is bipartite, specify the disjoint vertex sets.

15H. **16.**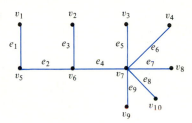

17. Figure 3.1.2 **18H.** Figure 3.1.5

19. Exercise 7 **20.** Exercise 9

21H. Exercise 11

22H. The graph $K_{m,n}$ called the **complete bipartite graph on m and n vertices**, has disjoint sets V_1 of m vertices and V_2 of n vertices. Every vertex in V_1 is joined to every vertex in V_2 by an edge. (There are no parallel edges.) Draw $K_{2,3}$, $K_{2,4}$, and $K_{3,3}$.

In Exercises 23–25, find a path of minimum length from v to w in the graph of Figure 3.1.7 that passes through each vertex exactly one time.

23H. $v = b$, $w = e$ **24.** $v = c$, $w = d$ **25.** $v = a$, $w = b$

26H. Draw the similarity graph that results from setting $S = 40$ in Example 3.1.6. How many classes are there?

27. Draw the similarity graph that results from setting $S = 50$ in Example 3.1.6. How many classes are there?

28. Suggest additional properties for Example 3.1.6 that might be useful in comparing essays.

29. How might one automate the selection of S to group data into classes using a similarity graph?

Exercises 30–32 refer to the following graph. The vertices represent offices. An edge connects two offices if there is a communication link between the two. Notice that any office can communicate with any other either directly through a communication link or by having others relay the message.

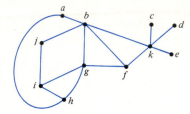

30H. Show, by giving an example, that communication among all offices is still possible even if some communication links are broken.

31. What is the maximum number of communication links that can be broken with communication among all offices still possible?

32. Show a configuration in which the maximum number of communication links are broken with communication among all offices still possible.

33H. In the graph the vertices represent cities and the numbers on the edges represent the costs of building the indicated roads. Find a least expensive road system that connects all the cities.

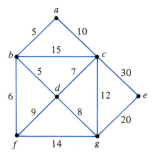

3.2 Paths and Cycles

If we think of the vertices in a graph as cities and the edges as roads, a path corresponds to a trip beginning at some city, passing through several cities, and terminating at some city. We begin by giving a formal definition of path.

Definition 3.2.1.   Let v_0 and v_n be vertices in a graph. A *path* from v_0 to v_n of length n is an alternating sequence of $n + 1$ vertices and n edges beginning with vertex v_0 and ending with vertex v_n,

$$(v_0, e_1, v_1, e_2, v_2, \ldots, v_{n-1}, e_n, v_n),$$

in which edge e_i is incident on vertices v_{i-1} and v_i for $i = 1, \ldots, n$.

Example 3.2.2.   In the graph of Figure 3.2.1,

$$(1, e_1, 2, e_2, 3, e_3, 4, e_4, 2) \qquad\qquad (3.2.1)$$

is a path of length 4 from vertex 1 to vertex 2.

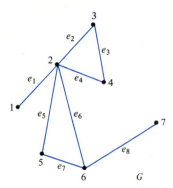

Figure 3.2.1

Example 3.2.3.   In the graph of Figure 3.2.1, the path (6) consisting solely of vertex 6 is a path of length 0 from vertex 6 to vertex 6.

In the absence of parallel edges, in denoting a path, we may suppress the edges. For example, the path (3.2.1) may also be written

$$(1, 2, 3, 4, 2).$$

A **connected graph** is a graph in which we can get from any vertex to any other vertex on a path. The formal definition follows.

Definition 3.2.4.   A graph G is *connected* if given any vertices v and w in G, there is a path from v to w.

Example 3.2.5.   The graph G of Figure 3.2.1 is connected since, given any vertices v and w in G, there is a path from v to w.

Example 3.2.6.   The graph G of Figure 3.2.2 is not connected since, for example, there is no path from vertex v_2 to vertex v_5.

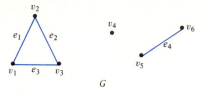

Figure 3.2.2

Example 3.2.7.   Let G be the graph whose vertex set consists of the 50 states of the United States. Put an edge between states v and w if v and w share a border. For example, there is an edge between California and Oregon and between Illinois and Missouri. There is no edge between Georgia and New York nor is there an edge between Utah and New Mexico. (Touching does not count; the states must share a border.) The graph G is not connected because there is no path from Hawaii to California (nor from Hawaii to any other state).

As we can see from Figures 3.2.1 and 3.2.2, a connected graph consists of one "piece," while a graph that is not connected consists of two or more "pieces." These "pieces" are **subgraphs** of the original graph and are called **components**. We give the formal definitions beginning with subgraph.

A subgraph G' of a graph G is obtained by selecting certain edges and vertices from G subject to the restriction that if we select an edge e in G that is incident on vertices v and w, we must include v and w in G'. The formal definition follows.

Definition 3.2.8.   If $G = (V, E)$ and $G' = (V', E')$ are graphs with $V' \subseteq V$ and $E' \subseteq E$, we call G' a *subgraph* of G.

Example 3.2.9.   The graph $G' = (V', E')$ of Figure 3.2.3 is a subgraph of the graph $G = (V, E)$ of Figure 3.2.4 since $V' \subseteq V$ and $E' \subseteq E$.

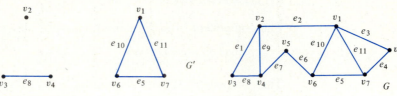

Figure 3.2.3   **Figure 3.2.4**

Example 3.2.10.   Find all subgraphs of the graph G of Figure 3.2.5 having at least one vertex.

   If we select no edges, we may select one or both vertices yielding the subgraphs G_1, G_2, and G_3 shown in Figure 3.2.6. If we select the one available edge e_1, we must select the two vertices on which e_1 is incident. In this case, we obtain the subgraph G_4 shown in Figure 3.2.6. Thus G has the four subgraphs shown in Figure 3.2.6.

   We can now define component.

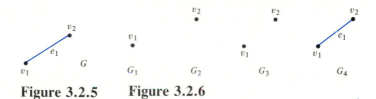

Figure 3.2.5   **Figure 3.2.6**

Definition 3.2.11.   Let G be a graph and let v be a vertex in G. The subgraph G' of G consisting of all edges and vertices in G that are contained in some path beginning at v is called the *component* of G containing v.

Example 3.2.12.   The graph G of Figure 3.2.1 has one component, namely itself. Indeed, a graph is connected if and only if it has exactly one component.

Example 3.2.13.   Let G be the graph of Figure 3.2.2. The component of G containing v_3 is the subgraph

$$G_1 = (V_1, E_1), \qquad V_1 = \{v_1, v_2, v_3\}, \qquad E_1 = \{e_1, e_2, e_3\}.$$

The component of G containing v_4 is the subgraph

$$G_2 = (V_2, E_2), \qquad V_2 = \{v_4\}, \qquad E_2 = \emptyset.$$

The component of G containing v_5 is the subgraph

$$G_3 = (V_3, E_3), \qquad V_3 = \{v_5, v_6\}, \qquad E_3 = \{e_4\}.$$

Notice that the definition of path allows repetitions of vertices or edges or both. In the path (3.2.1), vertex 2 appears twice.

Subclasses of paths are obtained by prohibiting duplicate vertices or edges or by making the vertices v_0 and v_n of Definition 3.2.1 identical.

Definition 3.2.14.   Let v and w be vertices in a graph G.

A *simple path* from v to w is a path from v to w with no repeated vertices.

A *cycle* (or *circuit*) is a path of nonzero length from v to v with no repeated edges.

A *simple cycle* is a cycle from v to v in which, except for the beginning and ending vertices that are both equal to v, there are no repeated vertices.

Example 3.2.15.   For the graph of Figure 3.2.1 we have

Path	Simple Path?	Cycle?	Simple Cycle?
(6, 5, 2, 4, 3, 2, 1)	No	No	No
(6, 5, 2, 4)	Yes	No	No
(2, 6, 5, 2, 4, 3, 2)	No	Yes	No
(5, 6, 2, 5)	No	Yes	Yes
(7)	Yes	No	No

We next reexamine the problem introduced in Section 3.1 of finding a cycle in a graph that traverses each edge exactly one time.

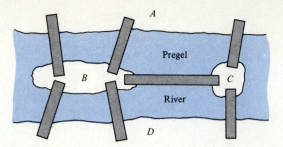

Figure 3.2.7

Example 3.2.16 Königsberg Bridge Problem.   The first paper in graph theory was Leonhard Euler's in 1736. The paper presented a general theory that included a solution to what is now called the Königsberg bridge problem.

Two islands lying in the Pregel River in Königsberg (now Kaliningrad in the Soviet Union) were connected to each other and the river banks by bridges as shown in Figure 3.2.7. The problem is to start at any location—*A*, *B*, *C*, or *D*; walk over each bridge exactly once; then return to the starting location.

The bridge configuration can be modeled as a graph as shown in Figure 3.2.8. The vertices represent the locations and the edges represent the bridges. The Königsberg bridge problem is now reduced to finding a cycle in the graph of Figure 3.2.8 that includes all the edges and all the vertices. In honor of Euler, a cycle in a graph *G* that includes all the edges and all the vertices of *G* is called an **Euler cycle**.† From the discussion of Section 3.1, we see that there is no Euler cycle in the graph of Figure 3.2.8 because there are an odd number of edges incident on vertex *A*. (In fact, in the graph of Figure 3.2.8, every vertex is incident on an odd number of edges.)

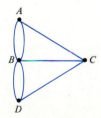

Figure 3.2.8

---

†For technical reasons, if *G* consists of one vertex v and no edges, we call the path (v) an Euler cycle for *G*.

Leonhard Euler (1707–1783)

Leonhard Euler was a Swiss mathematician who spent his professional life in St. Petersburg and Berlin. Euler was one of the most prolific mathematical authors of all time; he wrote over 800 papers. Long after his death, his papers were still appearing. He introduced "i" for the square root of -1 and the $f(x)$ notation for functions. Euler could carry out extensive calculations in his head. He once summed 17 terms in his head to settle an argument over the fiftieth decimal place in an answer. [*Photo courtesy of Archiv Für Kunst und Geschichte, Berlin*]

The solution to the existence of Euler cycles is nicely stated by introducing the degree of a vertex. The **degree of a vertex** v, $\delta(v)$, is the number of edges incident on v. (By definition, each loop on v contributes 2 to the degree of v.) In Section 3.1 we found that if a graph G has an Euler cycle, then every vertex in G has even degree. We can also prove that G is connected.

Theorem 3.2.17. *If a graph G has an Euler cycle, then G is connected and every vertex has even degree.*

PROOF. Suppose that G has an Euler cycle. We argued in Section 3.1 that every vertex in G has even degree. If v and w are vertices in G, the portion of the Euler cycle that takes us from v to w serves as a path from v to w. Therefore, G is connected. ■

The converse of Theorem 3.2.17 is also true. We state this result formally as Theorem 3.2.18.

Theorem 3.2.18. *If G is a connected graph and every vertex has even degree, then G has an Euler cycle.*

PROOF. See [Johnsonbaugh, Theorem 3.3.6]. ■

If G is a connected graph and every vertex has even degree and G has only a few edges, we can usually find an Euler cycle by inspection.

Example 3.2.19. Let G be the graph of Figure 3.2.9. Use Theorem 3.2.18 to verify that G has an Euler cycle. Find an Euler cycle for G.

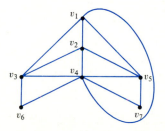

Figure 3.2.9

We find that

$$\delta(v_1) = \delta(v_2) = \delta(v_3) = \delta(v_5) = 4,$$
$$\delta(v_4) = 6, \qquad \delta(v_6) = \delta(v_7) = 2.$$

Since the degree of every vertex is even, by Theorem 3.2.18, G has an Euler cycle. By inspection, we find the Euler cycle

$$(v_6, v_4, v_7, v_5, v_1, v_3, v_4, v_1, v_2, v_5, v_4, v_2, v_3, v_6).$$

We conclude by proving a rather special result that we will use in Section 4.2.

Theorem 3.2.20.   *If a graph G contains a cycle from v to v, G contains a simple cycle from v to v.*

PROOF.   Let

$$C = (v_0, e_1, v_1, \ldots, e_i, v_i, e_{i+1}, \ldots, e_j,$$
$$v_j, e_{j+1}, v_{j+1}, \ldots, e_n, v_n)$$

be a cycle from v to v where $v = v_0 = v_n$ (see Figure 3.2.10). If C is not a simple cycle, then $v_i = v_j$, for some $i < j < n$. We can replace C by the cycle

$$C' = (v_0, e_1, v_1, \ldots, e_i, v_i, e_{j+1}, v_{j+1}, \ldots, e_n, v_n).$$

If C' is not a simple cycle from v to v, we repeat the previous procedure. Eventually we obtain a simple cycle from v to v.   ∎

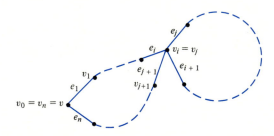

Figure 3.2.10

EXERCISES

In Exercises 1–9, tell whether the given path in the graph is
      (a) A simple path   (b) A cycle   (c) A simple cycle

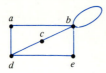

1H. (b, b)

2. (e, d, c, b)

3. (a, d, c, d, e)

4H. (d, c, b, e, d)

5. $(b, c, d, a, b, e, d, c, b)$

6. (b, c, d, e, b, b)

7H. (a, d, c, b, e)

8. (d)

9. (d, c, b)

10H. Find all the simple cycles in the graph.

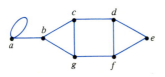

11.   Find all simple paths from a to e in the graph of Exercise 10.

12.   Find all connected subgraphs of the graph containing all of the vertices of the original graph and having as few edges as possible. Which are simple paths? Which are cycles? Which are simple cycles?

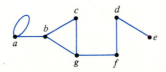

Find the degree of each vertex for the following graphs.

13H.

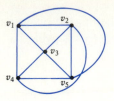

14.

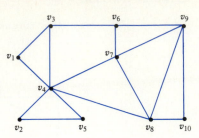

In Exercises 15–18, find all subgraphs having at least one vertex of the graph given.

15H.

16.

17.

*18H.

In Exercises 19–24, decide whether the graph has an Euler cycle. If the graph has an Euler cycle, exhibit one.

19H.   Exercise 12   **20.**   Exercise 13

21.   Exercise 14   **22H.** Figure 3.2.4

23.

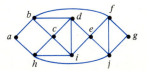

24.

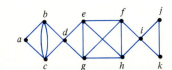

25H. The following graph is continued to an arbitrary, finite depth. Does the graph contain an Euler cycle? If the answer is yes, describe one.

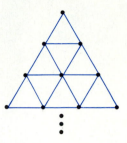

26. When does K_n contain an Euler cycle?

27. When does $K_{m,n}$ contain an Euler cycle?

28H. For which values of m and n does the graph contain an Euler cycle?

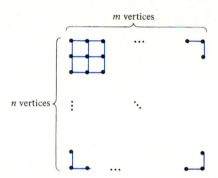

29. Give an example of a graph that contains a cycle which includes all the edges, but which is not an Euler cycle.

30. Let G be a connected graph. Suppose that an edge e is in a cycle. Show that G with e removed is still connected.

31H. Give an example of a connected graph such that the removal of any edge results in a graph that is not connected. (Assume that removing an edge does not remove any vertices.)

*32. Can a knight move around a chessboard and return to its original position making every move exactly once? (A move is

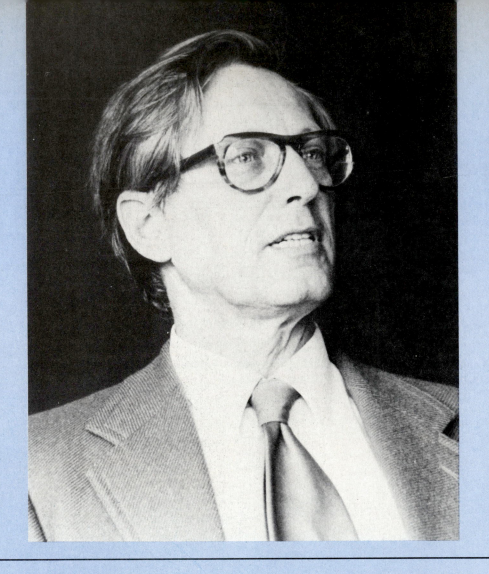

Nicolaas G. de Bruijn (1918–   )

Nicolaas G. de Bruijn is Emeritus Professor in the Department of Mathematics and Computer Science at Technological University Eindhoven in The Netherlands. In addition to holding academic positions at Technological University Eindhoven, Technological University Delft, and the University of Amsterdam, he was a consultant at Philips Research Laboratories (Eindhoven) from 1960 until 1984. He is the author of over 150 journal articles on geometry, number theory, analysis, combinatorics, computer science, logic, and mathematical languages. [*Photo by P. van Ern de Boas, courtesy of N. G. de Bruijn*]

considered to be made when the move is made in either
direction.)

33. A domino is a rectangle divided into two squares with each
square numbered one of $0, 1, \ldots, 6$. Two squares on a
single domino can have the same number. Can the distinct
dominoes be arranged in a circle so that touching dominoes
have adjacent squares with identical numbers?

*34H. Show that the maximum number of edges in a simple,
disconnected graph with n vertices is $(n - 1)(n - 2)/2$.

*35. Show that the maximum number of edges in a simple, bipartite
graph with n vertices is $\lfloor n^2/4 \rfloor$.

A vertex v in a connected graph G is an **articulation point** if the
removal of v and all edges incident on v disconnects G.

36H. Give an example of a graph with six vertices that has exactly
two articulation points.

37. Give an example of a graph with six vertices that has no
articulation points.

Let G be a directed graph and let v be a vertex in G. The **indegree**
of v, in(v), is the number of edges of the form (w, v). The **outdegree**
of v, out(v), is the number of edges of the form (v, w). A **directed
Euler cycle** in G is a sequence of edges of the form

$$(v_0, v_1), (v_1, v_2), \ldots, (v_{n-1}, v_n),$$

where $v_0 = v_n$, every edge in G occurs exactly one time, and all vertices appear.

It can be shown that a directed graph G contains a directed Euler
cycle if and only if the undirected graph obtained by ignoring the di-
rections of the edges of G is connected and in(v) = out(v) for every
vertex v in G.

A **de Bruijn sequence** for n (in 0's and 1's) is a sequence

$$a_1, \ldots, a_{2^n}$$

of 2^n bits having the property that if s is a bit string of length n, for
some m,

$$s = a_m a_{m+1} \cdots a_{m+n-1}. \qquad (3.2.2)$$

In (3.2.2), we define $a_{2^n + i} = a_i$ for $i = 1, \ldots, 2^n - 1$.

Sir William Rowan Hamilton *(1805–1865)*

Sir William Rowan Hamilton was one of Ireland's greatest scholars. He was professor of astronomy at the University of Dublin, where he published articles in physics and mathematics. Hamilton was a child prodigy. By age 8 he could read Latin, Greek, Hebrew, Italian, and French. In mathematics, Hamilton is most famous for inventing the quaternions, a generalization of the complex number system. The quaternions provided inspiration for the development of modern abstract algebra. In this connection, Hamilton introduced the term ''vector.'' [*Photo courtesy of the Ann Ronan Picture Library*]

38H. Verify that 00011101 is a de Bruijn sequence for $n = 3$.

39. Let G be a directed graph with vertices corresponding to all bit strings of length $n - 1$. A directed edge exists from vertex $x_1 \cdots x_{n-1}$ to $x_2 \cdots x_n$. Show that a directed Euler cycle in G corresponds to a de Bruijn sequence.

*40. Show that there is a de Bruijn sequence for every $n = 1, 2, \ldots$.

*41H. A **closed path** is a path from v to v. Show that a connected graph G is bipartite if and only if every closed path in G has even length.

†3.3 Hamiltonian Cycles and the Traveling Salesperson Problem

Sir William Rowan Hamilton marketed a puzzle in the mid-1800s in the form of a dodecahedron (see Figure 3.3.1). Each corner bore the name of a city and the problem was to start at any city, travel along the edges, visit each city exactly one time, and return to the initial city. The graph of the edges of the dodecahedron is given in Figure 3.3.2. We can solve Hamilton's puzzle if we can find a cycle in the graph of Figure 3.3.2 that contains each vertex exactly once (except for the starting and ending vertex that appears twice). See if you can find a solution before looking at a solution given in Figure 3.3.3.

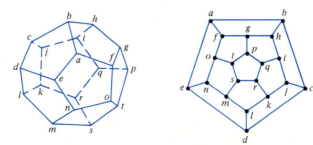

Figure 3.3.1   **Figure 3.3.2**

†This section can be omitted without loss of continuity.

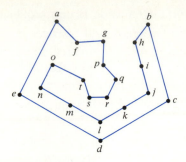

Figure 3.3.3

In honor of Hamilton, we call a cycle in a graph G that contains each vertex in G exactly once, except for the starting and ending vertex that appears twice, a **Hamiltonian cycle**.

Example 3.3.1. The cycle (a, b, c, d, e, f, g, a) is a Hamiltonian cycle for the graph of Figure 3.3.4.

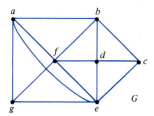

Figure 3.3.4

The problem of finding a Hamiltonian cycle in a graph sounds similar to the problem of finding an Euler cycle in a graph. An Euler cycle visits each edge once, whereas a Hamiltonian cycle visits each vertex once; however, the problems are actually quite distinct. For example, the graph G of Figure 3.3.4 does not have an Euler cycle since there are vertices of odd degree, yet Example 3.3.1 showed that G has a Hamiltonian cycle. Furthermore, unlike the situation for Euler cycles (see Theorems 3.2.17 and 3.2.18), no easily verified necessary and sufficient conditions are known for the existence of a Hamiltonian cycle in a graph.

The following examples show that sometimes we can argue that a graph does not contain a Hamiltonian cycle.

Example 3.3.2. Show that the graph of Figure 3.3.5 does not contain a Hamiltonian cycle.

Since there are five vertices, a Hamiltonian cycle must have five edges. Suppose that we could eliminate edges from the graph leaving just a Hamiltonian cycle. We would have to eliminate one edge incident at v_2 and one edge incident at v_4, since each vertex in a Hamiltonian cycle has degree 2. But this leaves only four edges—not enough for a Hamiltonian cycle of length 5. Therefore, the graph of Figure 3.3.5 does not contain a Hamiltonian cycle.

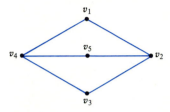

Figure 3.3.5

We must be careful not to count an eliminated edge more than once when using an argument like that in Example 3.3.2 to show that a graph does not have a Hamiltonian cycle. Notice in Example 3.3.2 (which refers to Figure 3.3.5) that if we eliminate one edge incident at v_2 and one edge incident at v_4, these edges are distinct. Therefore, we are correct in reasoning that we must eliminate two edges from the graph of Figure 3.3.5 to produce a Hamiltonian cycle.

As an example of double counting, consider the following *faulty* argument that purports to show that the graph of Figure 3.3.6 has no Hamiltonian cycle. Since there are five vertices, a Hamiltonian cycle must have five edges. Suppose that we could eliminate edges from the graph to produce a Hamiltonian cycle. We would have to eliminate two edges incident at c and one edge incident at each of a, b, d, and e. This

Figure 3.3.6

leaves two edges—not enough for a Hamiltonian cycle. Therefore, the graph of Figure 3.3.6 does not contain a Hamiltonian cycle. The error in this argument is that if we eliminate two edges incident at c (as we must do), we also eliminate edges incident at two of a, b, d, or e. We must not count the two eliminated edges incident at the two vertices again. Notice that the graph of Figure 3.3.6 does have a Hamiltonian cycle.

Example 3.3.3.   Show that the graph G of Figure 3.3.7 does not contain a Hamiltonian cycle.

Suppose that G has a Hamiltonian cycle H. The edges (a, b), (a, g), (b, c), and (c, k) must be in H since each vertex in a Hamiltonian cycle has degree 2. Thus edges (b, d) and (b, f) are not in H. Therefore, edges (g, d), (d, e), (e, f), and (f, k) are in H. The edges now known to be in H form a cycle C. Adding an additional edge to C will give some vertex in H degree greater than 2. This contradiction shows that G does not have a Hamiltonian cycle.

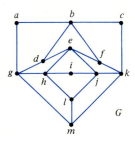

Figure 3.3.7

The **traveling salesperson problem** is related to the problem of finding a Hamiltonian cycle in a graph. (We referred briefly to a variant of the traveling salesperson problem in Section 3.1.) The problem is: Given a weighted graph G, find a minimum length Hamiltonian cycle in G. If we think of the vertices in a weighted graph as cities and the edge weights as distances, the traveling salesperson problem is to find a shortest route in which the salesperson can visit each city one time, starting and ending at the same city.

Example 3.3.4. The cycle $C = (a, b, c, d, a)$ is a Hamiltonian cycle for the graph G of Figure 3.3.8. Replacing any of the edges in C by either of the edges labeled 11 would increase the length of C; thus C is a minimum-length Hamiltonian cycle for G. Thus C solves the traveling salesperson problem for G.

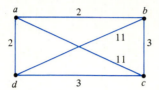

Figure 3.3.8

There are practical methods for finding Euler cycles, provided that they exist, in graphs with several hundred vertices. This is not true for either the Hamiltonian cycle problem or the traveling salesperson problem. All known methods for solving these problems are impractical for some graphs with even modest numbers of vertices. For this reason, methods that produce near-minimum-length cycles are often used for problems that ask for a solution to the traveling salesperson problem. Instant fame awaits the discoverer of a practical method for solving the Hamiltonian cycle problem or the traveling salesperson problem.

EXERCISES

Find a Hamiltonian cycle in each graph.

1H.

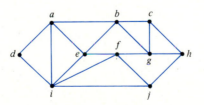

2.

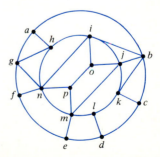

Show that none of the graphs contains a Hamiltonian cycle.

3H.

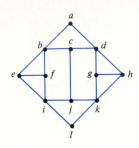

4.

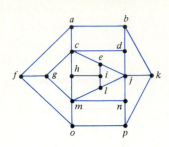

5.

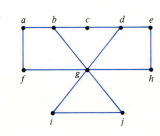

Determine whether or not each graph contains a Hamiltonian cycle. If there is a Hamiltonian cycle, exhibit it; otherwise, give an argument which shows that there is no Hamiltonian cycle.

6H.

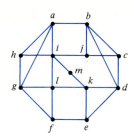

7.

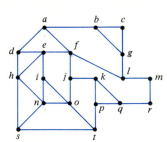

8.

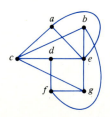

9H.   Give an example of a graph that has an Euler cycle but contains no Hamiltonian cycle.

10.   Give an example of a graph that has an Euler cycle which is also a Hamiltonian cycle.

11.   Give an example of a graph that has an Euler cycle and a Hamiltonian cycle which are not identical.

*12H. For which values of m and n does the graph of Exercise 28, Section 3.2, contain a Hamiltonian cycle?

13.   Show that if $n \geq 3$, the complete graph on n vertices K_n contains a Hamiltonian cycle.

14.   When does the complete bipartite graph $K_{m,n}$ contain a Hamiltonian cycle?

15H. Show that the cycle (e, b, a, c, d, e) provides a solution to the traveling salesperson problem for the graph shown.

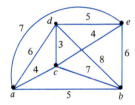

16.   Solve the traveling salesperson problem for the graph.

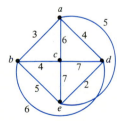

17.   Let G be a bipartite graph with disjoint vertex sets V_1 and V_2. Show that if G has a Hamiltonian cycle, V_1 and V_2 have the same number of elements.

18H. Let the vertices of a graph be the squares of an ordinary chessboard. Describe an edge set appropriate for modeling the problem of determining whether a knight can move around the

chessboard, return to its original position, and visit every
square exactly once. What does a solution correspond to in
graph terms?

3.4 Matrix Representations of Graphs

In Sections 3.1 to 3.3 we represented a graph by drawing it. Sometimes,
as for example in using a computer to analyze a graph, we need a more
formal representation. Our first method of representing a graph uses the
adjacency matrix.

Example 3.4.1 Adjacency Matrix.   Consider the graph of Figure 3.4.1. To obtain
the adjacency matrix of this graph, we first select an ordering of the
vertices, say a, b, c, d, e. Next, we label the rows and columns of a
matrix with the ordered vertices. The entry in this matrix is 1 if the
row and column vertices are adjacent and 0 otherwise. The adjacency
matrix for this graph is

$$\begin{array}{c@{}c} & \begin{array}{ccccc} a & b & c & d & e \end{array} \\ \begin{array}{c} a \\ b \\ c \\ d \\ e \end{array} & \left(\begin{array}{ccccc} 0 & 1 & 0 & 0 & 1 \\ 1 & 0 & 1 & 0 & 1 \\ 0 & 1 & 1 & 0 & 1 \\ 0 & 0 & 0 & 0 & 1 \\ 1 & 1 & 1 & 1 & 0 \end{array}\right). \end{array}$$

Notice that while the adjacency matrix allows us to represent loops,
it does not allow us to represent parallel edges. Also, note that if the

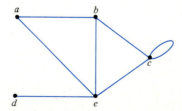

Figure 3.4.1

graph is a simple graph, we can obtain the degree of a vertex by summing its row or column.

The adjacency matrix is not a very efficient way to represent a graph. Since the matrix is symmetric about the main diagonal (the elements on a line from the upper left corner to the lower right corner), the information, except that on the main diagonal, appears twice.

Example 3.4.2. The adjacency matrix of the simple graph of Figure 3.4.2 is

$$
A = \begin{array}{c} \\ a \\ b \\ c \\ d \\ e \end{array}
\begin{array}{c} \begin{array}{ccccc} a & b & c & d & e \end{array} \\
\begin{pmatrix}
0 & 1 & 0 & 1 & 0 \\
1 & 0 & 1 & 0 & 1 \\
0 & 1 & 0 & 1 & 1 \\
1 & 0 & 1 & 0 & 0 \\
0 & 1 & 1 & 0 & 0
\end{pmatrix}
\end{array}.
$$

We will show that if A is the adjacency matrix of a simple graph G, the powers of A,

$$A, A^2, A^3, \dots,$$

count the number of paths of various lengths. More precisely, if the vertices of G are labeled $1, 2, \dots$, the ijth entry in the matrix A^n is equal to the number of paths from i to j of length n. For example, suppose that we square the matrix A of Example 3.4.2 to obtain

$$
A^2 = \begin{pmatrix}
0 & 1 & 0 & 1 & 0 \\
1 & 0 & 1 & 0 & 1 \\
0 & 1 & 0 & 1 & 1 \\
1 & 0 & 1 & 0 & 0 \\
0 & 1 & 1 & 0 & 0
\end{pmatrix}
\begin{pmatrix}
0 & 1 & 0 & 1 & 0 \\
1 & 0 & 1 & 0 & 1 \\
0 & 1 & 0 & 1 & 1 \\
1 & 0 & 1 & 0 & 0 \\
0 & 1 & 1 & 0 & 0
\end{pmatrix}
= \begin{array}{c} \\ a \\ b \\ c \\ d \\ e \end{array}
\begin{array}{c} \begin{array}{ccccc} a & b & c & d & e \end{array} \\
\begin{pmatrix}
2 & 0 & 2 & 0 & 1 \\
0 & 3 & 1 & 2 & 1 \\
2 & 1 & 3 & 0 & 1 \\
0 & 2 & 0 & 2 & 1 \\
1 & 1 & 1 & 1 & 2
\end{pmatrix}
\end{array}.
$$

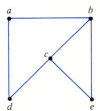

Figure 3.4.2

Consider the entry for row a, column c in A^2, obtained by multiplying pairwise the entries in row a by the entries in column c of the matrix A and summing:

$$a\ (0\ \ 1\ \ 0\ \ 1\ \ 0) \begin{pmatrix} 0 \\ 1 \\ 0 \\ 1 \\ 1 \end{pmatrix} \begin{matrix} \\ b \\ \\ d \\ \\ \end{matrix} = 0 \cdot 0 + 1 \cdot 1 + 0 \cdot 0 + 1 \cdot 1 + 0 \cdot 1$$

$$= 2.$$

The only way a nonzero product appears in this sum is if both entries to be multiplied are 1. This happens if there is a vertex v whose entry in row a is 1 and whose entry in column c is 1. In other words, there must be edges of the form (a, v) and (v, c). Such edges form a path (a, v, c) of length 2 from a to c and each such path increases the sum by 1. In this example, the sum is 2 because there are two paths

$$(a, b, c), \quad (a, d, c)$$

of length 2 from a to c. In general, the entry in row x and column y of the matrix A^2 is the number of paths of length 2 from vertex x to vertex y.

The entries on the main diagonal of A^2 give the degrees of the vertices (when the graph is a simple graph). Consider, for example, vertex c. The degree of c is 3 since c is incident on the three edges (c, b), (c, d), and (c, e). But each of these edges can be converted to a path of length 2 from c to c:

$$(c, b, c), \quad (c, d, c), \quad (c, e, c).$$

Similarly, a path of length 2 from c to c defines an edge incident on c. Thus the number of paths of length 2 from c to c is 3, the degree of c.

We now use induction to show that the entries in the nth power of an adjacency matrix give the number of paths of length n.

Theorem 3.4.3. *If A is the adjacency matrix of a simple graph, the ijth entry of A^n is the number of paths of length n from vertex i to vertex j, $n = 1, 2, \ldots$.*

PROOF.   We will use induction on n.

In case $n = 1$, A^1 is simply A. The ijth entry is 1 if there is an edge from i to j, which is a path of length 1, and 0 otherwise. Thus the theorem is true in case $n = 1$. The basis step has been verified.

Assume that the theorem is true for n. Now

$$A^{n+1} = A^n A$$

so that the ikth entry in A^{n+1} is obtained by multiplying pairwise the elements in the ith row of A^n by the elements in the kth column of A and summing:

$$\text{kth column of } A$$

$$\text{ith row of } A^n \; (s_1 \quad s_2 \quad \cdots \quad s_j \quad \cdots \quad s_m) \begin{pmatrix} t_1 \\ t_2 \\ \cdot \\ \cdot \\ \cdot \\ t_j \\ \cdot \\ \cdot \\ \cdot \\ t_m \end{pmatrix}$$

$$= s_1 t_1 + s_2 t_2 + \cdots + s_j t_j + \cdots + s_m t_m$$
$$= ik\text{th entry in } A^{n+1}.$$

By induction, s_j gives the number of paths of length n from i to j in the graph G. Now t_j is either 0 or 1. If t_j is 0, there is no edge from j to k, so there are $s_j t_j = 0$ paths of length $n + 1$ from i to k, where the last edge is (j, k). If t_j is 1, there is an edge from vertex j to vertex k (see Figure 3.4.3). Since there are s_j paths of length n from vertex i to vertex j, there are $s_j t_j = s_j$ paths of length $n + 1$ from i to k, where the last

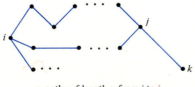

s_j paths of length n from i to j

Figure 3.4.3

edge is (j, k) (see Figure 3.4.3). Summing over all j, we will count all paths of length $n + 1$ from i to k. Thus the ikth entry in A^{n+1} gives the number of paths of length $n + 1$ from i to k and the inductive step is verified.

By the Principle of Mathematical Induction, the theorem is established.   ■

Example 3.4.4.   For the matrix A of Example 3.4.2 we find that

$$
A^4 = \begin{array}{c} \\ a \\ b \\ c \\ d \\ e \end{array}
\begin{array}{ccccc}
a & b & c & d & e \\
\left(\begin{array}{ccccc}
9 & 3 & 11 & 1 & 6 \\
3 & 15 & 7 & 11 & 8 \\
11 & 7 & 15 & 3 & 8 \\
1 & 11 & 3 & 9 & 6 \\
6 & 8 & 8 & 6 & 8
\end{array}\right)
\end{array}.
$$

The entry from row d, column e is 6, which means there are six paths of length 4 from d to e. By inspection, we find them to be

$$(d, a, d, c, e), \quad (d, c, d, c, e), \quad (d, a, b, c, e),$$
$$(d, c, e, c, e), \quad (d, c, e, b, e), \quad (d, c, b, c, e).$$

Another useful matrix representation of a graph is known as the **incidence matrix**.

Example 3.4.5 Incidence Matrix.   To obtain the incidence matrix of the graph in Figure 3.4.4, we label the rows with the vertices and the columns with the edges (in some arbitrary order). The entry for row v and column e is 1 if e is incident on v and 0 otherwise. Thus the incidence matrix for the graph of Figure 3.4.4 is

$$
\begin{array}{c} \\ v_1 \\ v_2 \\ v_3 \\ v_4 \\ v_5 \end{array}
\begin{array}{ccccccc}
e_1 & e_2 & e_3 & e_4 & e_5 & e_6 & e_7 \\
\left(\begin{array}{ccccccc}
1 & 1 & 1 & 0 & 0 & 0 & 0 \\
0 & 0 & 1 & 1 & 1 & 0 & 1 \\
0 & 0 & 0 & 0 & 0 & 1 & 0 \\
1 & 1 & 0 & 1 & 0 & 0 & 0 \\
0 & 0 & 0 & 0 & 1 & 1 & 0
\end{array}\right)
\end{array}.
$$

A column such as e_7 is understood to represent a loop.

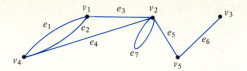

Figure 3.4.4

The incidence matrix allows us to represent both parallel edges and loops. Notice that in a graph without loops each column has two 1's and that the sum of a row gives the degree of the vertex identified with that row.

EXERCISES

In Exercises 1–6, write the adjacency matrix of each graph.

1H.

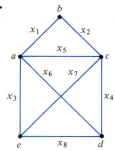

2.

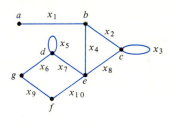

3.

4H. The graph of Figure 3.2.2

5. The complete graph on five vertices K_5

6. The complete bipartite graph $K_{2,3}$

In Exercises 7–12, write the incidence matrix of each graph.

7H. The graph of Exercise 1

8. The graph of Exercise 2

9.   The graph of Exercise 3

10H. The graph of Figure 3.2.1

11.   The complete graph on five vertices K_5

12.   The complete bipartite graph $K_{2,3}$

In Exercises 13–17, draw the graph represented by each adjacency matrix.

13H. c $\begin{array}{c} & a & b & c & d & e \\ a & \begin{pmatrix} 1 & 0 & 0 & 1 & 0 \\ b & 0 & 0 & 1 & 0 & 1 \\ c & 0 & 1 & 1 & 1 & 1 \\ d & 1 & 0 & 1 & 0 & 0 \\ e & 0 & 1 & 1 & 0 & 0 \end{pmatrix} \end{array}$.

14. c $\begin{array}{c} & a & b & c & d & e \\ a & \begin{pmatrix} 0 & 1 & 0 & 0 & 0 \\ b & 1 & 0 & 0 & 0 & 0 \\ c & 0 & 0 & 0 & 1 & 1 \\ d & 0 & 0 & 1 & 0 & 1 \\ e & 0 & 0 & 1 & 1 & 1 \end{pmatrix} \end{array}$.

15. c $\begin{array}{c} & a & b & c & d & e & f \\ a & \begin{pmatrix} 0 & 0 & 1 & 0 & 0 & 1 \\ b & 0 & 1 & 0 & 1 & 1 & 0 \\ c & 1 & 0 & 0 & 0 & 0 & 1 \\ d & 0 & 1 & 0 & 0 & 1 & 0 \\ e & 0 & 1 & 0 & 1 & 0 & 0 \\ f & 1 & 0 & 1 & 0 & 0 & 0 \end{pmatrix} \end{array}$.

16H. c $\begin{array}{c} & a & b & c & d & e & f \\ a & \begin{pmatrix} 1 & 1 & 1 & 1 & 0 & 1 \\ b & 1 & 0 & 1 & 1 & 1 & 0 \\ c & 1 & 1 & 0 & 1 & 1 & 1 \\ d & 1 & 1 & 1 & 0 & 1 & 1 \\ e & 0 & 1 & 1 & 1 & 0 & 1 \\ f & 1 & 0 & 1 & 1 & 1 & 0 \end{pmatrix} \end{array}$.

17.   The 7×7 matrix whose ijth entry is 1 if $i + 1$ divides $j + 1$ or $j + 1$ divides $i + 1$ and whose ijth entry is 0 otherwise.

18.   Write the adjacency matrices of the components of the graphs given by the adjacency matrices of Exercises 13–17.

19H. Compute the squares of the adjacency matrices of K_5 and the graphs of Exercises 1 and 3.

20.   Let A be the adjacency matrix for the graph of Exercise 1. What is the entry in row a, column d of A^5?

21.   Suppose that a graph has an adjacency matrix of the form

$$A = \left(\begin{array}{c|c} & A' \\ \hline A'' & \end{array} \right),$$

where all entries of the submatrices A' and A'' are 0. What must the graph look like?

22H. Repeat Exercise 21 with "adjacency" replaced by "incidence."

23.   How might the definition of adjacency matrix be changed to allow for the representation of parallel edges?

24. Let A be an adjacency matrix of a graph. Why is A^n symmetric about the main diagonal for every positive integer n?

In Exercises 25 and 26, draw the graphs represented by the incidence matrices.

25H. $\quad \begin{array}{c} a \\ b \\ c \\ d \\ e \end{array} \begin{pmatrix} 1 & 0 & 0 & 0 & 0 & 1 \\ 0 & 1 & 1 & 0 & 1 & 0 \\ 1 & 0 & 0 & 1 & 0 & 0 \\ 0 & 1 & 0 & 1 & 0 & 0 \\ 0 & 0 & 1 & 0 & 1 & 1 \end{pmatrix}$

26. $\quad \begin{array}{c} a \\ b \\ c \\ d \\ e \end{array} \begin{pmatrix} 0 & 1 & 0 & 0 & 1 & 1 \\ 0 & 1 & 1 & 0 & 1 & 0 \\ 0 & 0 & 0 & 0 & 0 & 1 \\ 1 & 0 & 0 & 1 & 0 & 0 \\ 1 & 0 & 0 & 1 & 0 & 0 \end{pmatrix}$

27. What must a graph look like if some row of its incidence matrix consists only of 0's?

28H. Let A be the adjacency matrix of a graph G with n vertices. Let

$$Y = A + A^2 + \cdots + A^{n-1}.$$

If some off-diagonal entry in the matrix Y is zero, what can you say about the graph G?

Exercises 29–32 refer to the adjacency matrix A of K_5.

29H. Let n be a positive integer. Explain why all the diagonal elements of A^n coincide and all the off-diagonal elements of A^n coincide.

Let d_n be the common value of the diagonal elements of A^n and let a_n be the common value of the off-diagonal elements of A^n.

*30. Show that

$$d_{n+1} = 4a_n; \qquad a_{n+1} = d_n + 3a_n; \qquad a_{n+1} = 3a_n + 4a_{n-1}.$$

*31. Use induction to show that

$$a_n = \tfrac{1}{5}[4^n + (-1)^{n+1}].$$

32H. Show that

$$d_n = \tfrac{4}{5}[4^{n-1} + (-1)^n].$$

*33. Derive results similar to Exercises 30–32 for the adjacency matrix A of the graph K_m.

*34. Let A be the adjacency matrix of the graph $K_{m,n}$. Find a formula for the entries in A^j.

3.5 Isomorphic Graphs

The following instructions are given to two persons who cannot see each other's paper: "Draw and label five vertices a, b, c, d, and e. Connect a and b, b and c, c and d, d and e, and a and e." The graphs G_1 and G_2 produced are shown in Figure 3.5.1. Surely, G_1 and G_2 are essentially the same even though they appear dissimilar. (The graphs G_1 and G_2 are not equal since the vertex set $\{a, b, c, d, e\}$ of G_1 is not equal to the vertex set $\{A, B, C, D, E\}$ of G_2.) In this section we discuss what it means for two graphs to be "essentially the same." Such graphs are said to be **isomorphic**. We give the formal definition (Definition 3.5.1) after a bit more discussion.

The adjacency matrix of graph G_1 in Figure 3.5.1 relative to the vertex ordering a, b, c, d, e,

$$
\begin{array}{c}
\\ a \\ b \\ c \\ d \\ e
\end{array}
\begin{array}{c}
\begin{array}{ccccc} a & b & c & d & e \end{array} \\
\left(\begin{array}{ccccc}
0 & 1 & 0 & 0 & 1 \\
1 & 0 & 1 & 0 & 0 \\
0 & 1 & 0 & 1 & 0 \\
0 & 0 & 1 & 0 & 1 \\
1 & 0 & 0 & 1 & 0
\end{array}\right),
\end{array}
$$

is equal to the adjacency matrix of graph G_2 in Figure 3.5.1 relative to the vertex ordering A, B, C, D, E,

$$
\begin{array}{c}
\\ A \\ B \\ C \\ D \\ E
\end{array}
\begin{array}{c}
\begin{array}{ccccc} A & B & C & D & E \end{array} \\
\left(\begin{array}{ccccc}
0 & 1 & 0 & 0 & 1 \\
1 & 0 & 1 & 0 & 0 \\
0 & 1 & 0 & 1 & 0 \\
0 & 0 & 1 & 0 & 1 \\
1 & 0 & 0 & 1 & 0
\end{array}\right).
\end{array}
\tag{3.5.1}
$$

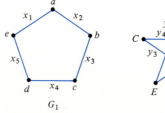

Figure 3.5.1

Simple graphs like G_1 and G_2 that have equal adjacency matrices (relative to some ordering of their vertices) are "essentially the same." We use this observation to *define* what it means for graphs to be isomorphic.

Definition 3.5.1.   Simple graphs† G_1 and G_2 are *isomorphic* if for some ordering of the vertices, their adjacency matrices are equal.

According to Definition 3.5.1, isomorphic simple graphs have equal adjacency matrices *for some ordering of the vertices*. In general, if we change the ordering of the vertices for one of the graphs, the adjacency matrices will not be equal. For example, the adjacency matrix of the graph G_1 of Figure 3.5.1 relative to the ordering e, a, d, b, c is

$$
\begin{array}{c c}
 & \begin{array}{ccccc} e & a & d & b & c \end{array} \\
\begin{array}{c} e \\ a \\ d \\ b \\ c \end{array} &
\left(\begin{array}{ccccc}
0 & 1 & 1 & 0 & 0 \\
1 & 0 & 0 & 1 & 0 \\
1 & 0 & 0 & 0 & 1 \\
0 & 1 & 0 & 0 & 1 \\
0 & 0 & 1 & 1 & 0
\end{array}\right),
\end{array}
$$

which is not equal to the adjacency matrix (3.5.1) of the graph G_2 of Figure 3.5.1. Nevertheless, the graphs G_1 and G_2 are isomorphic because there is *some* ordering of the vertices (namely a, b, c, d, e for G_1 and A, B, C, D, E for G_2) for which their adjacency matrices are equal.

One way to determine whether two simple graphs G_1 and G_2 are isomorphic is to order the vertices of one of the graphs, say G_1, and write its adjacency matrix. Then begin listing adjacency matrices of G_2 corresponding to the different orderings of the vertices of G_2. In case one of the adjacency matrices of G_2 is equal to the adjacency matrix of G_1, we know that G_1 and G_2 are isomorphic graphs. If none of the adjacency matrices of G_2 is equal to the adjacency matrix of G_1, G_1 and G_2 are not isomorphic graphs. The problem with this technique is that even if the number of vertices is modest, the number of orderings of vertices can be quite large, which will cause the method to be very time consuming. For example, 20 vertices can be ordered in 20! (which is approximately 2.4×10^{18}) ways (see Theorem 6.3.3). Every known method of testing graphs to see whether they are isomorphic is not much better than this "list all" technique, at least for certain "bad" graphs. At the same time, there are methods of testing to see whether graphs are

---

†To simplify the discussion, in this section we restrict our attention to simple graphs.

isomorphic that work well in practice; that is, they are much faster than the "list all" technique for "most pairs" of graphs.

The following is one way to show that two simple graphs G_1 and G_2 are *not* isomorphic. Find a property of G_1 that G_2 does *not* have, but which G_2 *would* have if G_1 and G_2 were isomorphic. Such a property is called an **invariant**. More precisely, a property P is an invariant, if whenever G_1 and G_2 are isomorphic graphs:

If G_1 has property P, G_2 also has property P.

By Definition 3.5.1, if simple graphs G_1 and G_2 are isomorphic, for some ordering of their vertices, their adjacency matrices are equal. If these matrices are $n \times n$, it follows that both graphs have n vertices. Therefore, if G_1 and G_2 are isomorphic, then G_1 and G_2 have the same number of vertices. Therefore, if n is a nonnegative integer, the property "has n vertices" is an invariant.

Example 3.5.2.   The graphs G_1 and G_2 in Figure 3.5.2 are not isomorphic since G_1 has four vertices and G_2 has three vertices and "has four vertices" is an invariant.

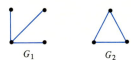

G_1 G_2

Figure 3.5.2

Suppose that simple graphs G_1 and G_2 are isomorphic. Let $v_1, \ldots, v_n$ denote the vertex ordering for G_1 and $w_1, \ldots, w_n$ denote the vertex ordering for G_2 that makes their adjacency matrices equal. The number of 1's in the first row is the number of edges on which v_1 or w_1 is incident. In general, the number of 1's in the row i is the number of edges on which v_i or w_i is incident. It follows that G_1 and G_2 have the same number of edges. Therefore, if e is a nonnegative integer, the property "has e edges" is an invariant.

Example 3.5.3.   The graphs G_1 and G_2 in Figure 3.5.3 are not isomorphic since G_1 has seven edges and G_2 has six edges and "has seven edges" is an invariant.

G_1　　　　　　　　　G_2

Figure 3.5.3

Example 3.5.4. We show that if k is a positive integer, "has a vertex of degree k" is an invariant.

Suppose that simple graphs G_1 and G_2 are isomorphic and that G_1 has a vertex of degree k. We must show that G_2 has a vertex of degree k. Let $v_1, \ldots, v_n$ denote the vertex ordering for G_1 and $w_1, \ldots, w_n$ denote the vertex ordering for G_2 that makes their adjacency matrices equal. Suppose that the degree of vertex v_i is k. The number of 1's in row i is equal to the number of edges on which v_i is incident. Thus there are k 1's in row i. Since the adjacency matrices of G_1 and G_2 are equal, the number of 1's in row i is also equal to the number of edges on which w_i is incident. Therefore, w_i has degree k and "has a vertex of degree k" is an invariant.

Example 3.5.5. Since "has a vertex of degree 3" is an invariant, the graphs G_1 and G_2 of Figure 3.5.4 are not isomorphic; G_1 has vertices (a and f) of degree 3, but G_2 does not have a vertex of degree 3. Notice that G_1 and G_2 have the same numbers of edges and vertices.

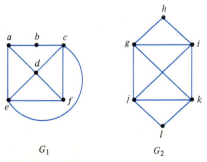

G_1　　　　　　　　　G_2

Figure 3.5.4

Another invariant that is sometimes useful is "has a simple cycle of length k." We leave the proof that this property is an invariant to the exercises (Exercise 11).

Example 3.5.6.   Since "has a simple cycle of length 3" is an invariant, the graphs G_1 and G_2 of Figure 3.5.5 are not isomorphic; the graph G_2 has a simple cycle of length 3, but all simple cycles in G_2 have length at least 4. Notice that G_1 and G_2 have the same numbers of edges and vertices and that every vertex in G_1 or G_2 has degree 4.

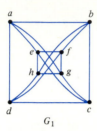 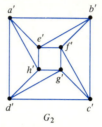

Figure 3.5.5

It would be easy to test whether a pair of graphs is isomorphic if we could find a small number of easily checked invariants that isomorphic graphs and *only* isomorphic graphs share. Unfortunately, no one has succeeded in finding such a set of invariants.

We conclude by giving an alternative characterization of isomorphic graphs. Suppose that G_1 and G_2 are isomorphic simple graphs. Let $v_1, \ldots, v_n$ denote a vertex ordering for G_1 and let $w_1, \ldots, w_n$ denote a vertex ordering for G_2 that makes their adjacency matrices equal. Define a function f from $\{v_1, \ldots, v_n\}$ to $\{w_1, \ldots, w_n\}$ by the rule $f(v_i) = w_i$. Then f is a one-to-one, onto function from the vertex set of G_1 to the vertex set of G_2. Moreover, if v_i and v_j are adjacent vertices in G_1, $f(v_i)$ and $f(v_j)$ are adjacent vertices in G_2. The reason is that if v_i and v_j are adjacent vertices, the entry in row i column j of the adjacency matrix is 1. Since the adjacency matrices for G_1 and G_2 are equal, there is an edge between w_i and w_j in G_2; that is, w_i and w_j are adjacent. Similarly, if w_i and w_j are adjacent vertices in G_2, v_i and v_j are adjacent vertices in G_1. Furthermore, the existence of a one-to-one, onto function f from the vertex set of a simple graph G_1 to the vertex set of a simple graph G_2 having the property "v_i and v_j are adjacent vertices in G_1 if and only if $f(v_i)$ and $f(v_j)$ are adjacent vertices in G_2" implies that G_1 is isomorphic to G_2. We formally state this result as Theorem 3.5.7.

Theorem 3.5.7. *Let G_1 and G_2 be simple graphs. The following are equivalent.*
(a) G_1 and G_2 are isomorphic.
(b) There is a one-to-one, onto function f from the vertex set of G_1 to the vertex set of G_2 satisfying: Vertices v_i and v_j are adjacent in G_1 if and only if the vertices $f(v_i)$ and $f(v_j)$ are adjacent in G_2.

PROOF. Just before the statement of the theorem, we showed that if (a) holds, then (b) holds.
The converse is given as Exercise 25. ■

If G_1 and G_2 are isomorphic simple graphs, we call the function f of Theorem 3.5.7b an **isomorphism** of G_1 and G_2.

Example 3.5.8. The graphs G_1 and G_2 of Figure 3.5.1 are isomorphic since their adjacency matrices relative to the vertex orderings a, b, c, d, e for G_1 and A, B, C, D, E for G_2 are equal. The discussion preceding Theorem 3.5.7 shows that the function which assigns the ith element in the vertex ordering for G_1, the ith element in the vertex ordering for G_2, is an isomorphism. Thus the function f defined by

$$f(a) = A, \quad f(b) = B, \quad f(c) = C, \quad f(d) = D, \quad f(e) = E$$

is an isomorphism of G_1 and G_2.

EXERCISES

In Exercises 1–10, determine whether the graphs G_1 and G_2 are isomorphic. If the graphs are isomorphic, give orderings of their vertices that produce equal adjacency matrices and specify the isomorphism function of Theorem 3.5.7b. If the graphs are not isomorphic, give an invariant that the graphs do not share.

1H.

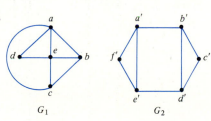

G_1 G_2

2.

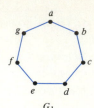 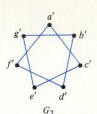

G_1 G_2

3.

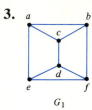 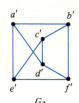

G_1 G_2

4H.

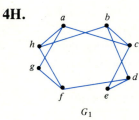 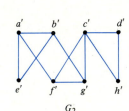

G_1 G_2

5.

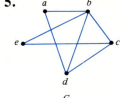 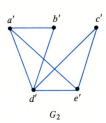

G_1 G_2

6.

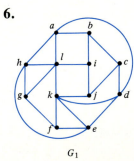 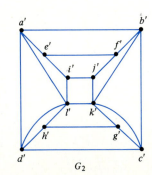

G_1 G_2

7H.

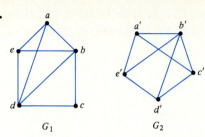

G_1 G_2

8.

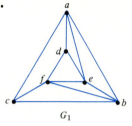

G_1 G_2

9.

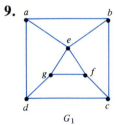 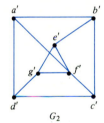

G_1 G_2

10H.

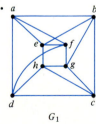 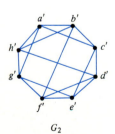

G_1 G_2

In Exercises 11–13, show that the property is an invariant.

11H. Has a simple cycle of length k

12. Has n vertices of degree k

13. Has an edge (v, w), where $\delta(v) = i$ and $\delta(w) = j$

14. Find an invariant not given in this section or in Exercises 11–13.

In Exercises 15–17, state whether the given property is an invariant. If the property is not an invariant, give a counterexample.

15H. Has an Euler cycle

16. Has a vertex inside some simple cycle

17. Is bipartite

18H. Draw all nonisomorphic simple graphs having three vertices.

19. Draw all nonisomorphic simple graphs having four vertices.

20. Draw all nonisomorphic, cycle-free, connected graphs having five vertices.

21H. Draw all nonisomorphic, cycle-free, connected graphs having six vertices.

The **complement** of a simple graph G is a simple graph $\overline{G}$ with the same vertices as G. An edge exists in $\overline{G}$ if and only if it does not exist in G.

22H. Draw the complement of the graph G_1 of Exercise 1.

23. Draw the complement of the graph G_2 of Exercise 1.

24. A simple graph G is **self-complementary** if G and $\overline{G}$ are isomorphic. Find a self-complementary graph having five vertices.

25. Complete the proof of Theorem 3.5.7 by assuming part (b) and deducing part (a).

†3.6 Instant Insanity

Instant Insanity is a puzzle consisting of four cubes each of whose faces is painted one of the four colors, red, white, blue, or green (see Figure 3.6.1). (There are different versions of the puzzle, depending on which

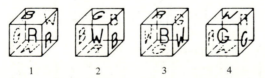

   1   2   3   4

Figure 3.6.1

†This section can be omitted without loss of continuity.

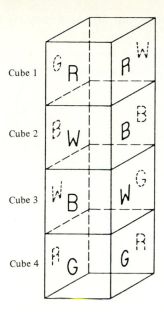

Figure 3.6.2

faces are painted which colors.) The problem is to stack the cubes, one on top of the other, so that whether viewed from front, back, left, or right, one sees all four colors (see Figure 3.6.2). Since over 300,000 different stacks are possible, a solution by hand by trial and error is impractical. We present a solution, using a graph model, that makes it possible to discover a solution, if there is one, in a few minutes!

First, notice that any particular stacking can be represented by two graphs, one representing the front/back colors and the other representing the left/right colors. For example, in Figure 3.6.3 we have represented the stacking of Figure 3.6.2. The vertices represent the colors, and an edge connects two vertices if the opposite faces have those colors. For example, in the front/back graph, the edge labeled 1 connects R and W, since the front and back faces of cube 1 are red and white. As another

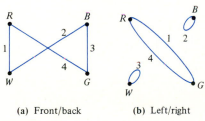

(a) Front/back   (b) Left/right

Figure 3.6.3

example, in the left/right graph, W has a loop, since both the left and right faces of cube 3 are white.

We can also construct a stacking from a pair of graphs such as those in Figure 3.6.3, which represent a solution of the Instant Insanity puzzle. Begin with the front/back graph. Cube 1 is to have red and white opposing faces. Arbitrarily assign one of these colors, say red, to the front. Then cube 1 has a white back face. The other edge incident on W is 2, so make cube 2's front face white. This gives cube 2 a blue back face. The other edge incident on B is 3, so make cube 3's front face blue. This gives cube 3 a green back face. The other edge incident on G is 4. Cube 4 then gets a green front face and a red back face. The front and back faces are now properly aligned. At this point, the left and right faces are randomly arranged; however, we will show how to correctly orient the left and right faces without altering the colors of the front and back faces.

Cube 1 is to have red and green opposing left and right faces. Arbitrarily assign one of these colors, say green, to the left. Then cube 1 has a red right face. Notice that by rotating the cube, we can obtain this left/right orientation without changing the colors of the front and back (see Figure 3.6.4). We can similarly orient cubes 2, 3, and 4. Notice that cubes 2 and 3 have the same colors on opposing sides. The stacking of Figure 3.6.2 has been reconstructed.

It is apparent from the discussion above that a solution to the Instant Insanity puzzle can be obtained if we can find two graphs like those of Figure 3.6.3. The properties needed are

<blockquote>Each vertex should have degree 2.   (3.6.1)</blockquote>

<blockquote>Each cube should be represented by an edge exactly once in each graph.   (3.6.2)</blockquote>

<blockquote>The two graphs should not have any edges in common.   (3.6.3)</blockquote>

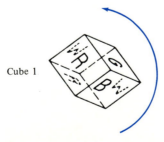

Cube 1

Figure 3.6.4

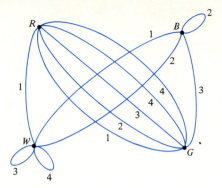

Figure 3.6.5

Property (3.6.1) assures us that each color can be used twice, once on the front (or left) and once on the back (or right). Property (3.6.2) assures us that each cube is used exactly once. Property (3.6.3) assures us that, after orienting the front and back sides, we can successfully orient the left and right sides.

To obtain a solution, we first draw a graph G that represents all of the faces of all of the cubes. The vertices of G represent the four colors and an edge labeled i connects two vertices (colors) if the opposing faces of cube i have those colors. In Figure 3.6.5 we have drawn the graph that represents the cubes of Figure 3.6.1. Then, by inspection, we find two subgraphs of G satisfying properties (3.6.1)–(3.6.3). Try your hand at the method by finding another solution to the puzzle represented by Figure 3.6.5.

Example 3.6.1.   Find a solution to the Instant Insanity puzzle of Figure 3.6.6.

We begin by trying to construct one subgraph having properties (3.6.1) and (3.6.2). We arbitrarily choose a vertex, say B, and choose

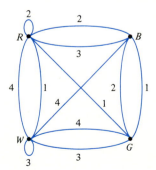

Figure 3.6.6

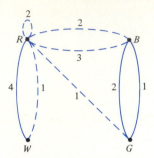

Figure 3.6.7

two edges incident on vertex B. Suppose that we select the two edges shown as solid lines in Figure 3.6.7. Now consider the problem of picking two edges incident on vertex R. We cannot select any edges incident on B or G since B and G must each have degree 2. Since each cube must appear in each subgraph exactly once, we cannot select any of the edges labeled 1 or 2 since we already have selected edges with these labels. Edges incident on R that cannot be selected are shown dashed in Figure 3.6.7. This leaves only the edge labeled 4. Since we need two edges incident on R, our initial selection of edges incident on B must be revised.

For our next attempt at choosing two edges incident on vertex B, let us select the edges labeled 2 and 3, as shown in Figure 3.6.8. Since this choice includes one edge incident on R, we must choose one additional edge incident on R. We have three possibilities for selecting the additional edge (shown as heavy solid lines in Figure 3.6.8). (The loop incident at R counts as two edges and so cannot be chosen.) If we select the edge labeled 1 incident on R and G, we

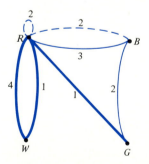

Figure 3.6.8

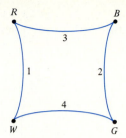

Figure 3.6.9

would need a loop at *W* labeled 4. Since there is no such loop, we do not select this edge. If we select the edge labeled 1 incident on *R* and *W*, we can then select the edge labeled 4 incident on *W* and *G* (see Figure 3.6.9). We have obtained one of the graphs.

We now look for a second graph having no edges in common with the graph just chosen. Let us again begin by picking two edges incident on *B*. Because we cannot reuse edges, our choices are limited to three edges (see Figure 3.6.10). Choosing the edges labeled 1 and 4 leads to the graph of Figure 3.6.11. The graphs of Figures 3.6.9 and 3.6.11 solve the Instant Insanity puzzle of Figure 3.6.6.

Figure 3.6.10

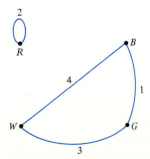

Figure 3.6.11

EXERCISES

Find solutions to the following Instant Insanity puzzles.

1H.

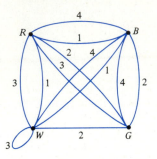

2.

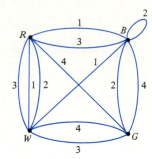

3.

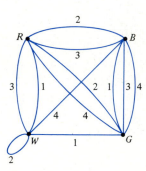

4H.

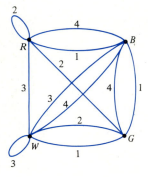

5.

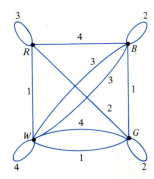

6.

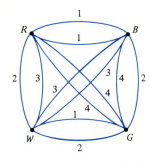

7H.   (a) Find all subgraphs of Figure 3.6.5 satisfying properties (3.6.1) and (3.6.2).

(b) Find all the solutions to the Instant Insanity puzzle of Figure 3.6.5.

8.   (a) Represent the Instant Insanity puzzle by a graph.

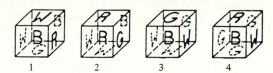

     (b) Find a solution to the puzzle.
     (c) Find all subgraphs of your graph of part (a) satisfying
         properties (3.6.1) and (3.6.2).
     (d) Use (c) to show that the puzzle has a unique solution.

9.   Show that the following Instant Insanity puzzle has no solution
     by giving an argument to show that no subgraph satisfies proper-
     ties (3.6.1) and (3.6.2). Notice that there is no solution even
     though each cube contains all four colors.

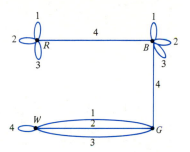

*10.   Give an example of an Instant Insanity puzzle satisfying:
     (a) There is no solution.
     (b) Each cube contains all four colors.
     (c) There is a subgraph satisfying properties (3.6.1) and
         (3.6.2).

*11H. Design an Instant Insanity puzzle that has exactly four
     solutions.

   Exercises 12–18 refer to a modified version of Instant Insanity
where a solution is defined to be a stacking which, when viewed from
the front, back, left, or right, shows one color. (The front, back, left,
and right are of different colors.)

12.   Give an argument that shows that if we graph the puzzle as in
     regular Instant Insanity, a solution to modified Instant Insanity

consists of two subgraphs of the form

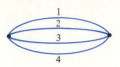

with no edges or vertices in common.

Find solutions to modified Instant Insanity for the following puzzles.

13H.

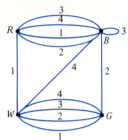

14.

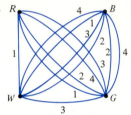

15.

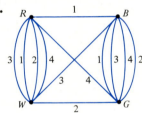

16. Graph of Exercise 6

17H. Show that for Figure 3.6.5, Instant Insanity, as given in the text, has a solution, but the modified version does not have a solution.

*18. Show that if modified Instant Insanity has a solution, the version given in the text must also have a solution.

3.7 Notes

Virtually any reference on discrete mathematics contains one or more chapters on graph theory. Books specifically on graph theory are [Berge; Busacker; Deo; Even, 1979; Harary; König; Ore]. [Deo] emphasizes applications. [Bellman] contains a unified treatment of several graph problems, including the traveling salesperson problem and several puzzles.

Euler's original paper on the Königsberg bridges, edited by J. R. Newman, was reprinted as [Euler].

[Duda] is an introductory text on pattern recognition. For an expository article about pattern recognition in both machines and animals, see [Gose].

In [Gardner, 1959], Hamiltonian circuits are related to the Tower of Hanoi puzzle.

So-called **branch-and-bound methods** (see [Horowitz, 1978]) often give solutions to the traveling salesperson problem more efficiently than will exhaustive search. [Parry] contains a computer program that uses a branch-and-bound method. For a good survey of results on the traveling salesperson problem, see [Bellmore].

Computer Exercises

1. Write a program that accepts as input any of
   (a) A listing of edges of a simple graph given as pairs of positive integers
   (b) The adjacency matrix of a simple graph
   (c) The incidence matrix of a simple graph

   and outputs the other two.

2. Write a program that accepts as input the edges of a graph and determines whether or not the graph contains an Euler cycle.

3. Write a program that randomly generates a 6 × 6 adjacency matrix. Have your program print the adjacency matrix, the number of edges, the number of loops, and the degree of each vertex.

4. Write a program that determines whether or not a graph is a bipartite graph. If it is a bipartite graph, the program should list the disjoint sets of vertices.

5. Write a program that accepts as input the edges of a graph and then draws the graph using a computer graphics display.

6. [*PROJECT*]   Prepare a report on some computer language developed especially to handle graphs. References may be found in [Deo, Section 11-10].

7. Write a program that finds an Euler cycle in a connected graph in which all vertices have even degree.

8. Write a program that lists all simple paths between two given vertices.

9. Write a program that determines whether a path is a simple path, a cycle, or a simple cycle.

10. Write a program that determines whether a graph is connected.

11. Write a program that finds the components of a graph.

12. Write a program that solves an arbitrary Instant Insanity puzzle.

Chapter Review

SECTION 3.1
Graph $G = (V, E)$ (undirected and directed)
Vertex
Edge
Edge e is incident on vertex v
Vertex v is incident on edge e
v and w are adjacent vertices
Parallel edges
Loop
Isolated vertex
Simple graph
Weighted graph
Weight of an edge
Length of a path in a weighted graph
Similarity graph
Dissimilarity function
Pattern recognition

SECTION 3.2
Path
Simple path
Cycle
Simple cycle
Connected graph

Chapter Self-Test

SECTION 3.1

1H. For the graph $G = (V, E)$, find V, E, all parallel edges, all
    loops, all isolated vertices, and state whether G is a simple
    graph. Also state on which vertices edge e_3 is incident and on
    which edges vertex v_2 is incident.

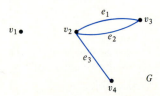

2H.   Explain why the graph does not have a path from a to a that passes through each edge exactly one time.

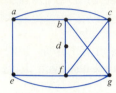

3H.   Draw $K_{2,5}$ the complete bipartite graph on 2 and 5 vertices.

4H.   A similarity graph G is constructed as in Example 3.1.6. We let v_i be the vertex corresponding to essay i. The values

$$s(v_1, v_2) = 18, \qquad s(v_1, v_3) = 24, \qquad s(v_1, v_4) = 11$$
$$s(v_1, v_5) = 34, \qquad s(v_2, v_3) = 41, \qquad s(v_2, v_4) = 66$$
$$s(v_2, v_5) = 84, \qquad s(v_3, v_4) = 12, \qquad s(v_3, v_5) = 23$$
$$s(v_4, v_5) = 46,$$

are obtained. Draw the similarity graph that results from setting $S = 20$. How many classes are there?

SECTION 3.2

5H.   Tell whether the path $(v_2, v_3, v_4, v_2, v_6, v_1, v_2)$ in the graph is a simple path, a cycle, a simple cycle, or none of these.

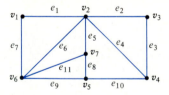

6H.   Draw all subgraphs containing exactly two edges of the graph

7H.   Find a connected subgraph of the graph of Exercise 5 containing all of the vertices of the original graph and having as few edges as possible.

8H.   Does the graph of Exercise 5 contain an Euler cycle? Explain.

SECTION 3.3

9H. Find a Hamiltonian cycle in the graph of Exercise 5.

10H. Show that the graph has no Hamiltonian cycle.

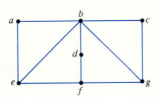

11H. Give an example of a complete bipartite graph $K_{m,n}$, m, $n > 1$, which has no Hamiltonian cycle.

12H. Show that the cycle $(a, b, c, d, e, f, g, h, i, j, a)$ provides a solution to the traveling salesperson problem for the graph shown.

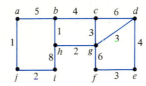

SECTION 3.4

13H. Write the adjacency matrix of the graph of Exercise 5.

14H. Write the incidence matrix of the graph of Exercise 5.

15H. If A is the adjacency matrix of the graph of Exercise 5, what does the entry in row v_2 and column v_3 of A^3 represent?

16H. Can a column of an incidence matrix consist only of zeros? Explain.

SECTION 3.5

In Exercises 17 and 18, determine whether the graphs G_1 and G_2 are isomorphic. If the graphs are isomorphic, give orderings of their vertices that produce equal adjacency matrices. If the graphs are not isomorphic, give an invariant that the graphs do not share.

17H.

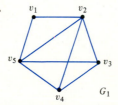

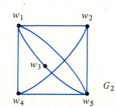

18H.

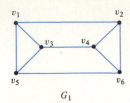

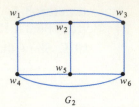

19H. Draw all nonisomorphic simple graphs having exactly five vertices and two edges.

20H. Draw all nonisomorphic, simple graphs having exactly five vertices, two components, and no cycles.

SECTION 3.6

21H. Represent the Instant Insanity puzzle by a graph.

22H. Find a solution to the puzzle of Exercise 21.

23H. Find all subgraphs of the graph of Exercise 21 satisfying properties (3.6.1) and (3.6.2).

24H. Use Exercise 23 to determine how many solutions the puzzle of Exercise 21 has.

I talk to the trees, but they don't listen to me.

—from *Paint Your Wagon*

4

Trees

Trees form one of the most widely used subclasses of graphs. Computer science, in particular, makes extensive use of trees. In computer science, trees are useful in organizing and relating data. For example, data in a data base can be structured and organized as a tree (see Example 4.1.6).

In Section 4.1, after giving some basic definitions, we look at an application that encodes text using minimal storage. In Section 4.2 we introduce additional terminology and present several alternative characterizations of trees. Binary trees and two special instances of binary trees useful in applications, decision trees and binary search trees, are treated in Section 4.3. Section 4.4, on isomorphic trees, extends the discussion of Section 3.5 on isomorphic graphs. Section 4.5 is concerned with game trees—structures that facilitate the generation of game-playing machines.

4.1 Introduction

Figure 4.1.1 shows the results of the last part of the 1985 NCAA basketball tournament, in which Villanova upset top-ranked and highly favored Georgetown to become the collegiate basketball champion. The

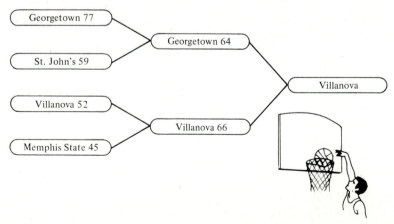

Figure 4.1.1

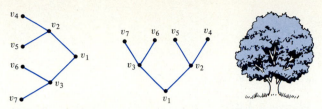

Figure 4.1.2   **Figure 4.1.3**

NCAA basketball tournament is a single-elimination tournament; that is, when a team loses, it is out of the tournament. Winning teams continue to play until only one team, the champion, remains. Figure 4.1.1 shows that initially Georgetown eliminated St. John's and Villanova eliminated Memphis State. The winners, Georgetown and Villanova, then played and Villanova eliminated Georgetown. Villanova, being the sole remaining team, won the tournament.

If we regard the single-elimination tournament of Figure 4.1.1 as a graph (see Figure 4.1.2), we obtain a **tree**. If we rotate Figure 4.1.2, it looks like a natural tree (see Figure 4.1.3). The formal definition follows.

Definition 4.1.1.   A (*free*) *tree* T is a simple graph satisfying: If v and w are vertices in T, there is a unique simple path from v to w.

A *rooted tree* is a tree in which a particular vertex is designated the root.

Example 4.1.2.   If we designate the winner as the root, the single-elimination tournament of Figure 4.1.1 (or Figure 4.1.2) is a rooted tree. Notice that if v and w are vertices in this graph, there is a unique simple path from v to w. For example, the unique simple path from v_2 to v_7 is (v_2, v_1, v_3, v_7).

In contrast to natural trees, which have their roots at the bottom, in graph theory rooted trees are typically drawn with their roots at the top. Figure 4.1.4 shows the way the tree of Figure 4.1.2 would be drawn (with v_1 as root). First, we place the root v_1 at the top. Under the root and on the same level, we place the vertices, v_2 and v_3, that can be reached from the root on a simple path of length 1. Under each of these vertices and on the same level, we place the vertices, v_4, v_5, v_6, and v_7, that can be reached from the root on a simple path of length 2. We

Figure 4.1.4

continue in this way until the entire tree is drawn. Since the simple path from the root to any given vertex is unique, each vertex is on a uniquely determined level. We call the level of the root level 0. The vertices under the root are said to be on level 1, and so on. Thus the **level of a vertex** v is the length of the simple path from the root to v. The **height** of a rooted tree is the maximum level number that occurs.

Example 4.1.3.   The vertices v_1, v_2, v_3, v_4, v_5, v_6, v_7 in the rooted tree of Figure 4.1.4 are on (respectively) levels 0, 1, 1, 2, 2, 2, 2. The height of the tree is 2.

Example 4.1.4.   If we designate e as the root in the tree T of Figure 4.1.5, we obtain the rooted tree T' shown in Figure 4.1.5. The vertices a, b, c, d, e, f, g, h, i, j are on (respectively) levels 2, 1, 2, 1, 0, 1, 1, 2, 2, 3. The height of T' is 3.

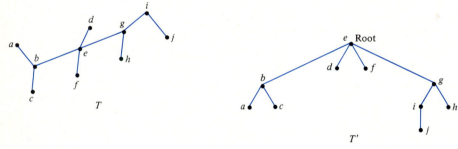

Figure 4.1.5

Example 4.1.5.   A rooted tree is often used to specify hierarchical relationships. When a tree is used in this way, if vertex a is on a level one less than the level of vertex b and a and b are adjacent, then a is "just above" b

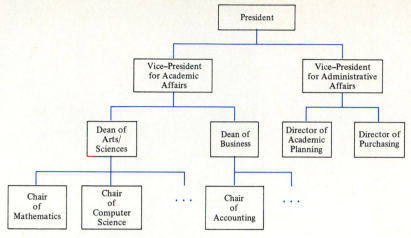

Figure 4.1.6

and a logical relationship exists between *a* and *b*: *a* dominates *b* or *b* is subordinate to *a* in some way. A partial example of such a tree, which is the administrative organizational chart of a hypothetical university, is given in Figure 4.1.6.

Example 4.1.6 Hierarchical Definition Trees. A **data base** is a collection of records that are manipulated by a computer. For example, an airline data base might contain records of passengers' reservations, flight schedules, equipment, and so on. Figure 4.1.7 is an example of a **hierarchical definition tree**. Such trees are used to show logical relationships among records in a data base. The tree of Figure 4.1.7 might be used as a model for setting up a data base to maintain records about books housed in several libraries.

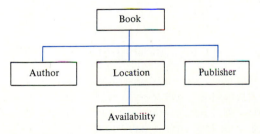

Figure 4.1.7

D. A. Huffman (1925–   )

D. A. Huffman published the paper [Huffman] describing what we now call Huffman codes in 1952. He received a bachelor's degree from Ohio State University in 1944. After serving as a radar maintenance officer during World War II, he returned to Ohio State, where he earned a master's degree. He then moved to the Massachusetts Institute of Technology, where he earned a doctoral degree in 1953. He remained at MIT as a faculty member until 1967, when he moved to the University of California at Santa Cruz. [*Photo courtesy of Dr. D. A. Huffman*]

Example 4.1.7 Huffman Codes.

The most common way to represent characters internally in a computer is by using fixed-length bit strings. For example, ASCII (American Standard Code for Information Interchange) represents each character by a string of seven bits. Examples are given in Table 4.1.1.

Table 4.1.1

Character	ASCII Code
A	100 0001
B	100 0010
C	100 0011
1	011 0001
2	011 0010
!	010 0001
*	010 1010

Huffman codes, which represent characters by variable-length bit strings, provide alternatives to ASCII and other fixed-length codes. The idea is to use short bit strings to represent the most frequently used characters and to use longer bit strings to represent less frequently used characters. In this way it is generally possible to represent strings of characters, such as text, programs, and so on, in less space than if ASCII were used. Because of limited memory, some hand-held computers have used Huffman codes (see [Williams]).

A Huffman code is most easily defined by a rooted tree (see Figure 4.1.8). To decode a bit string, we begin at the root and move down

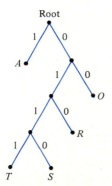

Figure 4.1.8

the tree until a character is encountered. The bit, 0 or 1, tells us whether to move right or left. As an example, let us decode the string

$$01010111. \tag{4.1.1}$$

We begin at the root. Since the first bit is 0, the first move is right. Next, we move left and then right. At this point, we encounter the first character R. To decode the next character, we begin again at the root. The next bit is 1, so we move left and encounter the next character A. The last bits 0111 decode as T. Therefore, the bit string (4.1.1) represents the word *RAT*.

Given a tree that defines a Huffman code, such as Figure 4.1.8, any bit string [e.g., (4.1.1)] can be uniquely decoded even though the characters are represented by variable-length bit strings. For the Huffman code defined by the tree of Figure 4.1.8, the character A is represented by a bit string of length 1, whereas S and T are represented by bit strings of length 4. (A is represented as 1, S is represented as 0110, and T is represented as 0111.)

It is possible to construct a Huffman code from a table giving the frequency of occurrence of the characters to be represented so that the code constructed represents strings of characters in minimal space, provided that the strings to be represented have character frequencies identical to the character frequencies in the table (see the discussion preceding Exercise 23).

EXERCISES

1H. Find the level of each vertex in the tree shown.

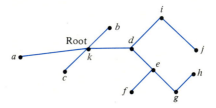

2. Find the height of the tree of Exercise 1.

3. Draw the tree T of Figure 4.1.5 as a rooted tree with a as root. What is the height of the resulting tree?

4H. Draw the tree T of Figure 4.1.5 as a rooted tree with b as root. What is the height of the resulting tree?

5.   Give an example similar to Example 4.1.5 of a tree that is used to specify hierarchical relationships.

6.   Give an example different from Example 4.1.6 of a hierarchical definition tree.

Decode each bit string using the Huffman code given.

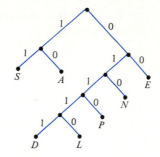

7H.   011000010

8.   01110100110

9.   01111001001110

10H. 1110011101001111

Encode each word using the preceding Huffman code.

11H. DEN   **12.** NEED   **13.** LEADEN   **14H.** PENNED

15.   What factors in addition to the amount of memory used should be considered when choosing a code, such as ASCII or a Huffman code, to represent characters in a computer?

16.   What techniques in addition to the use of Huffman codes might be used to save memory when storing text?

17H. Show that a tree is a bipartite graph.

18.   Explain how to color the vertices of a tree using only two colors so that each edge is incident on vertices of different colors.

The **eccentricity** of a vertex v in a tree T is the maximum length of a simple path that begins at v.

19.   Find the eccentricity of each vertex in the tree T of Figure 4.1.5.

A vertex v in a tree T is a **center** for T if the eccentricity of v is minimal.

20H. Find the center(s) of the tree T of Figure 4.1.5.

21. Define the radius r and the diameter d of a tree using the concepts of eccentricity and center. Is it always true, according to your definitions, that $2r = d$? Explain.

22. Give an example of a tree T that does not satisfy the property: If v and w are vertices in T, there is a unique path from v to w.

The following technique constructs a Huffman code from a table giving the frequency of occurrence of the characters to be represented so that the code represents strings of characters in minimal space, provided that the strings to be represented have character frequencies identical to the character frequencies in the table.

1. We begin by listing n sequences. The first sequence is the list of frequencies arranged in increasing order.
2. The sequence following

$$s_1, s_2, \ldots, s_i \qquad (4.1.2)$$

(s is in increasing order) is obtained by arranging

$$s_1 + s_2, s_3, \ldots, s_i \qquad (4.1.3)$$

in increasing order. Stop listing sequences when a one-element sequence is obtained.
3. We next construct a rooted tree with labeled vertices for each sequence beginning with the last sequence and working back through the list of sequences. The rooted tree that corresponds to the last sequence t_1 is

4. If the rooted tree T' corresponds to (4.1.3), a rooted tree that corresponds to the preceding sequence (4.1.2) is obtained from T' by replacing a terminal vertex in T' labeled $s_1 + s_2$ by

5. We stop when we obtain a rooted tree T that corresponds to the original sequence of frequencies. We replace the label f_i on the terminal vertex by the character that has frequency f_i. The rooted tree T defines an optimal Huffman code (see [Standish] for a proof of this fact).

23H. Use the preceding technique to construct an optimal Huffman code for the set of letters in the table.

Letter	Frequency
α	5
β	6
γ	6
δ	11
ϵ	20

24. Use the preceding technique to construct an optimal Huffman code for the set of letters in the table.

Letter	Frequency
I	7.5
U	20.0
B	2.5
S	27.5
C	5.0
H	10.0
M	2.5
P	25.0

25. Use the code developed in Exercise 24 to encode the following words (which have frequencies consistent with the table of Exercise 24): BUS, CUPS, MUSH, PUSS, SIP, PUSH, CUSS, HIP, PUP, PUPS, HIPS.

26H. Construct two optimal Huffman coding trees for the table of Exercise 23 of different heights.

4.2 Terminology and Characterizations of Trees

A portion of the family tree of the ancient Greek gods is shown in Figure 4.2.1. (Not all children are listed.) As shown, we can regard a family tree as a rooted tree. The vertices adjacent to a vertex v and on the next-lower level are the children of v. For example, Kronos's children are Zeus, Poseidon, Hades, and Ares. The terminology adapted from a family tree is used routinely for any rooted tree. The formal definitions follow.

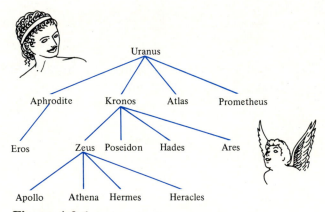

Figure 4.2.1

Definition 4.2.1.   Let T be a tree with root v_0. Suppose that x, y, and z are vertices in T and that $(v_0, v_1, \ldots, v_n)$ is a simple path in T. Then

(a) v_{n-1} is the *parent* of v_n.

(b) $v_0, \ldots, v_{n-1}$ are *ancestors* of v_n.

(c) v_n is a *child* of v_{n-1}.

(d) If x is an ancestor of y, then y is a *descendant* of x.

(e) If x and y are children of z, then x and y are *siblings*.

(f) If x has no children, x is a *terminal vertex* (or a *leaf*).

(g) If x is not a terminal vertex, x is an *internal* (or *branch*) *vertex*.

(h) The *subtree of T rooted at x* is the graph with vertex set V and edge set E, where V is x together with the descendants of x and

$$E = \{e \mid e \text{ is an edge on a simple path from } x \text{ to some vertex in } V\}.$$

Example 4.2.2. In the rooted tree of Figure 4.2.1,

    (a) The parent of Eros is Aphrodite.

    (b) The ancestors of Hermes are Zeus, Kronos, and Uranus.

    (c) The children of Zeus are Apollo, Athena, Hermes, and Heracles.

    (d) The descendants of Kronos are Zeus, Poseidon, Hades, Ares, Apollo, Athena, Hermes, and Heracles.

    (e) Aphrodite and Prometheus are siblings.

    (f) The terminal vertices are Eros, Apollo, Athena, Hermes, Heracles, Poseidon, Hades, Ares, Atlas, and Prometheus.

    (g) The internal vertices are Uranus, Aphrodite, Kronos, and Zeus.

    (h) The subtree rooted at Kronos is shown in Figure 4.2.2.

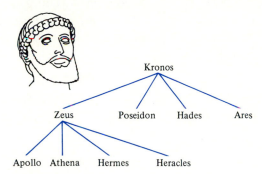

Figure 4.2.2

The remainder of this section is devoted to providing alternative characterizations of trees. Let T be a tree. We note that T is connected since there is a simple path from any vertex to any other vertex. Further, we can show that T does not contain a cycle. To see this, suppose that T contains a cycle C'. By Theorem 3.2.20, T contains a simple cycle (see Figure 4.2.3)

$$C = (v_0, \ldots , v_n),$$

$v_0 = v_n$. Since T is a simple graph, C cannot be a loop; so C contains

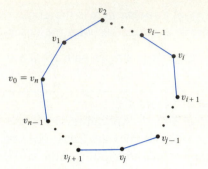

Figure 4.2.3

at least two distinct vertices v_i and v_j, $i < j$. Now

$$(v_i, v_{i+1}, \ldots, v_j), \qquad (v_i, v_{i-1}, \ldots, v_0, v_{n-1}, \ldots, v_j)$$

are distinct simple paths from v_i to v_j, which contradicts the definition of tree. Therefore, a tree cannot contain a cycle.

A graph with no cycles is called an **acyclic graph**. We just showed that a tree is a connected, acyclic graph. The converse is also true; every connected, acyclic graph is a tree. The next theorem gives this characterization of trees as well as others.

Theorem 4.2.3.   *Let T be a graph with n vertices. The following are equivalent.*
(a) T is a tree.
(b) T is connected and acyclic.
(c) T is connected and has n − 1 edges.
(d) T is acyclic and has n − 1 edges.

†PROOF.   To show that (a)–(d) are equivalent, we will prove four results: if (a), then (b); if (b), then (c); if (c), then (d); and, if (d), then (a).

[If (a), then (b).] The proof of this result was given before the statement of the theorem.

[If (b), then (c).] Suppose that T is connected and acyclic. We will prove that T has $n − 1$ edges by induction on n.

If $n = 1$, T consists of one vertex and zero edges, so the result is true if $n = 1$.

†The proof of Theorem 4.2.3 can be omitted on first reading.

Figure 4.2.4

Now suppose that the result holds for a connected, acyclic graph with n vertices. Let T be a connected, acyclic graph with $n + 1$ vertices. Choose a simple path P of maximum length. Since T is acyclic, P is not a cycle. Therefore, P contains a vertex v of degree 1 (see Figure 4.2.4). Let T^* be T with v and the edge incident on v removed. Then T^* is connected and acyclic, and because T^* contains n vertices, by the inductive hypothesis T^* contains $n - 1$ edges. Therefore, T contains n edges. The inductive argument is complete and this portion of the proof is complete.

[If (c), then (d).] Suppose that T is connected and has $n - 1$ edges. We must show that T is acyclic.

Suppose that T contains at least one cycle. Since removing an edge from a cycle does not disconnect a graph, we may remove edges, but no vertices, from cycle(s) in T until the resulting graph T^* is connected and acyclic. Now T^* is an acyclic, connected graph with n vertices. We may use our just proven result, (b) implies (c), to conclude that T^* has $n - 1$ edges. But now T has more than $n - 1$ edges. This is a contradiction. Therefore, T is acyclic. This portion of the proof is complete.

[If (d), then (a).] Suppose that T is acyclic and has $n - 1$ edges. We must show that T is a tree, that is, that T is a simple graph and that T has a unique simple path from any vertex to any other vertex.

The graph T cannot contain any loops because loops are cycles and T is acyclic. Similarly, T cannot contain distinct edges e_1 and e_2 incident on v and w because we would then have the cycle (v, e_1, w, e_2, v). Therefore, T is a simple graph.

Suppose, by way of contradiction, that T is not connected. Let

$$T_1, T_2, \ldots, T_k$$

be the components of T. Since T is not connected, $k > 1$. Suppose that T_i has n_i vertices. Each T_i is connected and acyclic, so we may use our previously proven result, (b) implies (c), to conclude that T_i has $n_i - 1$ edges. Now

$$n - 1 = (n_1 - 1) + (n_2 - 1)$$

$$+ \cdots + (n_k - 1) \qquad \text{(counting edges)}$$

$$< (n_1 + n_2 + \cdots + n_k) - 1 \qquad \text{(since } k > 1)$$

$$= n - 1, \qquad\qquad\qquad \text{(counting vertices)}$$

which is impossible. Therefore, T is connected.

Suppose that there are distinct simple paths P_1 and P_2 from a to b in T. Let c be the first vertex after a on P_1 that is also in the path P_2. (Since b is a vertex after a on P_1 that is also in P_2, such a vertex exists.) Let

$$(v_0, v_1, \ldots, v_{n-1}, v_n)$$

be the portion of P_1 from $a = v_0$ to $c = v_n$. Let

$$(w_0, w_1, \ldots, w_{m-1}, w_m)$$

be the portion of P_2 from $a = w_0$ to $c = w_m$. Now

$$(v_0, \ldots, v_n = w_m, w_{m-1}, \ldots, w_1, w_0) \qquad (4.2.1)$$

is a cycle in T, which is a contradiction. [In fact, (4.2.1) is a simple cycle since no vertices are repeated except for v_0 and w_0.] Thus there is a unique simple path from any vertex to any other vertex in T. Therefore, T is a tree. This completes the proof. ∎

EXERCISES

Answer the questions in Exercises 1–6 for the tree in Figure 4.2.1.

1H. Find the parent of Poseidon.

2. Find the ancestors of Eros.

3. Find the children of Uranus.

4H. Find the descendants of Zeus.

5. Find the siblings of Ares.

6. Draw the subtree rooted at Aphrodite.

Answer the questions in Exercises 7–15 for the two trees shown.

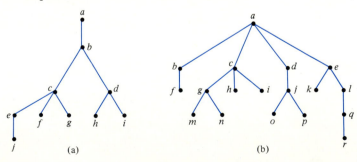

(a)   (b)

7H. Find the parents of c and of h.

8. Find the ancestors of c and of j.

9. Find the children of d and of e.

10H. Find the descendants of c and of e.

11. Find the siblings of f and of h.

12. Find the terminal vertices.

13H. Find the internal vertices.

14. Draw the subtree rooted at j.

15. Draw the subtree rooted at e.

16H. What can you say about two vertices in a rooted tree that have the same parent?

17. What can you say about two vertices in a rooted tree that have the same ancestors?

18. What can you say about a vertex in a rooted tree that has no ancestors?

19. What can you say about two vertices in a rooted tree that have a descendant in common?

20. What can you say about a vertex in a rooted tree that has no descendants?

In Exercises 21–25, draw a graph having the given properties or explain why no such graph exists.

21H. Six edges; eight vertices

22. Acyclic; four edges; six vertices

23. Tree; all vertices of degree 2

24H. Tree; six vertices having degrees 1, 1, 1, 1, 3, 3

25. Tree; four internal vertices; six terminal vertices

26H. Explain why if we allow cycles of length 0, a graph consisting of a single vertex and no edges is not acyclic.

27. Explain why if we allow cycles to repeat edges, a graph consisting of a single edge and two vertices is not acyclic.

28. The connected graph shown has a unique simple path from any vertex to any other vertex but it is not acyclic. Explain.

A **forest** is a simple graph with no cycles.

29H. Explain why a forest is a union of trees.

30. If a forest F consists of m trees and has n vertices, how many edges does F have?

31. If $P_1 = (v_0, \ldots, v_n)$ and $P_2 = (w_0, \ldots, w_m)$ are distinct simple paths from a to b in a simple graph G, is

$$(v_0, \ldots, v_n = w_m, w_{m-1}, \ldots, w_1, w_0)$$

necessarily a cycle? Explain. (This exercise is relevant to the last paragraph of the proof of Theorem 4.2.3.)

32H. Complete the following argument to show that a graph G with n vertices and fewer than $n - 1$ edges is not connected.

Suppose that G is connected. Add parallel edges until the resulting graph G^* has $n - 1$ edges. Therefore, G^* is acyclic (why?). This is a contradiction (why?). Thus G is not connected.

*33. Prove that the following are equivalent about a simple graph T.
   (a) T is a tree.
   (b) T is connected but the removal of any edge (but no vertices) from T disconnects T.
   (c) T is connected and if an edge is added between any two vertices, exactly one cycle is created.

†4.3 Binary Trees

The rooted tree in Figure 4.3.1 gives the rules for entering a national knitting contest. To use the tree, a prospective entrant begins at the root and, after answering each question, follows the appropriate edge. Eventually, the person will arrive at a terminal vertex that states what action to take. For example, suppose that Roger knitted boots for his dog from an original design and has already entered the boots in the local art museum's contest. Beginning at the root, the first question Roger must answer is: "Entered this contest already?" Since the answer to this question is no, Roger proceeds to the question "Original work?" Since the answer to this question is yes, Roger proceeds to the question "Submitted to another contest?" Since the answer to this question is yes, Roger

†This section can be omitted without loss of continuity.

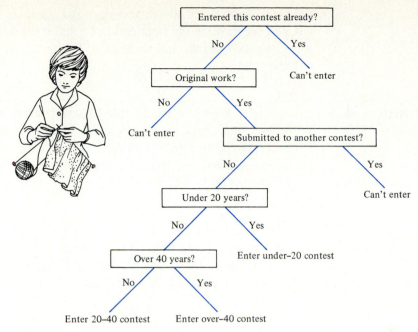

Figure 4.3.1

proceeds to the terminal node, "Can't enter." Roger learns that he is ineligible to enter this contest.

Elizabeth, 41 years of age, is considering submitting to the national knitting contest the cover for her discrete mathematics text that she knitted from an original design. Elizabeth has not entered her cover in any contest. Beginning at the root, the first question Elizabeth must answer is: "Entered this contest already?" Since the answer to this question is no, Elizabeth proceeds to the question "Original work?" Since the answer to this question is yes, Elizabeth proceeds to the question "Submitted to another contest?" Since the answer to this question is no, Elizabeth proceeds to the question "Under 20 years?" Since the answer to this question is no, Elizabeth proceeds to the question "Over 40 years?" Since the answer to this question is yes, Elizabeth proceeds to the terminal node, "Enter over-40 contest." Elizabeth learns that she is eligible to enter the over-40 contest.

The rooted tree of Figure 4.3.1 is an example of a **binary tree**. Every vertex in a binary tree has at most two children. Moreover, each child is designated as either a **left child** or a **right child**. In drawing a binary tree, a left child is drawn to the left and a right child is drawn to the right. In Figure 4.3.1, if the answer to the question at an internal vertex v is yes, we move to the right child of v and if the answer to the question is no, we move to the left child of v. We next give a formal definition of binary tree.

Definition 4.3.1.   A *binary tree* is a rooted tree in which each vertex has either no children, one child, or two children. If a vertex v has one child, that child is designated as either a left child or a right child (but not both). If a vertex v has two children, one child is designated a left child and the other child is designated a right child.

Example 4.3.2.   In the binary tree of Figure 4.3.2, vertex b is the left child of vertex a and vertex c is the right child of vertex a. Vertex d is the right child of vertex b; vertex b has no left child. Vertex e is the left child of vertex c; vertex c has no right child.

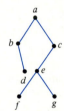

Figure 4.3.2

Example 4.3.3.   A tree that defines a Huffman code is a binary tree. For example, in the Huffman coding tree of Figure 4.1.8, moving from a vertex to a left child corresponds to using the bit 1 and moving from a vertex to a right child corresponds to using the bit 0.

Example 4.3.4 Decision Trees.   A binary tree such as that of Figure 4.3.1, in which the left and right children correspond to no and yes answers, is called a **decision tree**. In a decision tree, the internal vertices correspond to questions and the terminal vertices give the action to be taken.

Suppose that we have a set S whose elements can be ordered. For example, if S consists of numbers, we can use ordinary ordering defined on numbers, and if S consists of strings of alphabetic characters, we can use dictionary order. Binary trees are used extensively in computer science to store elements from an ordered set such as a set of numbers or a set of strings. If data item $d(v)$ is stored in vertex v and data item $d(w)$ is stored in vertex w, then if v is a left child (or right child) of w, some ordering relationship will be guaranteed to exist between $d(v)$ and $d(w)$. One example is a **binary search tree**.

Definition 4.3.5.   A *binary search tree* is a binary tree T in which data are associated with the vertices. The data are arranged so that, for *each* vertex v in T, each data item in the left subtree of v is less than the data item in v and each data item in the right subtree of v is greater than the data item in v.

Example 4.3.6.   The words

ONCE UPON A TIME THERE WAS AN OLD MAN   (4.3.1)

may be placed in a binary search tree as shown in Figure 4.3.3. Notice that for any vertex v, each data item in the left subtree of v is less than (i.e., precedes alphabetically) the data item in v and each data item in the right subtree of v is greater than the data item in v.

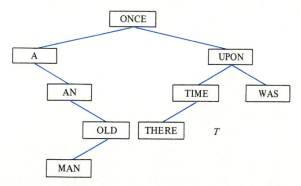

Figure 4.3.3

In general, there will be many ways to place data into a binary search tree. Figure 4.3.4 shows another binary search tree that stores the words (4.3.1).

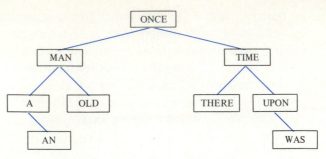

Figure 4.3.4

The binary search tree T of Figure 4.3.3 was constructed in the following way. We begin with an **empty tree**, that is, a tree with no vertices and no edges. We then inspect each of the words (4.3.1) *in the order in which they appear*, ONCE first, then UPON, then A, and so on. To start, we create a vertex and place the first word ONCE in this vertex. We designate this vertex the root. Thereafter, given a word in the list (4.3.1), we add a vertex v and an edge to the tree and place the word in the vertex v. To decide where to add the vertex and edge, we begin at the root. If the word to be added is less than (using dictionary order) the word at the root, we move to the left child and if the word to be added is greater than the word at the root, we move to the right child. If there is no child, we create one, put in an edge incident on the root and new vertex, and place the word in the new vertex. If there is a child v, we repeat this process. That is, we compare the word to be added with the word at v and move to the left child of v if the word to be added is less than the word at v; otherwise, we move to the right child of v. If there is no child to move to, we create one, put in an edge incident on v and the new vertex, and place the word in the new vertex. If there is a child to move to, we repeat this process. Eventually, we place the word in the tree. We then get the next word in the list, compare it with the root, move left or right, compare it with the new vertex, move left or right, and so on, and eventually store it in the tree. In this way, we store all of the words in the tree and thus create a binary search tree.

Example 4.3.7.   We show how the binary search tree in Figure 4.3.3 was created.
    We store the first word ONCE in the root (see Figure 4.3.5). The next word is UPON. We begin at the root. Since ONCE is at the root, we compare UPON with ONCE. Since UPON is greater than ONCE,

ONCE

Figure 4.3.5

we attempt to move to the right child. Since there is no right child, we create one and place UPON in the new vertex (see Figure 4.3.6).

The next word is A. We begin at the root. Since ONCE is at the root, we compare A with ONCE. Since A is less than ONCE, we attempt to move to the left child. Since there is no left child, we create one and place A in the new vertex (see Figure 4.3.7).

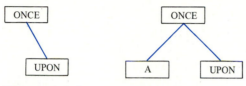

Figure 4.3.6   **Figure 4.3.7**

The next word is TIME. We begin at the root. Since ONCE is at the root, we compare TIME with ONCE. Since TIME is greater than ONCE, we attempt to move to the right child. This time we can move to the right child UPON. We next compare TIME with UPON. Since TIME is less than UPON, we attempt to move to the left child. Since there is no left child, we create one and place TIME in the new vertex (see Figure 4.3.8).

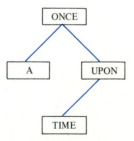

Figure 4.3.8

You should store the remaining words in the tree to verify that we eventually create the binary search tree shown in Figure 4.3.3.

Binary search trees are useful for locating data. That is, given a data item *D*, we can easily determine if *D* is in a binary search tree and if it is present, where it is located. To determine if a data item D is in a binary search tree, we would begin at the root. We would then repeatedly compare D with the data item at the current vertex. If D is equal to the data item at the current vertex, we have found D, so we stop. If D is less than the data item at the current vertex v, we move to v's left child and repeat this process. If D is greater than the data item at the current vertex v, we move to v's right child and repeat this process. If at any point the child to move to is missing, we conclude that D is not in the tree.

For example, to find OLD in the binary search tree of Figure 4.3.3, we would begin by comparing OLD with the data item at the root, ONCE. Since OLD is less than ONCE, we move to the left child A. Since OLD is greater than A, we move to the right child AN. Since OLD is greater than AN, we move to the right child OLD. At this point, we stop since we have found OLD.

You should work through the problem of finding OMIT in the binary search tree of Figure 4.3.3 to see what happens when the data item we wish to locate is not present.

The time spent searching for an item in a binary search tree is longest when the item is not present and we follow the longest path from the root to a terminal vertex. Thus the maximum time to search for an item in a binary search tree is approximately proportional to the height of the tree. Thus, if the height of a binary search tree is small, searching the tree will always be very fast. Many ways are known of minimizing the height of a binary search tree (see [Standish]).

By referring to Figure 4.3.9, we can compute the maximum number of vertices in a binary tree of height h. We see that for heights 0 to 3, the maximum number of vertices in a binary tree is as given in Table 4.3.1.

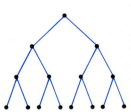

Level	Number of vertices on this level	Total number of vertices this level and above
0	$1 = 2^0$	$1 = 2^0$
1	$2 = 2^1$	$3 = 2^0 + 2^1$
2	$4 = 2^2$	$7 = 2^0 + 2^1 + 2^2$
3	$8 = 2^3$	$15 = 2^0 + 2^1 + 2^2 + 2^3$

Figure 4.3.9

Table 4.3.1

Height	Maximum Number of Vertices
0	1
1	3
2	7
3	15

It is apparent from Figure 4.3.9 that the maximum number of vertices on level h is 2^h. Thus the maximum number of vertices in a tree of height h is

$$1 + 2 + 2^2 + \cdots + 2^h = \frac{2^{h+1} - 1}{2 - 1} = 2^{h+1} - 1.$$

(We used Example 2.4.3, which gives the formula for the geometric sum.) For example, a binary tree of height 20 can contain up to $2^{20+1} - 1 = 2{,}097{,}149$ vertices. Thus it is possible to store over 2 million data items in a binary tree and find an item in the tree in at most 20 steps!

EXERCISES

1. Give an example of a real situation (like that of Figure 4.3.1) that can be modeled as a decision tree. Draw the decision tree.

Exercises 2–6 refer to the binary tree in Figure 4.3.2.

2H. Find the left child of e.

3. Find the right child of e.

4. Draw the binary tree that results from adding a right child h of f in Figure 4.3.2.

5H. Draw the binary tree that results from adding a left child i of g in Figure 4.3.2.

6. Draw the binary tree that results from adding as many children as possible to Figure 4.3.2.

7H. Place the words FOURSCORE AND SEVEN YEARS AGO OUR FOREFATHERS BROUGHT FORTH, in the order in which they appear, in a binary search tree.

8H. Explain how we would look for AGO in the binary search tree of Exercise 7.

9H. Explain how we would look for FOURTH in the binary search tree of Exercise 7.

10. Place the words OLD PROGRAMMERS NEVER DIE THEY JUST LOSE THEIR MEMORIES, in the order in which they appear, in a binary search tree.

11. Explain how we would look for AGO in the binary search tree of Exercise 10.

12. Explain how we would look for JUST in the binary search tree of Exercise 10.

13H. True or false? Let T be a binary tree. If for every vertex v in T the data item in v is greater than the data item in the left child of v and the data item in v is less than the data item in the right child of v, then T is a binary search tree. Explain.

A **full binary tree** is a binary tree in which every internal vertex has exactly two children.

14H. If T is a full binary tree with i internal vertices, how many terminal vertices does T have?

15. If T is a full binary tree with i internal vertices, how many vertices does T have?

16. If a single-elimination tournament has 70 contestants, how many matches are played?

In Exercises 17 and 18, draw a graph having the given properties or explain why no such graph exists.

17H. Full binary tree; four internal vertices; five terminal vertices

18. Full binary tree; height = 4; nine terminal vertices

A binary tree T is **balanced** if for every vertex v in T, the heights of the left and right subtrees of v differ by at most 1. (Here the height of an empty tree is defined to be -1.)

State whether each tree in Exercises 19–22 is balanced or not.

19H.

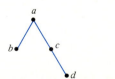

20.

21.
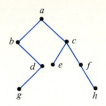

22H.
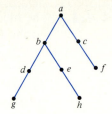

23. Draw a balanced binary tree with the fewest vertices having height 4.

*24. By using induction on h, show that if a binary tree of height h has t terminal vertices, then

$$\lg t \le h.$$

25. Draw a decision tree that can be used to determine who must file a federal tax return.

26. Draw a decision tree that gives a reasonable strategy for playing blackjack (see, e.g., [Ainslie]).

†4.4 Isomorphic Trees

In Section 3.5 we defined what it means for two graphs to be isomorphic. (You might want to review Section 3.5 before continuing.) In this section we discuss isomorphic trees, isomorphic rooted trees, and isomorphic binary trees.

Theorem 3.5.7 states that simple graphs G_1 and G_2 are isomorphic if and only if there is a one-to-one, onto function f from the vertex set of G_1 to the vertex set of G_2 that preserves the adjacency relation in the sense that vertices v_i and v_j are adjacent in G_1 if and only if the vertices $f(v_i)$ and $f(v_j)$ are adjacent in G_2. Since a (free) tree is a simple graph, trees T_1 and T_2 are isomorphic if and only if there is a one-to-one, onto function f from the vertex set of T_1 to the vertex set of T_2 that preserves the adjacency relation; that is, vertices v_i and v_j are adjacent in T_1 if and only if the vertices $f(v_i)$ and $f(v_j)$ are adjacent in T_2.

†This section can be omitted without loss of continuity.

Example 4.4.1. The function f from the vertex set of the tree T_1 shown in Figure 4.4.1 to the vertex set of the tree T_2 shown in Figure 4.4.2 defined by

$$f(a) = 1, \quad f(b) = 3, \quad f(c) = 2, \quad f(d) = 4, \quad f(e) = 5$$

is a one-to-one, onto function that preserves the adjacency relation. Thus the trees T_1 and T_2 are isomorphic.

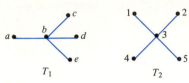

Figure 4.4.1   **Figure 4.4.2**

As in the case of graphs, we can show that two trees are not isomorphic if we can exhibit an invariant that the trees do not share.

Example 4.4.2. The trees T_1 and T_2 of Figure 4.4.3 are not isomorphic because T_2 has a vertex (x) of degree 3, but T_1 does not have a vertex of degree 3.

Figure 4.4.3

We can show that there are three nonisomorphic trees with five vertices. The three nonisomorphic trees are shown in Figures 4.4.1 and 4.4.3.

Theorem 4.4.3. *There are three nonisomorphic trees with five vertices.*

PROOF.   We will give an argument to show that any tree with five vertices is isomorphic to one of the trees in Figure 4.4.1 or 4.4.3.

If T is a tree with five vertices, by Theorem 4.2.3 T has four edges. If T had a vertex v of degree greater than 4, v would be incident on more than four edges. It follows that each vertex in T has degree at most 4.

We will first find all nonisomorphic trees with five vertices in which the maximum vertex degree that occurs is 4. We will next find all nonisomorphic trees with five vertices in which the maximum vertex degree that occurs is 3, and so on.

Let T be a tree with five vertices and suppose that T has a vertex v of degree 4. Then there are four edges incident on v and, because of Theorem 4.2.3, these are all the edges. It follows that in this case T is isomorphic to the tree in Figure 4.4.1.

Suppose that T is a tree with five vertices and the maximum vertex degree that occurs is 3. Let v be a vertex of degree 3. Then v is incident on three edges, as shown in Figure 4.4.4. The fourth edge cannot be incident on v since then v would have degree 4. Thus the fourth edge is incident on one of v_1, v_2, or v_3. Adding an edge incident on any of v_1, v_2, or v_3 gives a tree isomorphic to the tree T_2 of Figure 4.4.3.

Figure 4.4.4

Now suppose that T is a tree with five vertices and the maximum vertex degree that occurs is 2. Let v be a vertex of degree 2. Then v is incident on two edges as shown in Figure 4.4.5. A third edge cannot be incident on v; thus it must be incident on either v_1 or v_2. Adding the third edge gives the graph of Figure 4.4.6. For the same reason, the fourth edge cannot be incident on either of the vertices w_1 or w_2 of Figure 4.4.6. Adding the last edge gives a tree isomorphic to the tree T_1 of Figure 4.4.3.

Figure 4.4.5   **Figure 4.4.6**

Since a tree with five vertices must have a vertex of degree 2, we have found all nonisomorphic trees with five vertices. ∎

For two *rooted* trees T_1 and T_2 to be isomorphic, there must be a one-to-one, onto function f from T_1 to T_2 that preserves the adjacency relation and that preserves the root. The latter condition means that $f(\text{root of } T_1)$ = root of T_2. The formal definition follows.

Definition 4.4.4. Let T_1 be a rooted tree with root r_1 and let T_2 be a rooted tree with root r_2. The rooted trees T_1 and T_2 are *isomorphic* if there is a one-to-one, onto function f from the vertex set of T_1 to the vertex set of T_2 satisfying

(a) Vertices v_i and v_j are adjacent in T_1 if and only if the vertices $f(v_i)$ and $f(v_j)$ are adjacent in T_2.

(b) $f(r_1) = r_2$.

We call the function f an *isomorphism*.

Example 4.4.5. The rooted trees T_1 and T_2 in Figure 4.4.7 are isomorphic. An isomorphism is

$$f(v_1) = w_1, \quad f(v_2) = w_3, \quad f(v_3) = w_4, \quad f(v_4) = w_2,$$
$$f(v_5) = w_7, \quad f(v_6) = w_6, \quad f(v_7) = w_5.$$

The isomorphism of Example 4.4.5 is not unique. Can you find another isomorphism of the rooted trees of Figure 4.4.7?

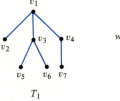

T_1 T_2

Figure 4.4.7

Example 4.4.6. The rooted trees T_1 and T_2 of Figure 4.4.8 are not isomorphic since the root of T_1 has degree 3 but the root of T_2 has degree 2. These trees are isomorphic as *free* trees. Each is isomorphic to the tree T_2 of Figure 4.4.3.

T_1 T_2

Figure 4.4.8

Arguing as in the proof of Theorem 4.4.3, we can show that there are four nonisomorphic rooted trees with four vertices.

Theorem 4.4.7. *There are four nonisomorphic rooted trees with four vertices. These four rooted trees are shown in Figure 4.4.9.*

PROOF. We first find all nonisomorphic rooted trees with four vertices in which the root has degree 3; we then find all nonisomorphic rooted trees with four vertices in which the root has degree 2; and so on. We note that the root of a rooted tree with four vertices cannot have degree greater than 3.

A rooted tree with four vertices in which the root has degree 3 must be isomorphic to the tree in Figure 4.4.9a.

A rooted tree with four vertices in which the root has degree 2 must be isomorphic to the tree in Figure 4.4.9b.

Let T be a rooted tree with four vertices in which the root has degree 1. Then the root is incident on one edge. The two remaining edges may be added in one of two ways (see Figure 4.4.9c and d). Therefore, all nonisomorphic rooted trees with four vertices are shown in Figure 4.4.9. ∎

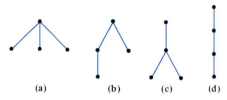

(a)   (b)   (c)   (d)

Figure 4.4.9

Binary trees are special kinds of rooted trees, thus an isomorphism of binary trees must preserve the adjacency relation and must preserve the roots. However, in binary trees a child is designated a left child or a right child. We require that an isomorphism of binary trees preserve the left and right children. The formal definition follows.

Definition 4.4.8. Let T_1 be a binary tree with root r_1 and let T_2 be a binary tree with root r_2. The binary trees T_1 and T_2 are *isomorphic* if there is a one-to-one, onto function f from the vertex set of T_1 to the vertex set of T_2 satisfying

(a) Vertices v_i and v_j are adjacent in T_1 if and only if the vertices $f(v_i)$ and $f(v_j)$ are adjacent in T_2.

(b) $f(r_1) = r_2$.

(c) v is a left child of w in T_1 if and only if $f(v)$ is a left child of $f(w)$ in T_2.

(d) v is a right child of w in T_1 if and only if $f(v)$ is a right child of $f(w)$ in T_2.

We call the function f an *isomorphism*.

Example 4.4.9. The binary trees T_1 and T_2 in Figure 4.4.10 are isomorphic. The isomorphism is $f(v_i) = w_i$ for $i = 1, \ldots, 4$.

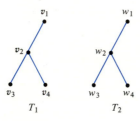

Figure 4.4.10

Example 4.4.10. The binary trees T_1 and T_2 in Figure 4.4.11 are not isomorphic. The root v_1 in T_1 has a right child, but the root w_1 in T_2 has no right child.

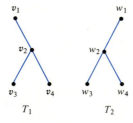

Figure 4.4.11

The trees T_1 and T_2 in Figure 4.4.11 *are* isomorphic as rooted trees and as free trees. As rooted trees, either of the trees of Figure 4.4.11 is isomorphic to the rooted tree T of Figure 4.4.9c.

Arguing similarly as in the proofs of Theorems 4.4.3 and 4.4.7, we can show that there are five nonisomorphic binary trees with three vertices.

Theorem 4.4.11. *There are five nonisomorphic binary trees with three vertices. These five binary trees are shown in Figure 4.4.12.*

PROOF. We first find all nonisomorphic binary trees with three vertices in which the root has degree 2. We then find all nonisomorphic binary trees with three vertices in which the root has degree 1. We note that the root of any binary tree cannot have degree greater than 2.

A binary tree with three vertices in which the root has degree 2 must be isomorphic to the tree in Figure 4.4.12a. In a binary tree with three vertices in which the root has degree 1, the root either has a left child and no right child or a right child and no left child. If the root has a left child, the child itself has either a left or a right child. We obtain the two binary trees in Figure 4.4.12b and c. Similarly, if the root has a right child, the child itself has either a left or a right child. We obtain the two binary trees in Figure 4.4.12d and e. Therefore, all nonisomorphic binary trees with three vertices are shown in Figure 4.4.12. ■

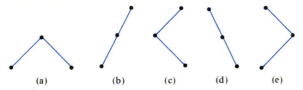

(a)   (b)   (c)   (d)   (e)

Figure 4.4.12

When discussing the graph isomorphism problem in Section 3.5, we remarked that there is no fast method of deciding whether two arbitrary graphs are isomorphic. The situation is different for trees. It is possible to determine whether two arbitrary trees are isomorphic very quickly even if the trees have enormous numbers of vertices.

In Theorem 4.4.3 we showed that there are three nonisomorphic free trees having five vertices. In Theorem 4.4.7 we showed that there are four nonisomorphic rooted trees having four vertices. In Theorem 4.4.11 we showed that there are five nonisomorphic binary trees having three vertices. You might have wondered if there are formulas for the number of nonisomorphic n-vertex trees of a particular type. There are formulas for the number of nonisomorphic n-vertex free trees, for the number of nonisomorphic n-vertex rooted trees, and for the number of nonisomorphic n-vertex binary trees. The formulas for the number of nonisomorphic free trees and for the number of nonisomorphic rooted trees with n vertices are quite complicated. Furthermore, the derivations of these formulas require techniques beyond those that we develop in this text. The formulas and proofs appear in [Deo, Section 10-3]. The formula for the number of binary trees with n vertices is quite simple. There are

$$\frac{1}{n+1} \cdot \frac{(2n)!}{n!n!}$$

binary trees with n vertices. In Section 7.1 we discuss how a derivation of this result might begin.

EXERCISES

In Exercises 1–6, determine whether each pair of free trees is isomorphic. If the pair is isomorphic, specify an isomorphism. If the pair is not isomorphic, give an invariant that one tree satisfies but the other does not.

1H.

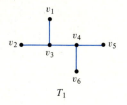 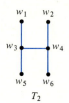

2. T_1 as in Exercise 1

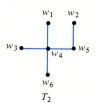

3.

 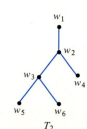

4H.

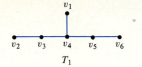

T_1

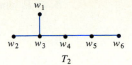

T_2

5.

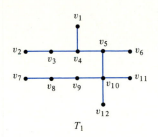

T_1

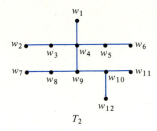

T_2

6.

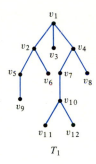

T_1

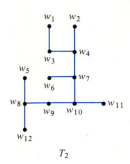

T_2

In Exercises 7–9, determine whether each pair of rooted trees is isomorphic. If the pair is isomorphic, specify an isomorphism. If the pair is not isomorphic, give an invariant that one tree satisfies but the other does not. Also, determine whether the trees are isomorphic as free trees.

7H.

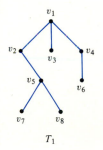

T_1

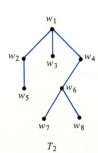

T_2

8. T_1 and T_2 as in Exercise 3

9.

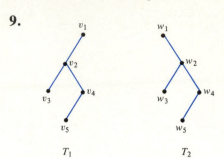

In Exercises 10–12, determine whether each pair of binary trees is isomorphic. If the pair is isomorphic, specify an isomorphism. If the pair is not isomorphic, give an invariant that one tree satisfies but the other does not. Also, determine whether the trees are isomorphic as free trees or as rooted trees.

10H. T_1 and T_2 as in Exercise 9

11.

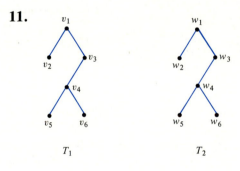

12.

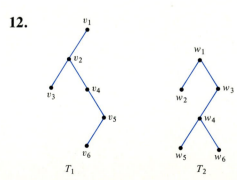

13H. Draw all nonisomorphic free trees having three vertices.

14. Draw all nonisomorphic free trees having four vertices.

15. Draw all nonisomorphic free trees having six vertices.

16H. Draw all nonisomorphic rooted trees having three vertices.

17. Draw all nonisomorphic rooted trees having five vertices.

18. Draw all nonisomorphic binary trees having two vertices.

19H. Draw all nonisomorphic binary trees having four vertices.

20. Draw all nonisomorphic full binary trees having seven vertices. (A full binary tree is a binary tree in which each internal vertex has two children.)

21. Draw all nonisomorphic full binary trees having nine vertices.

22. Report on the formulas for the number of nonisomorphic free trees and for the number of nonisomorphic rooted trees with n vertices (see [Deo]).

†4.5 Game Trees

Trees are useful in the analysis of games such as tic-tac-toe, chess, and checkers, in which players alternate moves. In this section we show how trees can be used to develop game-playing strategies. This kind of approach is used in the development of many computer programs that allow human beings to play against computers or even computers against computers.

As an example of the general approach, consider the game of nim. Initially, there are n piles, each containing a number of identical tokens. Players alternate moves. A move consists of removing one or more tokens from any one pile. The player who removes the last token loses. As a specific case, consider an initial distribution consisting of two piles: one containing three tokens and one containing two tokens. All possible move sequences can be listed in a **game tree** (see Figure 4.5.1). The first player is represented by a box and the second player is represented by a circle. Each vertex shows a particular position in the game. In our game, the initial position is shown as $\binom{3}{2}$. A path represents a sequence

---

†This section can be omitted without loss of continuity.

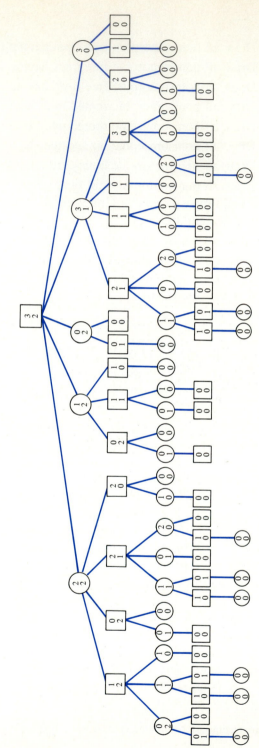

Figure 4.5.1

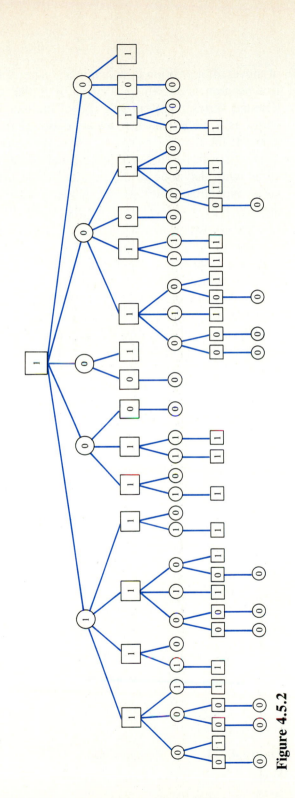

Figure 4.5.2

of moves. If a position is shown in a square, it is the first player's move; if a position is shown in a circle, it is the second player's move. A terminal vertex represents the end of the game. In nim, if the terminal vertex is a circle, the first player removed the last token and lost the game. If the terminal vertex is a box, the second player lost.

The analysis begins with the terminal vertices. We label each terminal vertex with the value of the position to the first player. If the terminal vertex is a circle, since the first player lost, this position is worthless to the first player and we assign it the value 0 (see Figure 4.5.2). If the terminal vertex is a box, since the first player won, this position is valuable to the first player and we label it with a value greater than 0, say 1 (see Figure 4.5.2). At this point, all terminal vertices have been assigned values.

Now consider the problem of assigning values to the internal vertices. Suppose, for example, that we have an internal box, all of whose children have been assigned a value. For example, if we have the situation shown in Figure 4.5.3, the first player (box) should move to the position represented by vertex B, since this position is the most valuable. In other words, box moves to a position represented by a child with the maximum value. We assign this maximum value to the box vertex.

Consider the situation from the second (circle) player's point of view. Suppose that we have the situation shown in Figure 4.5.4. Circle should move to the position represented by vertex C, since this position is least valuable to box and therefore most valuable to circle. In other words, circle moves to a position represented by a child with the minimum value. We assign this minimum value to the circle vertex. The process by which circle seeks the minimum of its children and box seeks the maximum of its children is called the **minimax procedure**.

Working upward from the terminal vertices and using the minimax procedure, we can assign values to all of the vertices in the game tree (see Figure 4.5.2). These numbers represent the value of the game, at any position, to the first player. Notice that the root in Figure 4.5.2,

Figure 4.5.3

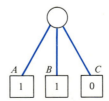

Figure 4.5.4

which represents the original position, has a value of 1. This means that the first player can always win the game by using an optimal strategy. This optimal strategy is contained in the game tree: The first player always moves to a position that maximizes the value of the children. No matter what the second player does, the first player can always move to a vertex having value 1. Ultimately, a terminal vertex having value 1 is reached where the first player wins the game.

Many interesting games, such as chess, have game trees so large that it is not feasible to use a computer to generate the entire tree. Nevertheless, the concept of a game tree is still useful for analyzing such games.

If the game tree is so large that it is not feasible to reach a terminal vertex, we do not evaluate vertices below some specified level. The evaluation is said to be an ***n*-level evaluation** if we evaluate vertices on levels 0 to n only. Since the vertices at the lowest level may not be terminal vertices, some method must be found to assign them a value. This is where the specifics of the game must be dealt with. An **evaluation function** E is constructed that assigns each possible game position P the value $E(P)$ of the position to the first player. After the vertices at the lowest level are assigned values by using the function E, the minimax procedure can be applied to generate the values of the other vertices. We illustrate these concepts with an example.

Example 4.5.1. Apply the minimax procedure to find the value of the root in tic-tac-toe using a two-level evaluation. Use the evaluation function E, which assigns a position the value

$$NX - NO,$$

where NX (respectively, NO) is the number of rows, columns, or diagonals containing an X (respectively, O) that X (respectively, O) might complete. For example, position P of Figure 4.5.5 has $NX = 2$, since X might complete the column or the diagonal, and $NO = 1$, since O can only complete a column. Therefore,

$$E(P) = 2 - 1 = 1.$$

Figure 4.5.5

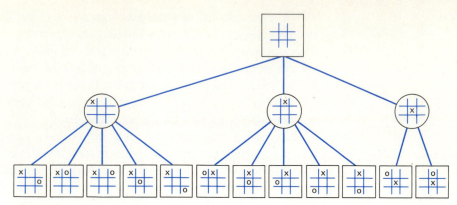

Figure 4.5.6

In Figure 4.5.6 we have drawn the game tree for tic-tac-toe to level 2. We have omitted symmetric positions. We first assign the vertices at level 2 the values given by E (see Figure 4.5.7). Next, we compute circle's values by minimizing over the children. Finally, we compute the value of the root by maximizing over the children. Using this analysis, the first move by the first player would be to the center square.

Evaluation of a game tree, or even a part of a game tree, can be a time-consuming task, so any technique that reduces the effort is welcomed. The most general technique is called **alpha-beta pruning**. In general, alpha-beta pruning allows us to bypass many vertices in a game tree, yet still find the value of a vertex. The value obtained is the same as if we had evaluated all the vertices.

As an example, consider the game tree in Figure 4.5.8. Suppose that we want to evaluate vertex A using a two-level evaluation. We evaluate children left to right. We begin at the lower left by evaluating the vertices, E, F, and G. The values shown are obtained from an evaluation

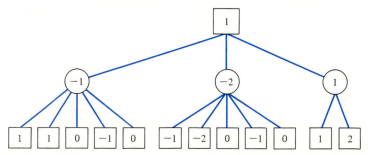

Figure 4.5.7

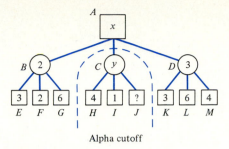

Figure 4.5.8

function. Vertex B is 2, the minimum of its children. At this point, we know that the value x of A must be at least 2, since the value of A is the maximum of its children; that is,

$$x \geq 2. \tag{4.5.1}$$

This lower bound for A is called an **alpha value** of A. The next vertices to be evaluated are H, I, and J. When I evaluates to 1, we know that the value y of C cannot exceed 1, since the value of C is the minimum of its children; that is,

$$y \leq 1. \tag{4.5.2}$$

It follows from (4.5.1) and (4.5.2) that whatever the value of y is, it will not affect the value of x; thus we need not concern ourselves further with the subtree headed by vertex C. We say that an **alpha cutoff** occurs. We next evaluate the children of D and then D itself. Finally, we find that the value of A is 3.

To summarize, an alpha cutoff occurs at a box vertex v when a grand-child w of v has a value less than or equal to the alpha value of v. The subtree whose root is the parent of w may be deleted (pruned). This deletion will not affect the value of v. An alpha value for a vertex v is only a lower bound for the value of v. The alpha value of a vertex is dependent on the current state of the evaluation and changes as the evaluation progresses.

Similarly, a **beta cutoff** occurs at a circle vertex v when a grandchild w of v has a value greater than or equal to the beta value of v. The subtree whose root is the parent of w may be pruned. This deletion will not affect the value of v. A **beta value** for a vertex v is only an upper bound for the value of v. The beta value of a vertex is dependent on the current state of the evaluation and changes as the evaluation progresses.

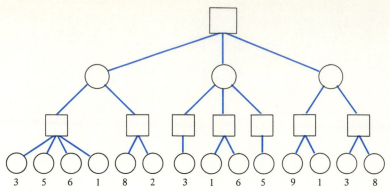

Figure 4.5.9

___

Example 4.5.2.   Evaluate the root of the tree of Figure 4.5.9 using the minimax procedure with alpha-beta pruning. Assume that children are evaluated left to right. For each vertex whose value is computed, write the value in the vertex. Place a check by the root of each subtree that is pruned. The value of each terminal vertex is written under the vertex.

We begin by evaluating vertices A, B, C, and D (see Figure 4.5.10). Next, we find that the value of E is 6. This results in a beta value of 6 for F. Next, we evaluate vertex G. Since its value is 8 and 8 exceeds the beta value of F, we obtain a beta cutoff and prune the subtree with root H. The value of F is 6. This results in an alpha value of 6 for I. Next, we evaluate vertices J and K. Since the value 3 of K is less than the alpha value 6 of I, an alpha cutoff occurs and the subtree

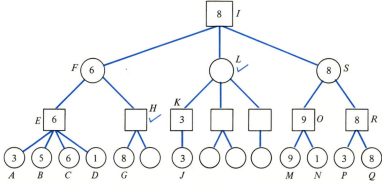

Figure 4.5.10

with root L may be pruned. Next, we evaluate M, N, O, P, Q, R, and S. No further pruning is possible. Finally, we determine that the root I has value 8.

It has been shown (see [Pearl]) that for game trees in which every parent has n children and in which the terminal values are randomly ordered, for a given amount of time, the alpha-beta procedure permits an n-level evaluation to a depth $\frac{4}{3}$ greater than the pure minimax procedure, which evaluates every vertex. Pearl also shows that for such game trees, the alpha-beta procedure is optimal.

Other techniques have been combined with alpha-beta pruning to facilitate the evaluation of a game tree. One idea is to order the children of the vertices· to be evaluated so that the most promising moves are examined first (see Exercises 19–21).

Some game-playing programs have been quite successful. A computer program beat the top human player in the world in backgammon (see [Berliner]). The best computer chess programs play at a level considerably above average human players. Success has been more elusive in card games. Although computers have been programmed to bid bridge hands at a competitive level, no computer program yet plays bridge hands (declarer or defender) at a very respectable level.

EXERCISES

1H. Draw the complete game tree for a version of nim in which the initial position consists of one pile of six tokens and a turn consists of taking one, two, or three tokens. Assign values to all vertices so that the resulting tree is analogous to Figure 4.5.2. Assume that the last player to take a token loses. Will the first or second player, playing an optimal strategy, always win? Describe an optimal strategy for the winning player.

2. Draw the complete game tree for nim in which the initial position consists of two piles of three tokens each. Omit symmetric positions. Assume that the last player to take a token loses. Assign values to all vertices so that the resulting tree is analogous to Figure 4.5.2. Will the first or second player, playing an optimal strategy, always win? Describe an optimal strategy for the winning player.

3.   Draw the complete game tree for nim in which the initial position consists of two piles, one containing three tokens and the other containing two tokens. Assume that the last player to take a token wins. Assign values to all vertices so that the resulting tree is analogous to Figure 4.5.2. Will the first or second player, playing an optimal strategy, always win? Describe an optimal strategy for the winning player.

4H. Draw the complete game tree for nim in which the initial position consists of two piles of three tokens each. Omit symmetric positions. Assume that the last player to take a token wins. Assign values to all vertices so that the resulting tree is analogous to Figure 4.5.2. Will the first or second player, playing an optimal strategy, always win? Describe an optimal strategy for the winning player.

5.   Draw the complete game tree for the version of nim described in Exercise 1. Assume that the last person to take a token wins. Assign values to all vertices so that the resulting tree is analogous to Figure 4.5.2. Will the first or second player, playing an optimal strategy, always win? Describe an optimal strategy for the winning player.

6.   Give an example of a (hypothetical) complete game tree in which a terminal vertex is 1 if the first player won and 0 if the first player lost having the following properties: There are more 0's than 1's among the terminal vertices, but the first player can always win by playing an optimal strategy.

Evaluate each vertex in each game tree. The values of the terminal vertices are given.

7H.

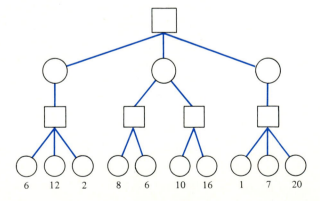

6   12   2   8   6   10   16   1   7   20

8.

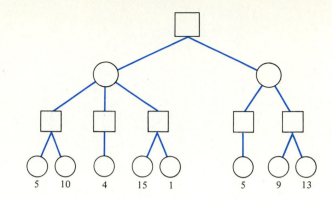

9.

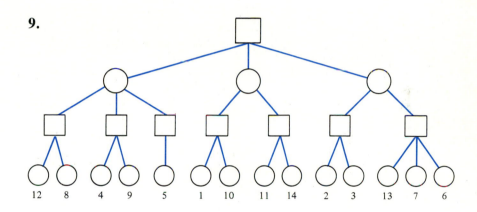

10H.

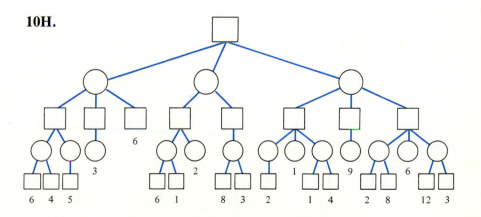

11.

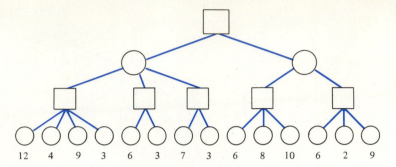

```
12   4   9   3   6   3   7   3   6   8   10   6   2   9
```

12H. Evaluate the root of each of the trees of Exercises 7–11 using the minimax procedure with alpha-beta pruning. Assume that children are evaluated left to right. For each vertex whose value is computed, write the value in the vertex. Place a check by the root of each subtree that is pruned. The value of each terminal vertex is written under the vertex.

In Exercises 13–16, determine the value of the tic-tac-toe position using the evaluation function of Example 4.5.1.

13H.

		O
	X	

14.

		O
	X	O
	X	

15.

O	X	O
	X	
	O	X

16H.

		O
	X	
		X

17. Assume that the first player moves to the center square in tic-tac-toe. Carry out a two-level evaluation of the game tree where the root has an X in the center square. Omit symmetric positions. Evaluate all the vertices using the evaluation function of Example 4.5.1. Where will O move?

*18. Would a program based on the evaluation function E of Example 4.5.1 that always performs a two-level evaluation of the game tree play a perfect game of tic-tac-toe? If not, can you alter E so that a program that always performs a two-level evaluation of the game tree will play a perfect game of tic-tac-toe?

The following approach often leads to more pruning than pure alpha-beta minimax. First, perform a two-level evaluation of the game tree. Evaluate children from left to right. At this point, all the children of the root will have values. Next, order the children of the root with the most promising moves to the left. Now, perform an *n*-level evaluation of the game tree using alpha-beta pruning. Evaluate children from left to right.

Carry out this procedure for *n* = 4 for each game tree of Exercises 19–21. Place a check by the root of each subtree that is pruned. The value of each vertex, as given by the evaluation function, is given under the vertex.

19H.

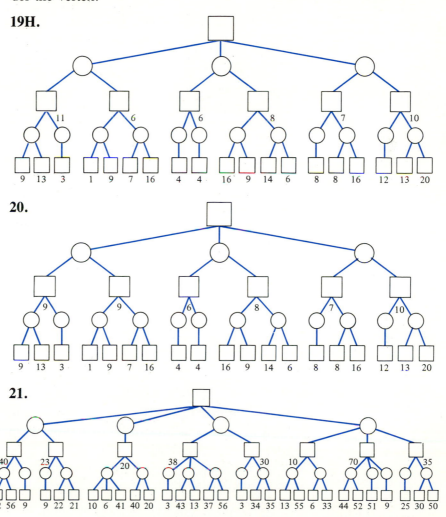

20.

21.

22.   [*Project*]   According to [Horowitz, 1976] and [Nilsson], the
      complete game tree for chess has over 10^{100} vertices. Report on
      how this estimate was obtained.

4.6 Notes

The following are recommended references on trees: [Berge; Busacker;
Deo; Harary; Horowitz, 1976; Knuth, 1973, Vol. 1; Lipschutz, 1976;
Liu, 1985; Ore; Standish; and Tucker].

See [Tsichritzis] for the use of trees in hierarchical data bases.

Decision trees are covered in [Standish].

The use of Huffman codes in hand-held computers is detailed in [Wil-
liams]. See [Standish] for additional information on Huffman codes.

Good references on game trees are [Horowitz, 1976; Nievergelt; Nilsson; and Slagle]. In [Frey], the minimax procedure is applied to a simple
game. Various methods to speed up the search of the game tree are
discussed and compared. Programs are given in BASIC. [Berlekamp]
contains a general theory of games as well as analyses of many specific
games.

Computer Exercises

1.   Write a program that tests if a graph is a tree.
2.   Write a program that given the adjacency matrix of a tree and a
     vertex v, draws the tree rooted at v using a computer graphics dis-
     play.
3.   Write a computer program to encode and decode strings of charac-
     ters according to the Huffman code given in the table.

Character	Huffman Code
(blank)	0
E	1100
T	1001
A	11111
O	11110
N	11100
R	11011
I	11010
S	10110
H	10101
D	111011
L	101111
F	101001
C	101000
M	100011
U	100010
G	100001
Y	100000
P	1110101
W	1011101
B	1011100
V	11101001
K	1110100011
X	1110100001
J	1110100000
Q	11101000101
Z	11101000100

4. Use your program of Exercise 3 to encode some sample text. Compare the number of bits used to encode your text with the number of bits necessary to encode your text in ASCII.

5. Write a program that given a frequency table for characters, constructs an optimum Huffman code.

6. Write a program that given a tree T, computes the eccentricity of each vertex in T and finds the center(s) of T.

7. Write a program that given a rooted tree and a vertex v,
(a) Finds the parent of v.

(b) Finds the ancestors of v.

(c) Finds the children of v.

(d) Finds the descendants of v.

(e) Finds the siblings of v.

(f) Determines whether v is a terminal vertex.

8. Write a program to generate the complete game tree for nim in which the initial position consists of two piles of four tokens each. Assume that the last player to take a token loses.

9. Implement the minimax procedure as a program.

10. Implement the minimax procedure with alpha-beta pruning as a program.

11. Implement the method of playing tic-tac-toe in Example 4.5.1 as a program.

12. [*Project*] Develop a computer program to play a game that has relatively simple rules. Suggested games are Othello, The Mill, Battleship, and Kalah. (See [Ainslie] and [Freeman] for rules and strategies.)

Chapter Review

SECTION 4.1

Free tree

Rooted tree

Level of a vertex in a rooted tree

Height of a rooted tree

Hierarchical definition tree

Data base

Huffman code

SECTION 4.2

Parent

Ancestor

Child

Descendant

Sibling

Terminal vertex

Internal vertex

Chapter Self-Test

SECTION 4.1

1H. Draw the free tree as a rooted tree with root c.

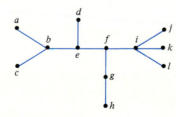

2H. Find the level of every vertex in the tree of Exercise 1 rooted at c.

3H. Find the height of the tree of Exercise 1 rooted at c.

4H. Decode the bit string 0010000111 using the Huffman code

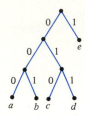

SECTION 4.2

5H. Draw the free tree of Exercise 1 as a rooted tree with root f. Find
(a) The parent of a.
(b) The children of b.
(c) The terminal vertices.
(d) The subtree rooted at e.

Answer true or false in Exercises 6–8 and explain your answer.

6H. If T is a tree with six vertices, T must have five edges.

7H. If T is a rooted tree with six vertices, the height of T is at most 5.

8H. An acyclic graph with eight vertices has seven edges.

SECTION 4.3

9H. Draw a binary tree with exactly two left children and one right child.

10H. Place the words WORD PROCESSING PRODUCES CLEAN MANUSCRIPTS BUT NOT NECESSARILY CLEAR PROSE, in the order in which they appear, in a binary search tree.

11H. Explain how we would look for NOT in the binary search tree of Exercise 10.

12H. Explain how we would look for MORE in the binary search tree of Exercise 10.

SECTION 4.4

Answer true or false in Exercises 13 and 14 and explain your answer.

13H. If T_1 and T_2 are isomorphic as rooted trees, then T_1 and T_2 are isomorphic as free trees.

14H. If T_1 and T_2 are rooted trees that are isomorphic as free trees, then T_1 and T_2 are isomorphic as rooted trees.

15H. Determine whether the free trees are isomorphic. If the trees are isomorphic, give an isomorphism. If the trees are not isomorphic, give an invariant the trees do not share.

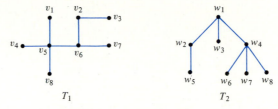

16H. Determine whether the rooted trees are isomorphic. If the trees are isomorphic, give an isomorphism. If the trees are not isomorphic, give an invariant the trees do not share.

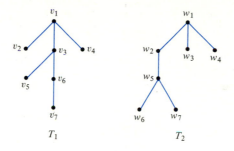

SECTION 4.5

17H. Find the value of the tic-tac-toe position using the evaluation function of Example 4.5.1.

18H. Give an evaluation function for a tic-tac-toe position different from that of Example 4.5.1. Attempt to discriminate more among the positions than does the evaluation function of Example 4.5.1.

19H. Evaluate each vertex in the game tree. The values of the terminal vertices are given.

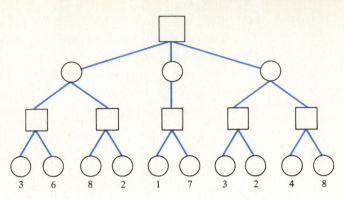

20H. Evaluate the root of the tree of Exercise 19 using the minimax procedure with alpha-beta pruning. Assume that the children are evaluated left to right. For each vertex whose value is computed, write the value in the vertex. Place a check by the root of each subtree that is pruned.

It's so simple.

Step 1: We find the worst play in the world—a sure fire flop.

Step 2: I raise a million bucks—there are a lot of little old ladies in the world.

Step 3: You go back to work on the books. Phoney lists of backers—one for the government, one for us. You can do it, Bloom, you're a wizard.

Step 4: We open on Broadway and before you can say

Step 5: We close on Broadway.

Step 6: We take our million bucks and we fly to Rio de Janeiro.

—from *The Producers*

Algorithms

An algorithm is a step-by-step method of solving some problem. The preceding excerpt from *The Producers* is an algorithm for making money from a play that flops on Broadway. Examples of algorithms can be found throughout history, going back at least as far as ancient Babylonia. Indeed, the word "algorithm" derives from the name of the ninth-century Arabic mathematician al-Khowārizmī. As emphasized by Donald Knuth and others, algorithms based upon sound mathematical principles play a central role in computer science. In order for a solution to a problem to be executed by a computer, the solution must be described as a sequence of precise steps.

In Section 5.1 we illustrate the concept of algorithm with several examples. In Section 5.1 and in all nonoptional sections of this book, we write algorithms in ordinary English. Optional Section 5.2 describes pseudocode—a method that is becoming a standard way to write algorithms. The greatest common divisor algorithm, an ancient Greek algorithm that is still much used, is the topic of Section 5.3. Complexity of algorithms (Section 5.4) refers to how long it takes to execute an algorithm and how much space is required. It is often not enough simply to have an algorithm to solve a problem; it is necessary to obtain an algorithm that solves the problem efficiently—quickly and without using too much space.

5.1 Introduction

Consider the step-by-step instructions to play a tape in a video cassette recorder (VCR):

1. Turn on the VCR.
2. Place the VCR selector switch in playback position.
3. Turn on the television set.
4. Select channel 3 on the television set.
5. Insert the cassette.
6. If the cassette has not been rewound, rewind it.
7. Press the play switch on the VCR.

Such a set of instructions furnishes an example of an **algorithm**.

Although it is possible to give a formal and mathematically precise definition of algorithm, we will consider an algorithm to be a method of solving a problem in a finite amount of time that can be described by a sequence of precise instructions. In the preceding example, steps 1 to 7 give an algorithm that solves the problem of playing a tape in a video cassette recorder. Theoretically, the instructions of an algorithm could be carried out by a machine because of the precision with which the steps are stated.

In this section we describe how we will write algorithms and present examples of algorithms. Algorithm 5.1.1, which finds the maximum of three numbers, illustrates our algorithmic notation.

Algorithm 5.1.1 Finding the Maximum of Three Numbers.   This algorithm finds the largest of the numbers a, b, and c.

Input: a, b, and c
Output: x, the maximum of a, b, and c

1. [Find the larger of a and b and call it x.] If $a > b$, then $x := a$; otherwise, $x := b$.
2. [Find the larger of x and c.] If $c > x$, then $x := c$.

The notation $y := z$ means "copy the value of z into y" or, equivalently, "replace the current value of y by the value of z." When $y := z$ is executed, the value of z is unchanged. We call $:=$ the **assignment operator**. For example, if y is 3 and z is 5, after $y := z$, y will be 5 and z will still be 5.

The meaning of line 1 of Algorithm 5.1.1 is that in case the condition "$a > b$" is true, we execute $x := a$, and in case the condition "$a > b$" is false, we execute $x := b$. In general, the instruction

$$\text{if } condition \text{ then } action \; 1 \text{ otherwise } action \; 2$$

means that if *condition* is true, execute *action 1*, and if *condition* is false, execute *action 2*. The meaning of line 2 of Algorithm 5.1.1 is that, in case the condition "$c > x$" is true, we execute $x := c$, and, in case the condition "$c > x$" is false, we do nothing. In general, the instruction

$$\text{if } condition \text{ then } action$$

means that if *condition* is true, execute *action* and, if *condition* is false, do nothing.

al-Khowārizmī (780–850?)

al-Khowārizmī used Hindu numerals, including zero, which were ultimately transmitted to Europe, becoming (incorrectly) known as Arabic numerals. He used the Arabic word *al-jabr* in the title of one of his books, from which the contemporary word *algebra* derives. He was also a geographer and calculated the circumference of the earth to be about twice its actual value. [*Photo courtesy of Süleymaniye Kütüphanesi, Istanbul, Turkey*]

Each algorithm has a title followed by a brief description of the algorithm. We then list the **input**, the values supplied to the algorithm, and the **output**, the values computed by the algorithm. Next are the steps, some of which may be numbered. The algorithm begins with the first line and proceeds sequentially through the lines unless there is an explicit instruction to do otherwise—such as "Go to line 2." Often a line will include a brief description of its purpose in brackets. In general, solutions to exercises that request algorithms should be written in the form illustrated by Algorithm 5.1.1.

We will show how Algorithm 5.1.1 executes for some specific input. Such a simulation is called a **trace**. First, suppose that

$$a = 1, \quad b = 5, \quad c = 3.$$

At line 1, $a > b$ ($1 > 5$) is false, so we set x to b (5). At line 2, $c > x$ ($3 > 5$) is false, so we do nothing. At this point x is 5, the largest of a, b, and c.

Suppose that

$$a = 6, \quad b = 1, \quad c = 9.$$

At line 1 of Algorithm 5.1.1, $a > b$ ($6 > 1$) is true, so we set x to 6. At line 2, $c > x$ ($9 > 6$) is true, so we set x to 9. At this point x is 9, the largest of a, b, and c.

For our next example of an algorithm, we generalize Algorithm 5.1.1 so that we can find the largest element in an arbitrarily long, but finite, sequence.

Algorithm 5.1.2   Finding the Largest Element in a Finite Sequence.
This algorithm finds the largest element among $s_1, s_2, \ldots, s_n$.

Input: A sequence $s_1, s_2, \ldots, s_n$
Output: x ($= \max \{s_1, \ldots, s_n\}$)

1. [Initialize x.] $x := s_1$.
2. [Step through the list looking for larger elements.] For $i := 2$ to n, execute line 3.
3.   If $s_i > x$, then (found a larger element) $x := s_i$.

Line 2 of Algorithm 5.1.2 means: Set i to 2 and execute line 3; then set i to 3 and execute line 3; then set i to 4 and execute line 3; . . . ; then set i to n and execute line 3; then stop.

Donald Knuth *(1938–   )*

Donald Knuth is professor of computer science at Stanford University. He became interested in computers while an undergraduate at Case Institute of Technology. While serving as student manager of the basketball team, he wrote a program that rated the players. The program proved successful; Case won the league championship. He is best known as the author of *The Art of Computing Programming* (see [1973 Vols. 1 and 3, 1981]), a projected seven-volume set of which three volumes have appeared. These books helped put computer algorithms on a sound mathematical footing. (His first publication was in *MAD* magazine.) An accomplished organist, he is also interested in computer-assisted typesetting. He won the prestigious Turing award from the Association for Computing Machinery in 1974. [*Photo © Thomas F. Black*]

We will sometimes insert **comments** in parentheses in the lines of algorithms. Comments are not executable instructions, but rather provide additional details that aid the reader in understanding the steps of algorithms. For example, in line 3 the text "(found a larger element)" is a comment.

We will provide a trace of Algorithm 5.1.2 for the input

$$s_1 = 5, \quad s_2 = 9, \quad s_3 = 7.$$

At line 1, we set x to 5. At line 2, we set i to 2 and execute line 3 and then we set i to 3 and execute line 3. First i is set to 2. The condition $s_2 > x$ ($9 > 5$) is true, so we set x to 9. Next, i is set to 3. The condition $s_3 > x$ ($7 > 9$) is false, so we do nothing. At this point x is 9, the largest of s_1, s_2, and s_3.

EXERCISES

1. Give an example, in the spirit of the video cassette recorder algorithm at the beginning of this section, of an algorithm in the real world.

In Exercises 2–4, trace Algorithm 5.1.1 for the given input.

2H. $a = 9, b = 6, c = 5$

3. $a = 6, b = 8, c = 10$

4. $a = 6, b = 4, c = 6$

In Exercises 5–7, trace Algorithm 5.1.2 for the given input.

5H. $s_1 = 5, s_2 = 7, s_3 = 9$

6. $s_1 = 5, s_2 = 12, s_3 = 5, s_4 = 9$

7. $s_1 = 8$

Unless instructed otherwise, write all algorithms in the style of Algorithms 5.1.1 and 5.1.2.

8H. Write an algorithm that outputs the smallest element among a, b, and c.

9. Write an algorithm that outputs the largest and smallest elements among a, b, c, and d.

10. Write an algorithm that outputs the largest and second largest elements among a, b, c, d, and e.

11H. Write an algorithm that outputs the smallest and second smallest elements among a, b, c, d, and e.

12. Write an algorithm that outputs the smallest element in the sequence

$$s_1, \ldots, s_n.$$

13. Write an algorithm that outputs the largest and second largest elements in the sequence

$$s_1, \ldots, s_n.$$

14H. Write an algorithm that outputs the smallest and second smallest elements in the sequence

$$s_1, \ldots, s_n.$$

15. Write an algorithm that outputs the largest and smallest elements in the sequence

$$s_1, \ldots, s_n.$$

16. Write an algorithm that outputs the index of the first occurrence of the largest element in the sequence

$$s_1, \ldots, s_n.$$

EXAMPLE: If the sequence were

$$6.2 \quad 8.9 \quad 4.2 \quad 8.9,$$

the algorithm would output the value 2.

17H. Write an algorithm that outputs the index of the last occurrence of the largest element in the sequence

$$s_1, \ldots, s_n.$$

EXAMPLE: If the sequence were

$$6.2 \quad 8.9 \quad 4.2 \quad 8.9,$$

the algorithm would output the value 4.

18. Write an algorithm that outputs the index of the first occurrence of the smallest element in the sequence

$$s_1, \ldots, s_n.$$

19. Write an algorithm that outputs the index of the last occurrence of the smallest element in the sequence

$$s_1, \ldots, s_n.$$

20H. Write an algorithm that outputs the index of the first occurrence of the value *key* in the sequence

$$s_1, \ldots, s_n.$$

If *key* is not in the sequence, the algorithm outputs the value 0.

EXAMPLE: If the sequence were

'MARY' 'JOE' 'MARK' 'RUDY',

and *key* were 'MARK', the algorithm would output the value 3.

21. Write an algorithm that outputs the index of the last occurrence of the value *key* in the sequence

$$s_1, \ldots, s_n.$$

If *key* is not in the sequence, the algorithm outputs the value 0.

22. Write an algorithm that outputs the index of the first item that is less than its predecessor in the sequence

$$s_1, \ldots, s_n.$$

If the items are in nondecreasing order (i.e., $s_i \le s_{i+1}$, for $i = 1, \ldots, n - 1$), the algorithm outputs the value 0.

EXAMPLE: If the sequence were

'AMY' 'BRUNO' 'ELIE' 'DAN' 'ZEKE',

the algorithm would output the value 4.

23H. Write an algorithm that outputs the index of the first item which is greater than its predecessor in the sequence

$$s_1, \ldots, s_n.$$

If the items are in nonincreasing order, the algorithm outputs the value 0.

24. Write an algorithm that reverses the sequence

$$s_1, \ldots, s_n.$$

EXAMPLE: If the sequence were

'AMY' 'BRUNO' 'ELIE',

the reversed sequence would be

'ELIE' 'BRUNO' 'AMY'.

25. Write an algorithm that outputs 1 if the positive integer $n > 1$ is prime and 0 if n is not prime.

26H. Write an algorithm that outputs the maximum sum of consecutive values in the numerical sequence

$$s_1, \ldots, s_n.$$

EXAMPLE: If the sequence were

27   6 -50 21 -3 14   16 -8 42   33 -21 9,

the algorithm outputs 115—the sum of

21   -3 14   16 -8 42   33.

If all the numbers in the sequence are negative, the maximum sum of consecutive values is defined to be 0.

27. Write the standard method of adding two positive decimal integers, taught in elementary schools, as an algorithm.

28. Write the standard method of multiplying two positive decimal integers, taught in elementary schools, as an algorithm.

29H. Write an algorithm that outputs the two's complement of a binary number. (Two's complement is defined before Exercise 65, Section 1.6.)

30. Write an algorithm that outputs the solutions of the quadratic equation

$$ax^2 + bx + c = 0.$$

†5.2 Formal Algorithmic Notation

In this section we introduce an alternative method of writing algorithms using **pseudocode**. This form of algorithm specification is called pseudocode because it resembles the actual code (programs) of languages such as Pascal, PL/I, or C. Pseudocode is becoming a standard way to write algorithms in mathematics and computer science. To fully under-

†This section can be omitted without loss of continuity.

stand and appreciate pseudocode, it is helpful to have written computer programs in a higher-level language such as Pascal, PL/I, or C. There are many varieties of pseudocode. We illustrate our version of pseudocode by rewriting Algorithm 5.1.1, which finds the maximum of three numbers.

Algorithm 5.2.1 Finding the Maximum of Three Numbers.   This algorithm finds the largest of the numbers a, b, and c.

1. **procedure** $max(a, b, c)$
2.   **if** $a > b$ **then**
3. $x := a$
4.   **else**
5. $x := b$ {x is the larger of a and b}
6.   **if** $c > x$ **then** {if c is larger than x, then update x}
7. $x := c$
8.   **return**(x)
9. **end** max

When using pseudocode, we will use the usual arithmetic operators $+$, $-$, $*$ (for multiplication), and $/$ as well as the relational operators $=$, $\neq$, $<$, $>$, $\leq$, and $\geq$ and the logical operators **and**, **or**, and **not**. We will use $=$ to denote the equality operator and $:=$ to denote the assignment operator. We will sometimes use less formal statements when to do otherwise would obscure the meaning. (*Example:* Choose an element x in S.) In general, solutions to exercises that request algorithms in pseudocode should be written in the form illustrated by Algorithm 5.2.1.

Our pseudocode algorithms will consist of a title, a brief description of the algorithm, and the procedures containing the steps of the algorithm. Algorithm 5.2.1 consists of a single procedure. To make it easy to refer to individual lines within an algorithm, we will sometimes number some of the lines. Algorithm 5.2.1 has nine numbered lines. The first line of a procedure will consist of the word **procedure**, then the name of the procedure, and then, in parentheses, the parameters of the procedure. The parameters describe the data, variables, arrays, and so on, that are available to the procedure. In Algorithm 5.2.1, the data available are the three numbers a, b, and c. The last line of a procedure consists of the word **end** followed by the name of the procedure. Between the **procedure** and **end** lines are the executable lines of the procedure. Lines 2–8 are the executable lines of Algorithm 5.2.1.

Comments will be enclosed in braces { }. The lines of a procedure, which are executed sequentially, are typically **assignment statements, conditional statements (if-then-else statements), while loops, for loops, return statements**, and combinations of these statements. We previously described assignment statements, conditional statements, and while loops (see Computer Notes for Section 1.1). The **return**(x) statement terminates the procedure and returns the value of x. The statement **return** [without (x)] simply terminates the procedure. If there is no return statement, the procedure terminates just before the **end** line.

A procedure that contains a **return**(x) statement is a function. The domain consists of all valid values for the parameters and the range is the set of all values that may be returned by the procedure.

One form of the for loop is

$$\textbf{for } var := init \textbf{ to } limit \textbf{ do} \qquad\qquad (5.2.1)$$
$$action$$

where var is a variable and $init$ and $limit$ are expressions that have integer values. To execute a for loop, var is first set to the value $init$. If $var \leq limit$, we execute $action$ and then add 1 to var. The process is then repeated. Repetition continues until $var > limit$. Notice that if $init > limit$, $action$ will not be executed at all. We will illustrate the for loop by rewriting Algorithm 5.1.2 that finds the largest value in a sequence.

Algorithm 5.2.2 Finding the Largest Element in a Finite Sequence.

This algorithm returns the largest value in the sequence

$$s(1), s(2), \ldots, s(n).$$

```
     procedure find_large(s, n)
1.     large := s(1)
2.     for i := 2 to n do
3.        if s(i) > large then {a larger value was found}
4.           large := s(i)
5.     return(large)
     end find_large
```

We will show how Algorithm 5.2.2 executes in some specific cases.

Example 5.2.3.

Suppose that $n = 1$ and s is the sequence $s(1) = 23$. At line 1 of Algorithm 5.2.2, we set $large$ to $s(1)$; in this case, we set $large$ to

23. Next, at line 2, i is set to 2. Now we test whether $i \leq n$; in this case, we test whether $2 \leq 1$. Since the condition is false, we do not execute the for loop. Thus, at line 5, we simply return *large* ($= 23$). We have indeed found the largest element in the sequence.

Example 5.2.4.   Suppose that $n = 4$ and s is the sequence

$$s(1) = -2, \quad s(2) = 6, \quad s(3) = 5, \quad s(4) = 6.$$

At line 1 of Algorithm 5.2.2, we set *large* to $s(1)$; in this case, we set *large* to -2. Next, at line 2, i is set to 2. Now we test whether $i \leq n$; in this case, we test whether $2 \leq 4$. Since this condition is true, we execute the for loop. At line 3, we test whether $s(i) > large$; in this case we test whether $s(2) = 6 > -2 = large$. Since the condition is true, we execute line 4; *large* is set to 6. Having reached the end of the for loop, i is incremented (so that i is now 3) and we return to line 2. We again test whether $i \leq n$; in this case, we test whether $3 \leq 4$. Since this condition is true, we again execute the for loop. At line 3, we test whether $s(i) > large$; in this case, we test whether $s(3) = 5 > 6 = large$. Since the condition is false, we skip line 4. Again, i is incremented and we return to line 2 with $i = 4$. We again test whether $i \leq n$; in this case, we test whether $4 \leq 4$. Since the condition is true, we again execute the for loop. At line 3, we test whether $s(i) > large$; in this case, we test whether $s(4) = 6 > 6 = large$. Since the condition is false, we skip line 4. Again, i is incremented and we return to line 2 with $i = 5$. We again test whether $i \leq n$; in this case, we test whether $5 \leq 4$. Since the condition is false, we do not execute the for loop. Thus, at line 5, we return *large* ($= 6$). We have found the largest element in the sequence.

Notice that the for loop (5.2.1) is logically the same as

$$var := init$$
while $var \leq limit$ **do**
    begin
    action
    $var := var + 1$
    end

A more general for loop has the form

for $var := init$ **to** $limit$ **by** $change$ **do**
    action

In this form of the for loop, if *change* is positive or zero, we execute *action* provided that $var \leq limit$, just as in the previous case (5.2.1). On the other hand, if *change* is negative, the condition for executing *action* is $var \geq limit$. In either case, after *action* is executed, *var* is replaced by $var + change$. If $change \geq 0$, the general for loop is logically the same as

$$
\begin{aligned}
&var := init \\
&\textbf{while } var \leq limit \textbf{ do} \\
&\quad \textbf{begin} \\
&\quad\quad action \\
&\quad\quad var := var + change \\
&\quad \textbf{end}
\end{aligned}
\qquad (5.2.2)
$$

If $change < 0$, the general for loop is logically the same as (5.2.2) except that the second line is changed to

$$\textbf{while } var \geq limit \textbf{ do}$$

The following algorithm illustrates the general for loop.

Algorithm 5.2.5 Finding the Index of the Last Largest Value in a Finite Sequence.

This algorithm returns the index of the largest value in the sequence

$$s(1), s(2), \ldots, s(n).$$

If the largest value occurs more than once, the index of the last occurrence is returned.

```
    procedure find_index_of_largest(s, n)
1.   large := s(n)
2.   index := n
3.   for i := n − 1 to 1 by − 1 do
4.      if s(i) > large then {a larger value was found, so update large
                                               and index}
           begin
5.            large := s(i)
6.            index := i
           end
7.   return(index)
    end find_index_of_largest
```

We will show how Algorithm 5.2.5 executes for a sample sequence.

Example 5.2.6. Suppose that $n = 4$ and s is the sequence

$$s(1) = -2, \quad s(2) = 6, \quad s(3) = 6, \quad s(4) = 5.$$

At lines 1 and 2 of Algorithm 5.2.5, we set *large* to $s(n)$ and *index* to n; in this case, we set *large* to 5 and *index* to 4. Next, at line 3, i is set to $n - 1 = 3$. Now, because *change* is negative, we test whether $i \geq 1$; in this case, we test whether $3 \geq 1$. Since this condition is true, we execute the for loop. At line 4, we test whether $s(i) > large$; in this case, we test whether $s(3) = 6 > 5 = large$. Since the condition is true, we execute lines 5 and 6; *large* is set to 6 and *index* is set to $i = 3$. Having reached the end of the for loop, *change* $= -1$ is added to i (so that i is now 2) and we return to line 3. We again test whether $i \geq 1$; in this case, we test whether $2 \geq 1$. Since this condition is true, we again execute the for loop. At line 4, we test whether $s(i) > large$; in this case, we test whether $s(2) = 6 > 6 = large$. Since the condition is false, we skip lines 5 and 6. Again, i is decremented and we return to line 3 with $i = 1$. We again test whether $i \geq 1$; in this case, we test whether $1 \geq 1$. Since the condition is true, we again execute the for loop. At line 4, we test whether $s(i) > large$; in this case, we test whether $s(1) = -2 > 6 = large$. Since the condition is false, we skip lines 5 and 6. Again, i is decremented and we return to line 3 with $i = 0$. We again test whether $i \geq 1$; in this case, we test whether $0 \geq 1$. Since the condition is false, we do not execute the for loop. Thus, at line 7, we return *index* ($= 3$), the index of the last largest value in the sequence s.

One statement we will not use is the **go to** statement, which causes an unconditional transfer to a designated line. By avoiding the go to statement and instead using the more restricted while, for, and if-then-else branching statements, we obtain **structured algorithms** as advocated by Dijkstra and others. Structured algorithms are easier to read, easier to prove correct, and often easier to construct.

In developing an algorithm, it is often a good idea to break the original problem into two or more subproblems. A procedure can be developed to solve each subproblem, after which these procedures can be combined to provide a solution to the original problem. Our final algorithms illustrate these ideas.

Suppose that we want an algorithm to find the least prime number that exceeds a given positive integer. That is, the problem is: Given a positive integer n, find the least prime p satisfying $p > n$. We might break this problem up into two subproblems. We could first develop an algorithm

Edsger W. Dijkstra (1930–)

Edsger W. Dijkstra was born in The Netherlands. He was an early proponent of programming as a science. So dedicated to programming was he that when he was married in 1957, he listed his profession as a programmer. However, the Dutch authorities said that there was no such profession and he had to change the entry to "theoretical physicist." He is well-known for early graph algorithms, one of which is known as Dijkstra's Algorithm. He won the prestigious Turing award from the Association for Computing Machinery in 1972. He was appointed to the Schlumberger Centennial Chair in Computer Science at the University of Texas at Austin in 1984. [*Photo courtesy of Dr. E. W. Dijkstra*]

to determine whether a positive integer is prime. We could then use this algorithm to find the least prime greater than a given positive integer.

Algorithm 5.2.7 tests whether an integer $m \geq 2$ is prime. We simply test whether any integer between 2 and $m - 1$ divides m. If we find an integer between 2 and $m - 1$ that divides m, m is not prime. If we fail to find an integer between 2 and $m - 1$ that divides m, m is prime. (Exercise 15 is to show that it suffices to check integers between 2 and $\sqrt{m}$ as possible divisors.) Algorithm 5.2.7 shows that we allow algorithms to return **true** or **false**.

Algorithm 5.2.7 Testing Whether a Positive Integer Is Prime. This algorithm returns **true** if the integer $m \geq 2$ is prime and **false** if m is not prime.

> **procedure** *is_prime*(m)
>   **for** $i := 2$ **to** $m - 1$ **do**
>   **if** $m \bmod i = 0$ **then** {i divides m}
>   **return**(**false**)
>   **return**(**true**)
> **end** *is prime*

Algorithm 5.2.8 that finds the least prime exceeding the positive integer n uses Algorithm 5.2.7. To invoke an algorithm such as Algorithm 5.2.7 that returns a value, we name it. To invoke an algorithm named say, *algor*, that does not return a value, we write

$$\textbf{call } algor(p_1, p_2, \ldots, p_k)$$

where $p_1, p_2, \ldots, p_k$ are the arguments passed to *algor*.

Algorithm 5.2.8 Finding a Prime Larger Than a Given Integer. This algorithm returns the smallest prime that exceeds the positive integer n.

> **procedure** *large_prime*(n)
> $m := n + 1$
>   **while not** *is_prime*(m)
> $m := m + 1$
>   **return**(m)
> **end** *large_prime*

Since the number of primes is known to be infinite, Algorithm 5.2.8 will eventually terminate.

EXERCISES

In Exercises 1–3, trace Algorithm 5.2.1 for the given values.

1H. $a = 9, b = 6, c = 5$
2. $a = 6, b = 8, c = 10$
3. $a = 6, b = 4, c = 6$

In Exercises 4–6, trace Algorithm 5.2.2 for the given values.

4H. $s(1) = 5, s(2) = 7, s(3) = 9$
5. $s(1) = 5, s(2) = 12, s(3) = 5, s(4) = 9$
6. $s(1) = 8$
7H. Trace Algorithm 5.2.5 for the values given in Exercises 4–6.

In Exercises 8–10, trace Algorithm 5.2.7 for the given value.

8H. $m = 9$ **9.** $m = 7$ **10.** $m = 35$

In Exercises 11–13, trace Algorithm 5.2.8 for the given value.

11H. $n = 8$ **12.** $n = 47$ **13.** $n = 135$

14H. Rewrite the algorithms of Exercises 8–30, Section 5.1, in pseudocode.
15. Show that the positive integer $m \geq 2$ is prime if and only if no integer between 2 and $\sqrt{m}$ divides m.

†5.3 Greatest Common Divisor Algorithm

An old and famous algorithm is the **Euclidean algorithm** for finding the greatest common divisor of two integers. The greatest common divisor of two integers m and n (not both zero) is the largest positive integer that divides both m and n. For example, the greatest common divisor of 4 and 6 is 2, and the greatest common divisor of 3 and 8 is 1. We use the notion of greatest common divisor when we check to see if a fraction m/n, where m and n are integers, is in lowest terms. If the greatest common divisor d of m and n is 1, m/n is in lowest terms; otherwise, we can reduce m/n. For example, 4/6 is not in lowest terms because

†This section can be omitted without loss of continuity.

the greatest common divisor of 4 and 6 is 2, not 1. (We can divide both 4 and 6 by 2). The fraction 3/8 is in lowest terms because the greatest common divisor of 3 and 8 is 1. After discussing the divisibility of integers, we examine the greatest common divisor in detail and present the Euclidean algorithm.

If a, b, and q are integers, $b \neq 0$, satisfying $a = bq$, we say that b **divides** a and write $b \mid a$. In this case, we call q the **quotient** and call b a **divisor** of a. If b does not divide a, we write $b \nmid a$.

Example 5.3.1. Since $21 = 3 \cdot 7$, 3 divides 21 and we write $3 \mid 21$. The quotient is 7.

Example 5.3.2. Since $-42 = -7 \cdot 6$, -7 divides -42 and we write $-7 \mid -42$. The quotient is 6.

Let m and n be integers that are not both zero. Among all the integers that divide both m and n, there is a largest divisor known as the **greatest common divisor** of m and n.

Definition 5.3.3. Let m and n be integers with not both m and n zero. A *common divisor* of m and n is an integer that divides both m and n. The *greatest common divisor*, written

$$\gcd(m, n),$$

is the largest common divisor of m and n.

Example 5.3.4. The positive divisors of 30 are

$$1, 2, 3, 5, 6, 10, 15, 30$$

and the positive divisors of 105 are

$$1, 3, 5, 7, 15, 21, 35, 105;$$

thus the positive common divisors of 30 and 105 are

$$1, 3, 5, 15.$$

It follows that the greatest common divisor of 30 and 105, $\gcd(30, 105)$, is 15.

Euclid (about 295 B.C.)

The Greek mathematician Euclid is best known for writing the *Elements*, the most famous mathematics text of all time. When Ptolemy, who was king of Egypt, had difficulty with geometry, Euclid is alleged to have said, ''In the real world there are two kinds of roads, roads for the common people to travel upon and roads reserved for the King to travel upon. In geometry there is no royal road.'' [*Photo courtesy of Brown Brothers*]

The properties of common divisors given in the following theorem will be useful in our subsequent work in this section.

Theorem 5.3.5. *Let m, n, and c be integers.*
(a) If c is a common divisor of m and n, then

$$c \mid (m + n).$$

(b) If c is a common divisor of m and n, then

$$c \mid (m - n).$$

(c) If c $\mid$ m, then c $\mid$ mn.

PROOF. (a) Let c be a common divisor of m and n. Since $c \mid m$,

$$m = cq_1, \tag{5.3.1}$$

for some integer q_1. Similarly, since $c \mid n$,

$$n = cq_2, \tag{5.3.2}$$

for some integer q_2. If we add equations (5.3.1) and (5.3.2), we obtain

$$m + n = cq_1 + cq_2 = c(q_1 + q_2).$$

Therefore, c divides $m + n$ (with quotient $q_1 + q_2$). We have proved part (a).

Part (b) is proved similarly to part (a) and is left as an exercise (see Exercise 31).

(c) Suppose that $c \mid m$. Then

$$m = cq$$

for some integer q. If we multiply both sides of this equation by n, we obtain

$$mn = c(nq).$$

Now, $c \mid mn$ (the quotient is nq). ■

We turn next to some terminology associated with division of integers.

Example 5.3.6.   Suppose that we divide 19,746 by 728:

$$
\begin{array}{r}
27 \\
728\overline{)19746} \\
1456 \\
\hline
5186 \\
5096 \\
\hline
90
\end{array}
$$

We call 27 the **quotient** and 90 the **remainder**. Since the remainder is nonzero, 728 does not divide 19,746 and we write $728 \nmid 19{,}746$.

The quotient, remainder, and original numbers in Example 5.3.6 are related by the equation

$$19{,}746 = 728 \cdot 27 + 90.$$

In general, if we divide the nonnegative integer a by the positive integer b, we obtain a quotient q and remainder r satisfying

$$a = bq + r, \qquad 0 \le r < b, \quad q \ge 0. \qquad (5.3.3)$$

Example 5.3.7.   The quotient q and the remainder r in (5.3.3) are illustrated for various values of a and b:

$$
\begin{array}{ll}
a = 22, \ b = 7, \ q = 3, \ r = 1; & 22 = 7 \cdot 3 + 1 \\
a = 24, \ b = 8, \ q = 3, \ r = 0; & 24 = 8 \cdot 3 + 0 \qquad (5.3.4) \\
a = 103, \ b = 21, \ q = 4, \ r = 19; & 103 = 21 \cdot 4 + 19 \\
a = 4895, \ b = 87, \ q = 56, \ r = 23; & 4895 = 87 \cdot 56 + 23 \\
a = 0, \ b = 47, \ q = 0, \ r = 0; & 0 = 47 \cdot 0 + 0. \qquad (5.3.5)
\end{array}
$$

In (5.3.4) and (5.3.5), the remainder r is zero and $b \mid a$. In all other cases, $b \nmid a$.

Now suppose that a is a nonnegative integer and b is a positive integer. We may divide a by b to obtain

$$a = bq + r, \qquad 0 \le r < b.$$

We show that the set of common divisors of a and b is equal to the set of common divisors of b and r. Let c be a common divisor of a and b. By Theorem 5.3.5c, $c \mid bq$. Since $c \mid a$ and $c \mid bq$, by Theorem 5.3.5b, $c \mid a - bq \ (= r)$. Thus c is a common divisor of b and r. Conversely, if c is a common divisor of b and r, then $c \mid bq$ and $c \mid bq + r \; (= a)$ and c is a common divisor of a and b. Thus the set of common divisors of a and b is equal to the set of common divisors of b and r. It follows that

$$\gcd(a, b) = \gcd(b, r).$$

We summarize this result as a theorem.

Theorem 5.3.8. *If a is a nonnegative integer, b is a positive integer, and*

$$a = bq + r, \qquad 0 \le r < b,$$

then

$$\gcd(a, b) = \gcd(b, r).$$

PROOF. The proof precedes the statement of the theorem. ∎

Example 5.3.9. If we divide 105 by 30, we obtain

$$105 = 30 \cdot 3 + 15.$$

The remainder is 15. By Theorem 5.3.8,

$$\gcd(105, 30) = \gcd(30, 15).$$

If we divide 30 by 15, we obtain

$$30 = 15 \cdot 2 + 0.$$

The remainder is 0. By Theorem 5.3.8,

$$\gcd(30, 15) = \gcd(15, 0).$$

By inspection, $\gcd(15, 0) = 15$. Therefore,

$$\gcd(105, 30) = \gcd(30, 15) = \gcd(15, 0) = 15.$$

In Example 5.3.4, we obtained the greatest common divisor of 105 and 30 by listing all of the divisors of 105 and 30. By using Theorem 5.3.8, two simple divisions produce the greatest common divisor. This computation illustrates the Euclidean algorithm.

In general, the Euclidean algorithm finds the greatest common divisor of a and b by repeatedly using Theorem 5.3.8 to replace the original problem of finding the greatest common divisor of a and b by the problem of finding the greatest common divisor of smaller numbers. We ultimately reduce the original problem to that of finding the greatest common divisor of two numbers, one of which is 0. Since gcd $(m, 0) = m$, we have solved the original problem. We now delineate this technique precisely.

Let r_0 and r_1 be nonnegative integers with r_1 nonzero. If we divide r_0 by r_1, we obtain

$$r_0 = r_1 q_2 + r_2, \qquad 0 \le r_2 < r_1.$$

By Theorem 5.3.8,

$$\gcd (r_0, r_1) = \gcd (r_1, r_2).$$

If $r_2 \ne 0$, we can divide r_1 by r_2 to obtain

$$r_1 = r_2 q_3 + r_3, \qquad 0 \le r_3 < r_2.$$

By Theorem 5.3.8,

$$\gcd (r_1, r_2) = \gcd (r_2, r_3).$$

We continue dividing r_i by r_{i+1} provided that $r_{i+1} \ne 0$. Since r_1, $r_2, \ldots$ are nonnegative integers and

$$r_1 > r_2 > r_3 > \ldots ,$$

eventually some r_i will be zero. Let r_n be the first zero remainder. Now

$$\gcd (r_0, r_1) = \gcd (r_1, r_2) = \gcd (r_2, r_3) = \cdots$$

$$= \gcd (r_{n-1}, r_n) = \gcd (r_{n-1}, 0).$$

The greatest common divisor of r_{n-1} and 0 is r_{n-1}; hence

$$\gcd (r_0, r_1) = \gcd (r_{n-1}, 0) = r_{n-1}.$$

Thus the greatest common divisor of r_0 and r_1 will be the last nonzero remainder.

We state the Euclidean algorithm as Algorithm 5.3.10 and in pseudocode as Algorithm 5.3.12 in the Computer Notes section.

Algorithm 5.3.10 Euclidean Algorithm.

This algorithm finds the greatest common divisor of the nonnegative integers a and b, where not both a and b are zero.

Input: a, b (nonnegative integers, not both zero)
Output: gcd (the greatest common divisor of a and b)

1. [Make b the nonzero integer.] If $b = 0$, swap a and b (i.e., execute *temp* := a, a := b, b := *temp*).
2. [Divide.] Divide a by b to obtain $a = bq + r$, $0 \leq r < b$.
3. [Terminate?] If $r = 0$, execute gcd := b and stop.
4. [Update and loop.] Execute a := b, b := r, and go to line 2.

Example 5.3.11.   We show how Algorithm 5.3.10 finds gcd (396, 504).

     Let $a = 504$ and $b = 396$. Since $b \neq 0$, we proceed right to line 2, where we divide a (504) by b (396) to obtain

$$504 = 396 \cdot 1 + 108.$$

We are now at line 3. Since $r = 108$ is not equal to zero, we move to line 4. We set a to 396 and b to 108 and return to line 2.
     At line 2, we divide a (396) by b (108) to obtain

$$396 = 108 \cdot 3 + 72.$$

We are again at line 3. Since $r = 72$ is not equal to zero, we move to line 4. We set a to 108 and b to 72 and return to line 2.
     At line 2, we divide a (108) by b (72) to obtain

$$108 = 72 \cdot 1 + 36.$$

We are again at line 3. Since $r = 36$ is not equal to zero, we move to line 4. We set a to 72 and b to 36 and return to line 2.
     At line 2, we divide a (72) by b (36) to obtain

$$72 = 36 \cdot 2 + 0.$$

Since $r = 0$, we set gcd to b ($= 36$) and stop. We find that 36 is the greatest common divisor of 396 and 504.

†COMPUTER NOTES FOR SECTION 5.3

Algorithm 5.3.12 is the formal statement of the Euclidean algorithm. In this algorithm we use the statement **while true do**, which causes the condition in the while loop always to be true. In this case, the while loop will execute until we leave the loop by a return statement.

†This section can be omitted without loss of continuity.

Algorithm 5.3.12 Euclidean Algorithm.   This algorithm finds the greatest common divisor of the nonnegative integers a and b, where not both a and b are zero.

```
procedure euclid(a, b)
   if b = 0 then {swap a and b so that b is nonzero}
      begin
      temp := b
      b := a
      a := temp
      end
   while true do
      begin
      divide a by b to obtain a = bq + r
      if r = 0 then
         return(b)
      a := b
      b := r
      end
end euclid
```

EXERCISES

In Exercises 1–10, show that b divides a by writing $a = bq$.

1H.	$a = 55, b = 11$	**2.**	$a = 14, b = 1$
3.	$a = 0, b = 23$	**4H.**	$a = 984, b = 24$
5.	$a = 90, b = -15$	**6.**	$a = -53, b = 1$
7H.	$a = -144, b = -12$	**8.**	$a = 0, b = -73$
9.	$a = -4242, b = 42$	**10H.**	$a = 71,412, b = -132$

In Exercises 11–20, find integers q and r so that $a = bq + r$, with $0 \le r < b$.

11H.	$a = 45, b = 6$	**12.**	$a = 106, b = 12$
13.	$a = 66, b = 11$	**14H.**	$a = 221, b = 17$
15.	$a = 0, b = 31$	**16.**	$a = 0, b = 47$

17H. $a = 8, b = 11$ **18.** $a = 33, b = 65$

19. $a = 67,942, b = 4209$ **20H.** $a = 490,256, b = 337$

Use the Euclidean algorithm to find the greatest common divisor of each pair of integers in Exercises 21–30.

21H. 60, 90 **22.** 110, 273

23. 220, 1400 **24H.** 315, 825

25. 20, 40 **26.** 331, 993

27H. 2091, 4807 **28.** 2475; 32,670

29. 67,942; 4209 **30H.** 490,256; 337

31H. Let m, n, and c be integers. Show that if c is a common divisor of m and n, then $c \mid (m - n)$.

32. Suppose that a, b, and c are positive integers. Show that if $a \mid b$ and $b \mid c$, then $a \mid c$.

33. Suppose that a, b, and c are positive integers. Show that $\gcd(a, b) = \gcd(a, a + b)$.

*34H. Using the notation in the text following Example 5.3.9, show that we may successively write

$$r_{n-1} = s_{n-3}r_{n-2} + t_{n-3}r_{n-3}$$

$$r_{n-1} = s_{n-4}r_{n-3} + t_{n-4}r_{n-4}$$

$$\cdot$$
$$\cdot$$
$$\cdot$$

$$r_{n-1} = s_0 r_1 + t_0 r_0$$

$$\gcd(r_0, r_1) = sr_1 + tr_0,$$

where the s's and t's are integers.

*35. Use the method of Exercise 34 to write each greatest common divisor of Exercises 21–30 in the form $ta + sb$.

*36. Show that if p is a prime number and a and b are positive integers and $p \mid ab$, then $p \mid a$ or $p \mid b$.

37H. Give an example of positive integers a, b, and c where $a \mid bc$, $a \nmid b$, and $a \nmid c$.

5.4 Analysis of Algorithms

A computer program, even though derived from a correct algorithm, might be useless for certain types of input because the time needed to run the program or the storage needed to hold the data, program variables, and so on, is too great. **Analysis of an algorithm** refers to the process of deriving estimates for the time and space needed to execute the algorithm. **Complexity of an algorithm** refers to the amount of time and space required to execute the algorithm. In this section we deal with the problem of estimating the time required to execute algorithms.

Determining the performance parameters of a computer program is a difficult task and depends on a number of factors such as the computer that is being used, the way the data are represented, and how the program is translated into machine instructions. Although precise estimates of the execution time of a program must take such factors into account, useful information can be obtained by analyzing the time complexity of the underlying algorithm.

The time needed to execute an algorithm is a function of the input. Usually, it is difficult to obtain an explicit formula for this function and we settle for less. Instead of dealing directly with the input, we use parameters that characterize the *size* of the input. We can ask for the minimum time needed to execute the algorithm among all inputs of size n. This time is called the **best-case time** for inputs of size n. We can also ask for the maximum time needed to execute the algorithm among all inputs of size n. This time is called the **worst-case time** for inputs of size n. (Another important case is **average-case time**—the average time needed to execute the algorithm over some finite set of inputs all of size n. In this book we discuss only best-case and worst-case times.)

We could measure the time required by an algorithm by counting the number of instructions executed. Alternatively, we could use a cruder time estimate such as the number of times each loop is executed. If the principal activity of an algorithm is making comparisons, as might happen in a sorting routine, we might count the number of comparisons. Usually, we are interested in general estimates since, as we have already observed, the actual performance of a program implementation of an algorithm is dependent on many factors.

Example 5.4.1.   A reasonable definition of the size of input for Algorithm 5.1.2 that finds the largest value in a finite sequence is the number of elements in the input sequence. A reasonable definition of the execution time is the number of iterations of the loop at line 2. With these definitions, the worst-case and best-case times for Algorithm 5.1.2 for input of size n are each $n - 1$ since the loop is always executed $n - 1$ times.

Often we are less interested in the exact best-case or worst-case time required for an algorithm to execute than we are in how the best-case or worst-case time grows as the size of the input increases. For example, suppose that the worst-case time of an algorithm is

$$t(n) = 60n^2 + 5n + 1 \qquad (5.4.1)$$

for input of size n. For large n, the term $60n^2$ is approximately equal to $t(n)$ (see Table 5.4.1). In this sense, $t(n)$ grows like $60n^2$.

If (5.4.1) measures the worst-case time for input of size n in seconds, then

$$T(n) = n^2 + \frac{5}{60} n + \frac{1}{60}$$

measures the worst-case time for input of size n in minutes. Now this change of units does not affect how the worst-case time grows as the size of the input increases but only the units in which we measure the worst-case time for input of size n. Thus when we describe how the best-case or worst-case time grows as the size of the input increases, we not only seek the dominant term [e.g., $60n^2$ in (5.4.1)], but we also may ignore constant coefficients. Under these assumptions, $t(n)$ grows like n^2 as n increases. We say that $t(n)$ is of **order at most** n^2 and write

$$t(n) = O(n^2),$$

which is read "$t(n)$ is **big oh** of n^2." The formal definition follows.

Table 5.4.1

n	$t(n) = 60n^2 + 5n + 1$	$60n^2$
10	6,051	6,000
100	600,501	600,000
1000	60,005,001	60,000,000
10000	6,000,050,001	6,000,000,000

Definition 5.4.2.   Let f and g be functions with domain $\{1, 2, 3, \ldots\}$. We write

$$f(n) = O(g(n))$$

and say that $f(n)$ is of *order at most* $g(n)$ if there exists a positive constant C such that

$$|f(n)| \leq C|g(n)|$$

for all but finitely many positive integers n.

Example 5.4.3.   Since

$$60n^2 + 5n + 1 \leq 60n^2 + 60n + 60$$
$$\leq 60n^2 + 60n^2 + 60n^2$$
$$= 180n^2 \quad \text{for } n \geq 1,$$

we may take $C = 180$ in Definition 5.4.2 to obtain

$$60n^2 + 5n + 1 = O(n^2).$$

Example 5.4.4.   Since $\lg n < n$ for $n \geq 1$ (see Figure 5.4.1),

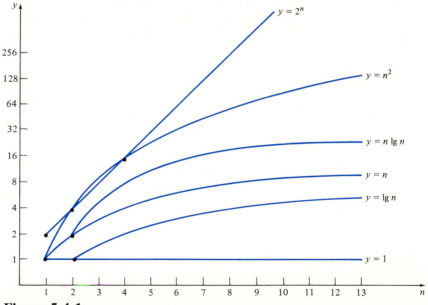

Figure 5.4.1

$$2n + 3 \lg n < 2n + 3n = 5n \qquad \text{for } n \geq 1;$$

thus

$$2n + 3 \lg n = O(n).$$

Example 5.4.5. If we replace each integer $1, 2, \ldots, n$ by n in the sum $1 + 2 + \cdots + n$, the sum does not decrease and we have

$$1 + 2 + \cdots + n \leq n + n + \cdots + n = n \cdot n = n^2 \quad (5.4.2)$$

for $n \geq 1$. It follows that

$$1 + 2 + \cdots + n = O(n^2).$$

Example 5.4.6. If k is a positive integer and, as in Example 5.4.5, we replace each integer $1, 2, \ldots, n$ by n, we have

$$1^k + 2^k + \cdots + n^k \leq n^k + n^k + \cdots + n^k = n \cdot n^k = n^{k+1}$$

for $n \geq 1$; hence

$$1^k + 2^k + \cdots + n^k = O(n^{k+1}).$$

Example 5.4.7. Since[†]

$$|3n^3 + 6n^2 - 4n + 2| \leq 3n^3 + 6n^2 + 4n + 2$$

$$\leq 6n^3 + 6n^3 + 6n^3 + 6n^3$$

$$= 24n^3 \qquad \text{for } n \geq 1,$$

it follows that

$$3n^3 + 6n^2 - 4n + 2 = O(n^3).$$

The method of Example 5.4.7 can be used to show that a polynomial in n of degree k is $O(n^k)$.

Theorem 5.4.8. *Let*

$$a_k n^k + a_{k-1} n^{k-1} + \cdots + a_1 n + a_0$$

[†]We are using the fact that for real numbers a and b, $|a + b| \leq |a| + |b|$.

be a polynomial in n of degree k. Then

$$a_k n^k + a_{k-1} n^{k-1} + \cdots + a_1 n + a_0 = O(n^k).$$

PROOF. Let

$$C = \max \{|a_k|, |a_{k-1}|, \ldots, |a_1|, |a_0|\}.$$

Then

$$|a_k n^k + a_{k-1} n^{k-1} + \cdots + a_1 n + a_0|$$

$$\leq |a_k| n^k + |a_{k-1}| n^{k-1} + \cdots + |a_1| n + |a_0|$$

$$\leq C n^k + C n^{k-1} + \cdots + C n + C$$

$$\leq C n^k + C n^k + \cdots + C n^k + C n^k$$

$$= (k+1) C n^k.$$

Therefore,

$$a_k n^k + a_{k-1} n^{k-1} + \cdots + a_1 n + a_0 = O(n^k). \quad \blacksquare$$

Example 5.4.9. Since $3n^4 - 6n^2 + 4n$ is a polynomial in n of degree 4,

$$3n^4 - 6n^2 + 4n = O(n^4).$$

We next define what it means for the best-case or worst-case time of an algorithm to be of order at most $g(n)$.

Definition 5.4.10. If an algorithm requires $t(n)$ units of time to terminate in the best case for an input of size n and

$$t(n) = O(g(n)),$$

we say that the *best-case time required by the algorithm is of order at most $g(n)$* or that the *best-case time required by the algorithm is $O(g(n))$*.

If an algorithm requires $t(n)$ units of time to terminate in the worst case for an input of size n and

$$t(n) = O(g(n)),$$

we say that the *worst-case time required by the algorithm is of order at most $g(n)$* or that the *worst-case time required by the algorithm is $O(g(n))$*.

Example 5.4.11.   Suppose that an algorithm is known to take

$$3n^4 - 6n^2 + 4n$$

units of time to terminate in the worst case for inputs of size n. We showed in Example 5.4.9 that

$$3n^4 - 6n^2 + 4n = O(n^4).$$

Thus the worst-case time required by this algorithm is $O(n^4)$.

Example 5.4.12.   Find a "big oh" notation in terms of n for the number of times the statement $x := x + 1$ is executed.

1. For $i := 1$ to n, execute line 2.
2.   For $j := 1$ to i, execute line 3.
3. $x := x + 1$

First, i is set to 1 and, as j runs from 1 to 1, line 3 is executed one time. Next, i is set to 2 and, as j runs from 1 to 2, line 3 is executed two times, and so on. Thus the total number of times line 3 is executed is (see Example 5.4.5)

$$1 + 2 + \cdots + n = O(n^2).$$

Thus a "big oh" notation for the number of times the statement $x := x + 1$ is executed is $O(n^2)$.

It can be shown by a method similar to that of Example 5.4.12 that the best-case and worst-case times required by Algorithm 5.4.13 that multiplies two $n \times n$ matrices are each $O(n^3)$.

Algorithm 5.4.13 Matrix Multiplication.   This algorithm computes $C = AB$, where A and B are $n \times n$ matrices. (In line 5, $*$ denotes multiplication.)

Input: The $n \times n$ matrices A and B
Output: The $n \times n$ matrix C, which is equal to the product AB

1. For $i := 1$ to n, execute line 2.
2.   For $j := 1$ to n, execute lines 3 and 4.
3. $C_{ij} := 0$ (Initialize C_{ij} to zero.)
4.   For $k := 1$ to n, execute line 5.
5. $C_{ij} := C_{ij} + A_{ik} * B_{kj}$ (Compute $\sum_{k=1}^{n} A_{ik}B_{kj}$.)

Example 5.4.14.   Find a "big oh" notation in terms of n for the number of times the statement $x := x + 1$ is executed.

1. $j := n$
2. If $j < 1$, then stop.
3. For $i := 1$ to j, execute line 4.
4. $x := x + 1$
5. $j := \lfloor j/2 \rfloor$
6. Go to line 2.

Suppose that $j \geq 1$. After j is set to n, we execute line 3 for the first time and execute line 4 n times. Next, j is replaced by $\lfloor n/2 \rfloor$; hence $j \leq n/2$. If $j \geq 1$, we will execute line 4 at most $n/2$ additional times. Now j is replaced $\lfloor j/2 \rfloor$; hence $j \leq n/4$. If $j \geq 1$, we will execute line 4 at most $n/4$ additional times. If we let k denote the number of times we execute line 3, the number of times we execute line 4 is at most

$$n + \frac{n}{2} + \frac{n}{4} + \cdots + \frac{n}{2^{k-1}}.$$

This geometric sum (see Example 2.4.3) is equal to

$$\frac{n\left(1 - \frac{1}{2^k}\right)}{1 - \frac{1}{2}}.$$

Since

$$\frac{n\left(1 - \frac{1}{2^k}\right)}{1 - \frac{1}{2}} = 2n\left(1 - \frac{1}{2^k}\right) \leq 2n = O(n),$$

a "big oh" notation for the number of times we execute $x := x + 1$ is $O(n)$.

If the worst-case time required by an algorithm is $O(f(n))$ and the algorithm uses $t(n)$ time units in the worst case, for some constant C

$$t(n) \leq Cf(n) \tag{5.4.3}$$

for all but finitely many n. In this sense, $Cf(n)$ gives only an upper estimate for the exact worst-case time $t(n)$. (Similarly, the best-case "big oh" notation gives only an upper estimate of the exact best-case time.) If we are told that the worst-case run time of algorithm A is $O(n)$ and the worst-case run time of algorithm B is $O(n^2)$, we may feel that algorithm A is superior. However, since we are provided only upper estimates, we cannot be sure. It may be that algorithm B is also $O(n)$. Normally, one tries to choose the "best" function $f(n)$ to describe the best-case or worst-case time $O(f(n))$ of an algorithm.

The constant C in (5.4.3) may also be important. Even if algorithm A requires exactly $C_1 n$ time units in the worst case and algorithm B requires exactly $C_2 n^2$ time units in the worst case, for certain sizes of inputs algorithm B may be superior. For example, suppose that for an input of size n, algorithm A requires $300n$ units of time in the worst case and algorithm B requires $5n^2$ units of time in the worst case. For an input size of $n = 5$, algorithm A requires 1500 units of time in the worst case and algorithm B requires 125 units of time in the worst case. We see that algorithm B executes its worst-case input of size 5 faster than algorithm A executes its worst-case input of size 5. Of course, for sufficiently large inputs, algorithm A is considerably faster than algorithm B in the worst case. Despite these cautionary remarks, the "big oh" notation is very useful.

Certain forms occur so often that they are given special names: $O(1)$ is called **constant**; $O(n)$ is called **linear**; $O(n^2)$ is called **quadratic**; $O(n^3)$ is called **cubic**; $O(n^p)$, where p is a fixed positive integer, is called **polynomial**; $O(p^n)$, where p is a fixed integer greater than 1, is called **exponential**; and $O(n!)$ is called **factorial**.

It is important to develop some feeling for the relative sizes of frequently encountered "big oh" forms. In Figure 5.4.1, we have graphed some of these functions. (In Figure 5.4.1, lg denotes the logarithm to the base 2.) Another way to develop some appreciation for the relative sizes of these functions $f(n)$ is to determine how long it would take a computer to execute $f(n)$ steps. For this purpose, let us assume that we have a computer that can execute one step in 1 microsecond (10^{-6} seconds). Table 5.4.2 shows the execution times, under this assumption, for various n. Notice that it is feasible to implement an algorithm that requires 2^n steps for only very small values of n. Algorithms requiring n^2 or n^3 steps also become infeasible, but for relatively larger values of n.

EXERCISES

Select the best "big oh" notation from among $O(1)$, $O(\lg n)$, $O(n)$, $O(n \lg n)$, $O(n^2)$, $O(n^3)$, $O(2^n)$, or $O(n!)$ for each of the expressions in Exercises 1–17.

Table 5.4.2 Time to Execute an Algorithm if One Step Takes 1 Microsecond to Execute

Number of Steps to Termination for Input of Size n	*Time to Execute if n =*								
	3	*6*	*9*	*12*	*50*	*100*	*1000*	10^5	10^6
1	10^{-6} sec	10^{-6} sec	10^{-6} sec	10^{-6} sec	10^{-6} sec	10^{-6} sec	10^{-6} sec	10^{-6} sec	10^{-6} sec
$\lg \lg n$	10^{-6} sec	10^{-6} sec	2×10^{-6} sec	2×10^{-6} sec	2×10^{-6} sec	3×10^{-6} sec	3×10^{-6} sec	4×10^{-6} sec	4×10^{-6} sec
$\lg n$	2×10^{-6} sec	3×10^{-6} sec	3×10^{-6} sec	4×10^{-6} sec	6×10^{-6} sec	7×10^{-6} sec	10^{-5} sec	2×10^{-5} sec	2×10^{-5} sec
n	3×10^{-6} sec	6×10^{-6} sec	9×10^{-6} sec	10^{-5} sec	5×10^{-5} sec	10^{-4} sec	10^{-3} sec	0.1 sec	1 sec
$n \lg n$	5×10^{-6} sec	2×10^{-5} sec	3×10^{-5} sec	4×10^{-5} sec	3×10^{-4} sec	7×10^{-4} sec	10^{-2} sec	2 sec	20 sec
n^2	9×10^{-6} sec	4×10^{-5} sec	8×10^{-5} sec	10^{-4} sec	3×10^{-3} sec	0.01 sec	1 sec	3 hrs	12 days
n^3	3×10^{-5} sec	2×10^{-4} sec	7×10^{-4} sec	2×10^{-3} sec	0.13 sec	1 sec	16.7 min	32 yr	$31{,}710$ yr
2^n	8×10^{-6} sec	6×10^{-5} sec	5×10^{-4} sec	4×10^{-3} sec	36 yr	4×10^{16} yr	3×10^{287} yr	3×10^{30089} yr	3×10^{301016} yr

1H. $6n + 1$ **2.** $2n^2 + 1$

3. $6n^3 - 12n^2 - 1$ **4H.** 3

5. $2 + 4 + \cdots + 2n$ **6.** $(6n - 1)^2$

7H. $\dfrac{(n + 1)(n - 3)}{n + 2}$ **8.** $2 + 4 + 8 + \cdots + 2^n$

9. $n^2 + 2^n$ **10H.** $1^2 + 2^2 + \cdots + n^2$

11. $\dfrac{n + 1}{n^2}$ **12.** 3^n

13H. $n + \lg n$ **14.** $n^2 + n \lg n$

15. $\dfrac{\lg n}{n}$ **16H.** $3n^3 - 4n^2 + 6n \lg n$

17. $10 + \lg n$

In Exercises 18–29, select the best "big oh" notation from among $O(1)$, $O(\lg n)$, $O(n)$, $O(n \lg n)$, $O(n^2)$, $O(n^3)$, $O(2^n)$, or $O(n!)$ for the number of times the statement $x := x + 1$ is executed.

18H.   For $i := 1$ to n, execute line 1.
     1. $x := x + 1$

19.   For $i := 1$ to $2n$, execute line 1.
     1. $x := x + 1$

20. $i := 1$
     1.   If $i > 2n$, then stop.
          $x := x + 1$
          $i := i + 2$
          Go to line 1.

21H.   For $i := 1$ to n, execute line 1.
     1.   For $j := 1$ to n, execute line 2.
     2. $x := x + 1$

22.   For $i := 1$ to $2n$, execute line 1.
     1.   For $j := 1$ to n, execute line 2.
     2. $x := x + 1$

23.   For $i := 1$ to n, execute line 1.
     1.   For $j := 1$ to $\lfloor i/2 \rfloor$, execute line 2.
     2. $x := x + 1$

24H.   For $i := 1$ to n, execute line 1.
     1.   For $j := 1$ to n, execute line 2.
     2.   For $k := 1$ to j, execute line 3.
     3. $x := x + 1$

25.   For $i := 1$ to n, execute line 1.
   1.   For $j := 1$ to n, execute line 2.
   2.   For $k := 1$ to i, execute line 3.
   3. $x := x + 1$

26.   For $i := 1$ to n, execute line 1.
   1.   For $j := 1$ to i, execute line 2.
   2.   For $k := 1$ to j, execute line 3.
   3. $x := x + 1$

27H. $j := n$
   1.   If $j < 1$, then stop.
            For $i := 1$ to j, execute line 2.
   2. $x := x + 1$
            $j := \lfloor j/3 \rfloor$
            Go to line 1.

28. $i := n$
   1.   If $i < 1$, then stop.
            $x := x + 1$
            $i := \lfloor i/2 \rfloor$
            Go to line 1.

29. $i := n$
   1.   If $i < 1$, then stop.
            For $j = 1$ to n, execute line 2.
   2. $x := x + 1$
            $i := \lfloor i/2 \rfloor$
            Go to line 1.

30H. Write an algorithm that adds two $n \times n$ matrices and find a "big oh" notation for its worst-case run time.

31. Write an algorithm that performs scalar multiplication (of a number and an $n \times n$ matrix) and find a "big oh" notation for its worst-case run time.

32. Show that $n! = O(n^n)$.

33H. Show that $2^n = O(n!)$.

34. Show that if

$$f(n) = O(h(n)) \text{ and } g(n) = O(h(n)),$$

then

$$f(n) + g(n) = O(h(n)) \text{ and } cf(n) = O(h(n))$$

for any number c.

35. Find "big oh" notations for the worst-case run times of the algorithms of Exercises 12–24, Section 5.1.

36. Find a "big oh" notation for the worst-case time of Binary Search, Algorithm 5.4.15, that searches a sorted sequence for a particular item.

Algorithm 5.4.15 Binary Search.   Given the sequence

$$s(1), s(2), \ldots, s(n)$$

sorted in increasing order [i.e., $s(1) \le s(2) \le \cdots \le s(n)$] and a value *key*, this algorithm finds an occurrence of *key* and returns its index. If *key* is not found, the algorithm returns the value 0.

> **procedure** *binary_search*(*s*, *n*, *key*)
>   *left* := 1   {*left* and *right* mark the boundaries of the }
>   *right* := *n*   {portion of the sequence currently being searched}
>   **while** *left* ≤ *right* **do**
>   **begin**
>   *middle* := ⌊(*left* + *right*)/2⌋ {divide sequence in two parts}
>   **if** *key* = *s*(*middle*) **then** {found}
>   **return**(*middle*)
>   **if** *key* < *s*(*middle*) **then** {if present, *key* is in the left half}
>   *right* := *middle* − 1
>   **else** {if present, *key* is in the right half}
>   *left* := *middle* + 1
>   **end**
>   **return**(0) {not found}
> **end** *binary_search*

37. Find a "big oh" notation for the best-case time of Binary Search, Algorithm 5.4.15.

5.5 Notes

Most general references on computer science contain some discussion of algorithms. Advanced texts on algorithms are [Aho; Baase; Horowitz, 1978; Knuth 1973 Vols. 1 and 3, 1981; Nievergelt; and Reingold].

[McNaughton] contains a very thorough discussion on an introductory level of what an algorithm is. Knuth's expository article about algorithms ([Knuth, 1977]) and his article about the role of algorithms in the mathematical sciences ([Knuth, 1985]) are also recommended.

Computer Exercises

1. Implement Algorithm 5.1.2 as a program.
2. Implement the algorithms of Exercises 12–30, Section 5.1, as programs.
3. Implement Algorithm 5.2.5 as a program.
4. Implement Algorithm 5.2.7 as a program.
5. Implement Algorithm 5.2.8 as a program.
6. Implement Algorithm 5.3.10 as a program.
7. Implement binary search (Algorithm 5.4.15) as a program. Measure the time used by the program for various *key*s and for various values of n. Compare these results with the theoretical estimate for the worst case $O(\lg n)$.

Chapter Review

SECTION 5.1
Algorithm
Assignment statement: $x := y$
If-then statement: if *condition* then *action*
If-then-otherwise statement: if *condition* then *action1* otherwise *action2*
Input
Output
Trace
Comments: (—)

SECTION 5.2
Pseudocode
Procedure

Comments: {—}
If-then statement: **if** *condition* **then** *action*
If-then-else statement: **if** *condition* **then** *action1* **else** *action2*
While loop: **while** *condition* **do**
For loops: **for** *var* := *init* **to** *limit* **do**
          for *var* := *init* **to** *limit* **by** *change* **do**
Return statements: **return**, **return**(*val*)
begin end
Structured algorithm

SECTION 5.3
Euclidean algorithm
Greatest common divisor
b divides a: $b \mid a$
b is a divisor of a
Quotient
Remainder

SECTION 5.4
Analysis of algorithms
Complexity of algorithms
Worst-case time of an algorithm
Best-case time of an algorithm
Average-case time of an algorithm
"Big oh" notation: $f(n) = O(g(n))$
"Big oh" forms: $O(1)$, constant
          $O(n)$, linear
          $O(n^2)$, quadratic
          $O(n^3)$, cubic
          $O(n^p)$, polynomial
          $O(p^n)$, exponential
          $O(n!)$, factorial

Chapter Self-Test

SECTION 5.1
1H. Trace Algorithm 5.1.2 for the input

$$s_1 = 7, \quad s_2 = 9, \quad s_3 = 17, \quad s_4 = 7.$$

2H.   Write an algorithm that receives as input the distinct numbers a, b, and c, and assigns the values a, b, and c to the variables x, y, and z so that

$$x < y < z.$$

3H.   Write an algorithm that receives as input the $n \times n$ matrix A and outputs the transpose A^T.

4H.   Write an algorithm that receives as input the sequence

$$s_1, \ldots, s_n$$

sorted in nondecreasing order (i.e., the sequence satisfies $s_i \leq s_{i+1}$ for $i = 1, \ldots, n - 1$) and prints all values that appear more than once.

EXAMPLE: If the sequence were

$$1 \quad 1 \quad 1 \quad 5 \quad 8 \quad 8 \quad 9 \quad 12$$

the output would be

$$1 \quad 8.$$

SECTION 5.2

How many times will the statement $x := x + 1$ be executed in Exercises 5 and 6?

5H.   **for** $i := 1$ **to** 10 **by** 2 **do**
$\quad x := x + 1$

6H.   **procedure** *exercise6*
$\quad i := 5$
$\quad$ **while** $i > 1$ **do**
$\quad\quad$ **begin**
$\quad\quad$ **if** $i < 3$ **then**
$\quad\quad\quad$ **return**
$\quad\quad x := x + 1$
$\quad\quad i := i - 1$
$\quad\quad$ **end**
$\quad$ **end** *exercise6*

7H.   Write the algorithm of Exercise 4 in pseudocode.

8H.   Write an algorithm in pseudocode that returns the number of occurrences of the value *key* in the sequence

$$s(1), \ldots, s(n).$$

SECTION 5.3

9H. If $a = 333$ and $b = 24$, find integers q and r so that $a = bq + r$, with $0 \leq r < b$.

10H. Using the Euclidean algorithm, find the greatest common divisor of the integers 396 and 480.

11H. Using the Euclidean algorithm, find the greatest common divisor of the integers 2390 and 4326.

12H. Fill in the blank to make a true statement: If a and b are integers satisfying $a > b > 0$ and $a = bq + r$, $0 \leq r < b$, then gcd $(a, b) = $ _____.

SECTION 5.4

Select the best "big oh" notation from among $O(1)$, $O(n)$, $O(n^2)$, $O(n^3)$, $O(n^4)$, $O(2^n)$, or $O(n!)$ for each of the expressions in Exercises 13 and 14.

13H. $4n^3 + 2n - 5$ **14H.** $1^3 + 2^3 + \cdots + n^3$

15H. Select the best "big oh" notation from among $O(1)$, $O(n)$, $O(n^2)$, $O(n^3)$, $O(2^n)$, or $O(n!)$ for the number of times the line $x := x + 1$ is executed.

> For $i := 1$ to n, execute line 1.
> 1.   For $j := i$ to n, execute line 2.
> 2. $x := x + 1$

16H. Write an algorithm that tests whether two $n \times n$ matrices are equal and find a "big oh" notation for its worst-case run time.

6

There's only so many hands in a deck o'cards.

—from *Shane*

Permutations, Combinations, and the Pigeonhole Principle

In many discrete problems, we are confronted with the problem of counting. For example, in Section 5.4 we saw that in order to estimate the run time of an algorithm, we needed to count the number of times certain steps or loops were executed. Counting also plays a crucial role in probability theory. Because of the importance of counting, a variety of useful aids, some quite sophisticated, have been developed. Sections 6.1 to 6.4 present the fundamental principles used in counting. These techniques can be used to derive the Binomial Theorem (Section 6.5). In optional Section 6.6 we present algorithms for generating all possible orderings and for generating unordered selections from a finite set of elements. The Pigeonhole Principle, which often allows us to assert the existence of an object with certain properties, is developed in Section 6.7.

6.1 First Counting Principle

The menu for Ron's Greasy Spoon Restaurant is shown in Figure 6.1.1. As you can see, it features two appetizers, three main courses, and four beverages. How many different dinners consist of one main course and one beverage?

If we list all possible dinners consisting of one main course and one beverage:

HT, HM, HC, HR, CT, CM, CC, CR, FT, FM, FC, FR,

we see that there are 12 different dinners. (The dinner consisting of a main course whose first letter is X and a beverage whose first letter is Y is denoted XY. For example, CR refers to the dinner consisting of a cheeseburger and root beer.) Notice that there are three main courses and four beverages and $12 = 3 \cdot 4$.

There are 24 possible dinners consisting of one appetizer, one main course, and one beverage:

NHT, NHM, NHC, NHR, NCT, NCM, NCC, NCR,

Ron's Greasy Spoon Restaurant

*** APPETIZERS ***

Nachos *1.15*

Salad . . . . *.90*

*** MAIN COURSES ***

Hamburger *2.25*

Cheeseburger *2.45*

Fish Filet *2.15*

*** BEVERAGES ***

Tea . . . . *.30*

Milk . . . . *.45*

Cola . . . . *.35*

Root Beer . . . . *.35*

Figure 6.1.1

NFT, NFM, NFC, NFR, SHT, SHM, SHC, SHR,
SCT, SCM, SCC, SCR, SFT, SFM, SFC, SFR.

(The dinner consisting of an appetizer whose first letter is X, a main course whose first letter is Y, and a beverage whose first letter is Z is denoted XYZ.) Notice that there are two appetizers, three main courses, and four beverages and $24 = 2 \cdot 3 \cdot 4$.

In each of these examples, we found that the total number of dinners was equal to the product of numbers of each of the courses. These examples illustrate the **First Counting Principle**.

First Counting Principle. *If an activity can be constructed in t successive steps and step 1 can be done in n_1 ways; step 2 can then be done in n_2 ways; . . .; and step t can then be done in n_t ways, then the number of different possible activities is $n_1 \cdot n_2 \cdot \cdots \cdot n_t$.*

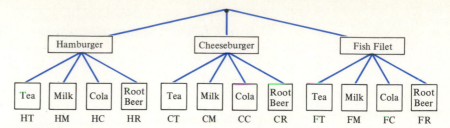

Figure 6.1.2

In the problem of counting the number of dinners consisting of one main course and one beverage, the first step is "select the main course" and the second step is "select the beverage." Thus $n_1 = 3$ and $n_2 = 4$ and, by the First Counting Principle, the total number of dinners is $3 \cdot 4 = 12$.

We may summarize the First Counting Principle by saying that we multiply together the numbers of ways of doing each step when an activity is constructed in successive steps. The reason that we multiply can be seen by drawing a tree that represents the various choices that can be made. In Figure 6.1.2 we have drawn the tree that enumerates the dinners from Ron's Greasy Spoon Restaurant consisting of one main course and one beverage. Each path from the root to a terminal vertex corresponds to one dinner. The first level corresponds to choosing a main course and the second level corresponds to choosing a beverage. Since each of the three vertices on level 1 has four children, the number of terminal vertices, which is the same as the number of dinners, is the product $3 \cdot 4$.

Example 6.1.1.   How many dinners are available from Ron's Greasy Spoon Restaurant consisting of one main course and an *optional* beverage?

We may construct a dinner consisting of one main course and an optional beverage by a two step process. The first step is "select the main course" and the second step is "select an optional beverage." There are $n_1 = 3$ ways to select the main course (hamburger, cheeseburger, fish filet) and $n_2 = 5$ ways to select the optional beverage (tea, milk, cola, root beer, none). By the First Counting Principle, there are $3 \cdot 5 = 15$ dinners. As confirmation, we list the 15 dinners:

$$HT, HM, HC, HR, HN, CT, CM,$$
$$CC, CR, CN, FT, FM, FC, FR, FN.$$

The tree that enumerates these dinners is shown in Figure 6.1.3.

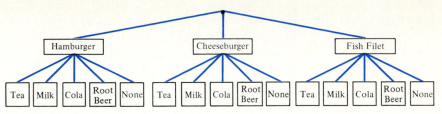

Figure 6.1.3

Example 6.1.2. (a) How many strings of length 4 can be formed using the letters ABCDE if repetitions are not allowed?
(b) How many strings of part (a) begin with the letter B?
(c) How many strings of part (a) do not begin with the letter B?

(a) We use the First Counting Principle. A string of length 4 can be constructed in four successive steps: Choose the first letter, choose the second letter, choose the third letter, and choose the fourth letter. The first letter can be selected in five ways. Having selected the first letter, the second letter can be selected in four ways. Having selected the second letter, the third letter can be selected in three ways. Having selected the third letter, the fourth letter can be selected in two ways. By the First Counting Principle, there are

$$5 \cdot 4 \cdot 3 \cdot 2 = 120$$

strings.

(b) The strings that begin with the letter B can be constructed in four successive steps: Choose the first letter, choose the second letter, choose the third letter, and choose the fourth letter. The first letter (B) can be chosen in one way, the second letter in four ways, the third letter in three ways, and the fourth letter in two ways. Thus, by the First Counting Principle, there are

$$1 \cdot 4 \cdot 3 \cdot 2 = 24$$

strings that start with the letter B.

(c) Part (a) shows that there are 120 strings of length 4 that can be formed using the letters ABCDE and part (b) shows that 24 of these start with the letter B. It follows that there are

$$120 - 24 = 96$$

strings that do not begin with the letter B.

Example 6.1.3. How many strings of length 3 can be formed using the letters ABCDE if repetitions are allowed?

We may construct a string of length 3 by a three-step process. The first step is "select the first letter," the second step is "select the second letter," and the third step is "select the third letter." Since we allow repetitions, $n_1 = n_2 = n_3 = 5$. By the First Counting Principle, the total number of strings of length 3 is $5 \cdot 5 \cdot 5 = 125$.

Example 6.1.4. In a digital picture, we wish to encode the amount of light at each point as an eight-bit string. How many values are possible at one point?

An eight-bit encoding can be constructed in eight successive steps: Select the first bit, select the second bit, . . . , select the eighth bit. Since there are two ways to select each bit, by the First Counting Principle the total number of eight-bit encodings is

$$2 \cdot 2 \cdot 2 \cdot 2 \cdot 2 \cdot 2 \cdot 2 \cdot 2 = 2^8 = 256.$$

We next give a proof using the First Counting Principle that a set with n elements has 2^n subsets. We previously gave a proof of this result using mathematical induction (Theorem 2.4.5).

Example 6.1.5. Use the First Counting Principle to show that a set $\{x_1, \ldots, x_n\}$ containing n elements has 2^n subsets.

A subset can be constructed in n successive steps: Pick or do not pick x_1, pick or do not pick x_2, . . . , pick or do not pick x_n. Each step can be done in two ways. Thus the number of possible subsets is

$$\underbrace{2 \cdot 2 \cdot \ldots \cdot 2}_{n \text{ factors}} = 2^n.$$

Example 6.1.6. A six-person committee composed of Alice, Ben, Connie, Dolph, Egbert, and Francisco is to select a chairperson, secretary, and treasurer.

(a) In how many ways can this be done?

(b) In how many ways can this be done if Egbert must hold one of the offices?

(c) In how many ways can this be done if both Dolph and Francisco must hold office?

(a) We use the First Counting Principle. The officers can be selected in three successive steps: Select the chairperson, select the secretary, select the treasurer. The chairperson can be selected in six ways. Having selected the chairperson, the secretary can be selected in five ways. Having selected the chairperson and secretary, the treasurer can be selected in four ways. Therefore, the total number of possibilities is

$$6 \cdot 5 \cdot 4 = 120.$$

(b) Let us consider the activity of assigning Egbert and two others to offices to be made up of three successive steps: Assign Egbert an office, fill the highest remaining office, fill the last office. There are three ways to assign Egbert an office. Having assigned Egbert, there are five ways to fill the highest remaining office. Having assigned Egbert and filled the highest remaining office, there are four ways to fill the last office. By the First Counting Principle, there are

$$3 \cdot 5 \cdot 4 = 60$$

possibilities.

(c) Let us consider the activity of assigning Dolph, Francisco, and one other person to offices to be made up of three successive steps: Assign Dolph, assign Francisco, fill the remaining office. There are three ways to assign Dolph. Having assigned Dolph, there are two ways to assign Francisco. Having assigned Dolph and Francisco, there are four ways to fill the remaining office. By the First Counting Principle, there are

$$3 \cdot 2 \cdot 4 = 24$$

possibilities.

EXERCISES

Find the number of dinners at Ron's Greasy Spoon Restaurant (Figure 6.1.1) satisfying the conditions of Exercises 1–6.

1H. One appetizer and one beverage

2. One appetizer and one main course

3. An optional appetizer and one main course

4H. An optional appetizer, one main course, and one beverage

5. One appetizer, one main course, and an optional beverage

6. An optional appetizer, one main course, and an optional beverage

7H. A man has eight shirts, four pairs of pants, and five pairs of shoes. How many different outfits are possible?

8. The options available on a particular model of a car are five interior colors, six exterior colors, two types of seats, three types of engines, and three types of radios. How many different possibilities are available to the consumer?

In Exercises 9–13, two dice are rolled, one blue and one red.

9H. How many outcomes are possible?

10. How many outcomes have the blue die showing 2?

11. How many outcomes have exactly one die showing 2?

12H. How many outcomes have at least one die showing 2?

13. How many outcomes have neither die showing 2?

In Exercises 14–17, suppose there are 10 roads from Oz to Mid Earth and five roads from Mid Earth to Fantasy Island.

14H. How many routes are there from Oz to Fantasy Island passing through Mid Earth?

15. How many round trips are there of the form Oz–Mid Earth–Fantasy Island–Mid Earth–Oz?

16. How many round trips are there of the form Oz–Mid Earth–Fantasy Island–Mid Earth–Oz in which on the return trip we do not reverse the original route from Oz to Fantasy Island?

17H. How many trips are there of the form Oz–Mid Earth–Fantasy Island–Mid Earth?

18H. How many different car license plates can be constructed if the licenses contain three letters followed by two digits if repetitions are allowed?

19. How many different car license plates can be constructed if the licenses contain three letters followed by two digits if repetitions are not allowed?

20. How many eight-bit strings begin with 1?

21H. How many eight-bit strings begin 1100?

22. How many eight-bit strings begin and end with 1?

23. How many eight-bit strings read the same from either end?
        EXAMPLE: 01111110. (Such strings are called *palindromes*.)

In Exercises 24–27, a six-person committee composed of Alice, Ben, Connie, Dolph, Egbert, and Francisco is to select a chairperson, secretary, and treasurer.

24H. How many selections exclude Connie?

25. How many selections are there in which neither Ben nor Francisco is an officer?

26. How many selections are there in which both Ben and Francisco are officers?

27H. How many selections are there in which Dolph is an officer and Francisco is not an officer?

In Exercises 28–35, the letters ABCDE are to be used to form strings of length 3.

28H. How many strings can be formed if we allow repetitions?

29. How many strings can be formed if we do not allow repetitions?

30. How many strings begin with A, allowing repetitions?

31H. How many strings begin with A if repetitions are not allowed?

32. How many strings do not contain the letter A, allowing repetitions?

33. How many strings do not contain the letter A if repetitions are not allowed?

34H. How many strings contain the letter A, allowing repetitions?

35. How many strings contain the letter A if repetitions are not allowed?

36H. Show that there are 24 orientations of a cube.

37. Number the cubes of an Instant Insanity puzzle 1, 2, 3, and 4. Show that the number of stackings in which the cubes are stacked 1, 2, 3, and 4, reading from bottom to top, is 331,776.

38. In how many ways can the months of the birthdays of five people be distinct?

39H. In how many ways can the months of the birthdays of six people be distinct?

40. How many possibilities are there for the months of the birthdays of five people?

41. In how many ways can at least two people from among five have their birthdays in the same month?

42H. In one version of BASIC, a variable name consists of a string of one or two alphanumeric characters beginning with a letter (excepting FN, IF, ON, OR, and TO), optionally followed by one of %, !, #, or $. (An **alphanumeric** character is one of A to Z or 0 to 9.) How many BASIC variable names are there?

43. A valid FORTRAN identifier consists of a string of one to six alphanumeric characters beginning with a letter. How many valid FORTRAN identifiers are there?

44. A password-free file specification in the TRSDOS operating system consists of a name followed by an optional extension. The name is a string of one to eight alphanumeric characters beginning with a letter. The extension is a slash (/) followed by a string of one to three alphanumeric characters beginning with a letter. EXAMPLE: EXERCISE/CH1. How many password-free file specifications are there in TRSDOS?

45H. If X is an n-element set and Y is an m-element set, how many functions are there from X to Y?

46. How many terms are there in the expansion of

$$(x + y)(a + b + c)(e + f + g)(h + i)?$$

*47. How many subsets of a $(2n + 1)$-element set have n elements or less?

48H. Find a formula for the number of edges in $K_{m,n}$, the complete bipartite graph on m and n vertices.

49. There are 10 copies of one book and one copy of each of 10 other books. In how many ways can we select 10 books?

6.2   Second Counting Principle

Two dice are rolled, one blue and one red. How many outcomes give the sum of 7 or the sum of 11?

There are six ways to obtain the sum of 7:

$$(1, 6), (2, 5), (3, 4), (4, 3), (5, 2), (6, 1),$$

where (x, y) means that the blue die shows x and the red die shows y. There are two ways to obtain the sum of 11:

$$(5, 6), (6, 5).$$

The set of pairs of dice giving the sum of 7 is disjoint from the set of pairs of dice giving the sum of 11; thus we may add 6 and 2 to obtain 8—the number of outcomes that give the sum of 7 or the sum of 11.

The **Second Counting Principle** tells us when to add to compute the total number of possibilities.

Second Counting Principle. *Suppose that $X_1, \ldots, X_t$ are sets and that the ith set X_i has n_i elements. If $\{X_1, \ldots, X_t\}$ is a pairwise disjoint family (i.e., if $i \neq j$, $X_i \cap X_j = \emptyset$), the number of possible elements that can be selected from X_1 or X_2 or . . . or X_t is*

$$n_1 + n_2 + \cdots + n_t.$$

(Equivalently, the union $X_1 \cup X_2 \cup \cdots \cup X_t$ contains $n_1 + n_2 + \cdots + n_t$ elements.)

In the preceding example concerning the dice, we could let X_1 denote the set of outcomes that give the sum of 7 and X_2 denote the set of outcomes that give the sum of 11. Since X_1 is disjoint from X_2, according to the Second Counting Principle, the number of outcomes that give either the sum of 7 or the sum of 11, which is the number of elements in $X_1 \cup X_2$, is $6 + 2 = 8$.

We may summarize the Second Counting Principle by saying that we add the numbers of elements in each subset when the elements being counted can be decomposed into disjoint subsets.

The First Counting Principle involves multiplication, whereas the Second Counting Principle involves adding. If we are counting objects that are constructed in successive steps, we use the First Counting Principle, which involves multiplication. If we have disjoint sets of objects and we want to know the total number of objects, we use the Second Counting Principle, which involves addition. It is important to recognize when to apply each principle. This skill comes from practice and carefully thinking about each problem.

We illustrate the Second Counting Principle with several more examples.

Example 6.2.1.   How many eight-bit strings begin either 101 or 111?

An eight-bit string that begins 101 can be constructed in five successive steps: Select the fourth bit, select the fifth bit, . . . , select the eighth bit. Since each of the five bits can be selected in two ways, by the First Counting Principle, there are

$$2 \cdot 2 \cdot 2 \cdot 2 \cdot 2 = 2^5 = 32$$

eight-bit strings that begin 101. The same argument can be used to show that there are 32 eight-bit strings which begin 111. Since the set of eight-bit strings that begin 101 is disjoint from the set of eight-bit strings that begin 111, by the Second Counting Principle, there are 32 + 32 = 64 eight-bit strings that begin either 101 or 111.

Example 6.2.2.   In how many ways can we select two books from different subjects from among five distinct computer science books, three distinct mathematics books, and two distinct art books?

Using the First Counting Principle, we find that we can select two books, one from computer science and one from mathematics in 5 · 3 = 15 ways. Similarly, we can select two books, one from computer science and one from art in 5 · 2 = 10 ways, and we can select two books, one from mathematics and one from art in 3 · 2 = 6 ways. Since these sets of selections are pairwise disjoint, we may use the Second Counting Principle to conclude that there are

$$15 + 10 + 6 = 31$$

ways of selecting two books from different subjects from among the computer science, mathematics, and art books.

Example 6.2.3.   A six-person committee composed of Alice, Ben, Connie, Dolph, Egbert, and Francisco is to select a chairperson, secretary, and treasurer. In how many ways can this be done if either Alice or Ben must be chairperson?

If Alice is chairperson, there are five ways to choose a secretary and four ways to choose a treasurer. By the First Counting Principle, there are 5 · 4 = 20 ways to select a chairperson, secretary, and treasurer if Alice is chairperson. Similarly, there are 20 ways to select a chairperson, secretary, and treasurer if Ben is chairperson. Since these cases are disjoint, by the Second Counting Principle, there are

$$20 + 20 = 40$$

possibilities.

EXERCISES

1H. How many eight-bit strings begin either 10 or 11?

2. How many eight-bit strings end either 111 or 001?

3. How many eight-bit strings begin and end with the same bit?

In Exercises 4–9, two dice are rolled, one blue and one red.

4H. How many outcomes give the sum of 4?

5. How many outcomes give the sum of 8?

6. How many outcomes give a sum of 4 or 8?

7H. How many outcomes are doubles? (A double occurs when both dice show the same number.)

8. How many outcomes give an odd sum?

9. How many outcomes are doubles or give an odd sum?

Exercises 10–12 refer to a set of five distinct computer science books, three distinct mathematics books, four distinct business books, and two distinct art books.

10H. In how many ways can we select two books each from a different subject?

11. In how many ways can we select three books each from a different subject?

12. In how many ways can we select four books each from a different subject?

In Exercises 13–15, a six-person committee composed of Alice, Ben, Connie, Dolph, Egbert, and Francisco is to select a chairperson, secretary, and treasurer.

13H. How many selections are there in which either Dolph is chairperson or he is not an officer?

14. How many selections are there in which Ben is either chairperson or treasurer?

15.   How many selections are there in which both Egbert and Francisco are officers or Egbert is not an officer?

16H. If X and Y are not disjoint subsets, we cannot add $|X|$ to $|Y|$ to compute the number of elements in $X \cup Y$. By referring to a Venn diagram or otherwise, justify the formula

$$|X \cup Y| = |X| + |Y| - |X \cap Y|,$$

which is valid for arbitrary sets X and Y.

Use the result of Exercise 16 to solve Exercises 17–22.

17H. How many eight-bit strings either begin 100 or have the fourth bit 1?

18.   How many eight-bit strings either start with a 1 or end with a 1?

In Exercises 19–21, a six-person committee composed of Alice, Ben, Connie, Dolph, Egbert, and Francisco is to select a chairperson, secretary, and treasurer.

19.   How many selections are there in which either Ben is chairperson or Alice is secretary?

20H. How many selections are there in which either Connie is chairperson or Alice is an officer?

21.   How many selections are there in which either Ben is chairperson or Alice is secretary?

22.   Two dice are rolled, one blue and one red. How many outcomes have either the blue die 3 or an even sum?

6.3 Permutations

Four candidates, Zeke, Yung, Xeno, and Wilma, are running for the same office. So that the positions of the names on the ballot will not influence the voters, it is necessary to print ballots with the names listed in every possible order. How many distinct ballots will there be?

We can use the First Counting Principle. A ballot can be constructed in four successive steps: Select the first name to be listed, select the second name to be listed, select the third name to be listed, select the fourth name to be listed. The first name can be selected in four ways. Having selected the first name, the second name can be selected in three ways. Having selected the second name, the third name can be selected in two ways. Having selected the third name, the fourth name can be selected in one way. By the First Counting Principle, the number of ballots is

$$4 \cdot 3 \cdot 2 \cdot 1 = 24.$$

An ordering of objects, such as the names on the ballot, is called a **permutation**.

Definition 6.3.1.   A *permutation* of n distinct elements $x_1, \ldots, x_n$, is an ordering of the n elements $x_1, \ldots, x_n$.

Example 6.3.2.   There are six permutations of three elements. If the elements are denoted A, B, C, the six permutations are

$$ABC, \ ACB, \ BAC, \ BCA, \ CAB, \ CBA.$$

We found that there are 24 ways to order four candidates on a ballot; thus there are 24 permutations of four objects. The method that we used to count the number of distinct ballots containing four names may be used to derive a formula for the number of permutations of n elements.

Theorem 6.3.3.   *There are n! permutations of n elements.*

PROOF.   We use the First Counting Principle. A permutation of n elements can be constructed in n successive steps: Select the first element, select the second element, . . . , select the last element. The first element can be selected in n ways. Having selected the first element, the second element can be selected in $n - 1$ ways. Having selected the second element, the third element can be selected in $n - 2$ ways, and so on. By the First Counting Principle, there are

$$n(n - 1)(n - 2) \cdots 2 \cdot 1 = n!$$

permutations of n elements.   ∎

Example 6.3.4.   There are

$$10! = 10 \cdot 9 \cdot 8 \cdot 7 \cdot 6 \cdot 5 \cdot 4 \cdot 3 \cdot 2 \cdot 1 = 3,628,800$$

permutations of 10 elements.

Example 6.3.5.   How many permutations of the letters *ABCDEF* contain the substring *DEF*?

To guarantee the presence of the pattern *DEF* in the substring, these three letters must be kept together in this order. The remaining letters *A*, *B*, and *C* can be placed arbitrarily. We can think of constructing permutations of the letters *ABCDEF* that contain the pattern *DEF* by permuting four tokens—one labeled *DEF* and the others labeled *A*, *B*, and *C* (see Figure 6.3.1). By Theorem 6.3.3, there are 4! permutations of four objects. Thus the number of permutations of the letters *ABCDEF* that contain the substring *DEF* is

$$4! = 24.$$

| DEF | | A | | B | | C |

Figure 6.3.1

Example 6.3.6.   How many permutations of the letters *ABCDEF* contain the letters *DEF* together in any order?

We can solve the problem by a two-step procedure: Select an ordering of the letters *DEF*, construct a permutation of *ABCDEF* containing the given ordering of the letters *DEF*. By Theorem 6.3.3, the first step can be done in 3! = 6 ways and, according to Example 6.3.5, the second step can be done in 24 ways. By the First Counting Principle, the number of permutations of the letters *ABCDEF* containing the letters *DEF* together in any order is

$$6 \cdot 24 = 144.$$

Sometimes we want to consider an ordering of *r* elements selected from *n* available elements. Such an ordering is called an **r-permutation**.

Definition 6.3.7.   An *r-permutation* of *n* (distinct) elements $x_1, \ldots, x_n$, is an ordering of an *r*-element subset of $\{x_1, \ldots, x_n.\}$ The number of *r*-permutations of a set of *n* distinct elements is denoted $P(n, r)$.

Example 6.3.8. Examples of 2-permutations of a, b, c are

$$ab, ba, ca.$$

If $r = n$ in Definition 6.3.7, we obtain an ordering of all n elements. Thus an n-permutation of n elements is what we previously called, simply, a permutation. Theorem 6.3.3 tells us that $P(n, n) = n!$. The number $P(n, r)$ of r-permutations of an n-element set when $r < n$ may be derived as in the proof of Theorem 6.3.3.

Theorem 6.3.9. *The number of r-permutations of a set of n distinct objects is*

$$P(n, r) = n(n - 1)(n - 2) \cdots (n - r + 1), \qquad r \leq n.$$

PROOF. We are to count the number of ways to order r elements selected from an n-element set. The first element can be selected in n ways. Having selected the first element, the second element can be selected in $n - 1$ ways. We continue selecting elements until having selected the $(r - 1)$st element, we select the rth element. This last element can be chosen in $n - r + 1$ ways. By the First Counting Principle, the number of r-permutations of a set of n distinct objects is

$$n(n - 1)(n - 2) \cdots (n - r + 1). \qquad \blacksquare$$

Example 6.3.10. According to Theorem 6.3.9, the number of 2-permutations of $X = \{a, b, c\}$ is

$$P(3, 2) = 3 \cdot 2 = 6.$$

These six 2-permutations are

$$ab, ac, ba, bc, ca, cb.$$

Example 6.3.11. In how many ways can we select a chairperson, vice-chairperson, secretary, and treasurer from a group of 10 persons?

We need to count the number of orderings of four persons selected from a group of 10, since an ordering picks (uniquely) a chairperson (first pick), a vice-chairperson (second pick), a secretary (third pick), and a treasurer (fourth pick). By Theorem 6.3.9, the solution is

$$P(10, 4) = 10 \cdot 9 \cdot 8 \cdot 7 = 5040.$$

We could also have solved Example 6.3.11 by appealing directly to the First Counting Principle.

We may also write $P(n, r)$ in terms of factorials (see Example 2.3.7):

$$P(n, r) = n(n - 1) \cdots (n - r + 1)$$

$$= \frac{n(n - 1) \cdots (n - r + 1)(n - r) \cdots 2 \cdot 1}{(n - r) \cdots 2 \cdot 1} \qquad (6.3.1)$$

$$= \frac{n!}{(n - r)!}.$$

Example 6.3.12.   Using (6.3.1), we may rewrite the solution of Example 6.3.11 as

$$P(10, 4) = \frac{10!}{(10 - 4)!} = \frac{10!}{6!}.$$

Example 6.3.13.   In how many ways can five distinct Martians and five distinct Jovians wait in line if no two Jovians stand together?

We can line up the Martians and Jovians by a two-step process: Line up the Martians, line up the Jovians. The Martians can line up in 5! ways. Having lined up the Martians (e.g., in positions M_1–M_5), since no two Jovians can stand together, the Jovians have six possible positions to stand (indicated by blanks):

$$— M_1 — M_2 — M_3 — M_4 — M_5 —.$$

Thus the Jovians can stand in $P(6, 5) = 6 \cdot 5 \cdot 4 \cdot 3 \cdot 2$ ways. By the First Counting Principle, the number of ways five Martians and five Jovians can wait in line if no two Jovians stand together is

$$5!(6 \cdot 5 \cdot 4 \cdot 3 \cdot 2) = 86,400.$$

EXERCISES

1H.   How many permutations are there of a, b, c, d?

2.   List the permutations of a, b, c, d.

3.   How many 3-permutations are there of a, b, c, d?

4H.   List the 3-permutations of a, b, c, d.

5.   How many permutations are there of 11 distinct objects?

6. How many 5-permutations are there of 11 distinct objects?

7H. In how many ways can we select a chairperson, vice-chairperson, and recorder from a group of 11 persons?

8. In how many ways can we select a chairperson, vice-chairperson, secretary, and treasurer from a group of 12 persons?

9. In how many different ways can 12 horses finish in the order: Win, Place, Show?

In Exercises 10–16, determine how many strings can be formed by ordering the letters *ABCDE* subject to the conditions given.

10H. Contains the substring *ACE*.

11. Contains the letters *ACE* together in any order.

12. Contains the substrings *DB* and *AE*.

13H. Contains either the substring *AE* or the substring *EA*.

14. *A* appears before *D*. EXAMPLES: *BCAED*, *BCADE*.

*15. Contains neither of the substrings *AB*, *CD*.

*16H. Contains neither of the substrings *AB*, *BE*.

17H. In how many ways can five distinct Martians and eight distinct Jovians wait in line if no two Martians stand together?

18. In how many ways can five distinct Martians, ten distinct Vesuvians, and eight distinct Jovians wait in line if no two Martians stand together?

19. In how many ways can six persons be seated around a circular table? (If a seating is obtained from another seating by having everyone move *n* seats clockwise, the seatings are considered identical.)

20H. In how many ways can six distinct keys be put on a ring? (Turning the ring over does not count as a different arrangement.)

21. In how many ways can five distinct Martians and five distinct Jovians wait in line?

22. In how many ways can five distinct Martians and five distinct Jovians be seated at a circular table?

23H. In how many ways can five distinct Martians and five distinct Jovians be seated at a circular table if no two Martians sit together?

24. In how many ways can five distinct Martians and eight distinct Jovians be seated at a circular table if no two Martians sit together?

Exercises 25–30 refer to a set of five distinct computer science books, three distinct mathematics books, and two distinct art books.

25H. In how many ways can these books be ordered?

26. In how many ways can these books be ordered if all five computer science books are on the left and both art books are on the right?

27. In how many ways can these books be ordered if all five computer science books are on the left?

28H. In how many ways can these books be ordered if all books of the same discipline are grouped together?

29. In how many ways can these books be ordered if the two art books are not together?

30. In how many ways can these books be ordered if no two computer science books are together?

6.4 Combinations

A group of five students, Mary, Boris, Rosa, Ahmad, and Nguyen, has decided to talk with the Mathematics Department chairperson about having the Mathematics Department offer more courses in discrete mathematics. The chairperson has said that she will speak with three of the students. In how many ways can these five students choose three of their group to talk with the chairperson?

In solving this problem, we must *not* take order into account. (For example, it will make no difference whether the chairperson talks with Mary, Ahmad, and Nguyen or to Nguyen, Mary, and Ahmad.) By simply listing the possibilities, we see that there are 10 ways that the five students can choose three of their group to talk to the chairperson:

MBR, MBA, MRA, BRA, MBN, MRN, BRN, MAN, BAN, RAN.

A selection of objects without regard to order is called a **combination**. A permutation *takes order into account*, whereas a combination *does not take order into account*.

Definition 6.4.1.   Given a set $X = \{x_1, \ldots, x_n\}$ containing n (distinct) elements,

(a) An *r-combination* of X is an unordered selection of r elements of X (i.e., an r-element subset of X).

(b) The number of r-combinations of a set of n distinct elements is denoted $C(n, r)$ or $\binom{n}{r}$.

Example 6.4.2.   In the example preceding Definition 6.4.1, we found that there were ten ways for five students to select three representatives to see the Mathematics Department chairperson; thus there are ten 3-combinations of a 5-element set. We write

$$C(5, 3) = 10.$$

Example 6.4.3.   There are three 2-combinations of $\{a, b, c\}$, namely

$$\{a, b\}, \{a, c\}, \{b, c\}.$$

We write

$$C(3, 2) = 3.$$

We next derive a formula for $C(n, r)$ by counting the number of r-permutations of an n-element set in two ways. The first way simply uses the formula $P(n, r)$ from Section 6.3. The second way of counting the number of r-permutations of an n-element set involves $C(n, r)$. Equating the two values will enable us to derive a formula for $C(n, r)$.

We can construct r-permutations of an n-element X set in two successive steps: First, select an r-combination of X (an unordered subset of r items), and then order it. For example, to construct a 2-permutation of $\{a, b, c, d\}$, we can first select a 2-combination and then order it. The tree in Figure 6.4.1 shows how all 2-permutations of $\{a, b, c, d\}$ are

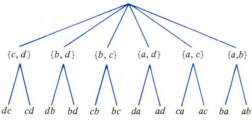

Figure 6.4.1

obtained in this way. The First Counting Principle tells us that the number of r-permutations is the product of the number of r-combinations and the number of orderings of r elements. That is,

$$P(n, r) = C(n, r)r!.$$

Therefore,

$$C(n, r) = \frac{P(n, r)}{r!}.$$

Our next theorem states this result and gives some alternative ways to write $C(n, r)$.

Theorem 6.4.4.   *The number of r-combinations of a set of n distinct objects is*

$$C(n, r) = \frac{P(n, r)}{r!}$$

$$= \frac{n(n - 1) \cdots (n - r + 1)}{r!}$$

$$= \frac{n!}{(n - r)!r!}, \qquad r \le n.$$

PROOF.   The proof of the first equation is given before the statement of the theorem. The other forms of the equation follow from Theorem 6.3.9 and equation (6.3.1). ■

Example 6.4.5.   Since $C(3, 2) = P(3, 2)/2! = 3 \cdot 2/(2 \cdot 1) = 3$, there are three 2-combinations of a 3-element set. The three 2-combinations of $\{a, b, c\}$ are shown in Example 6.4.3.

Example 6.4.6.   In how many ways can we select a committee of three from a group of 10 distinct persons?

Since a committee is an unordered group of people, the answer is

$$C(10, 3) = \frac{10 \cdot 9 \cdot 8}{3!} = 120.$$

Example 6.4.7. In how many ways can we select a committee of two women and three men from a group of five distinct women and six distinct men?

As in Example 6.4.6, we find that the two women can be selected in

$$C(5, 2) = 10$$

ways and that the three men can be selected in

$$C(6, 3) = 20$$

ways. The committee can be constructed in two successive steps: Select the women, select the men. By the First Counting Principle, the total number of committees is

$$10 \cdot 20 = 200.$$

Example 6.4.8. How many eight-bit strings contain exactly four 1's?

An eight-bit string containing four 1's is uniquely determined once we tell which bits are 1. But this can be done in

$$C(8, 4) = 70$$

ways.

Example 6.4.9

(a) How many (unordered) five-card poker hands, selected from an ordinary 52-card deck, are there?

(b) How many poker hands contain cards all of the same suit?

(c) How many poker hands contain three cards of one denomination and two cards of a second denomination?

(a) The answer is given by the combination formula

$$C(52, 5) = 2,598,960.$$

(b) A hand containing cards all of the same suit can be constructed in two successive steps: Select a suit, select five cards from the chosen suit. The first step can be done in four ways and the second step can be done in $C(13, 5)$ ways. By the First Counting Principle, the answer is

$$4 \cdot C(13, 5) = 5148.$$

(c) A hand containing three cards of one denomination and two cards of a second denomination can be constructed in four successive steps: Select the first denomination, select the second denomination, select three cards of the first denomination, select two cards of the second denomination. The first denomination can be chosen in 13 ways. Having selected the first denomination, the second denomination can be chosen in 12 ways. We can select three cards of the first denomination in $C(4, 3)$ ways and we can select two cards of the second denomination in $C(4, 2)$ ways. By the First Counting Principle, the answer is

$$13 \cdot 12 \cdot C(4, 3) \cdot C(4, 2) = 3744.$$

We conclude this section by considering orderings of objects where repetitions are allowed.

Example 6.4.10.   How many strings can be formed using the following letters?

$$MISSISSIPPI$$

Because of the duplication of letters, the answer is not 11!, but some number less than 11!.

Let us consider the problem of filling 11 blanks,

$$\_\ \_\ \_\ \_\ \_\ \_\ \_\ \_\ \_\ \_\ \_\ ,$$

with the letters given. There are $C(11, 2)$ ways to choose positions for the two P's. Having selected the positions for the P's, there are $C(9, 4)$ ways to choose positions for the four S's. Having selected the positions for the S's, there are $C(5, 4)$ ways to choose positions for the four I's. Having made these selections, there is one position left to be filled by the M. By the First Counting Principle, the number of ways of ordering the letters is

$$C(11, 2)C(9, 4)C(5, 4) = \frac{11!}{2!9!} \frac{9!}{4!5!} \frac{5!}{4!1!}$$

$$= \frac{11!}{2!4!4!1!}$$

$$= 34,650.$$

The solution to Example 6.4.10 assumes a nice form. The number 11 that appears in the numerator is the total number of letters. The values in the denominator give the numbers of duplicates of each letter. The method can be used to establish a general formula.

Theorem 6.4.11. *Suppose that a sequence S of n items has n_1 identical objects of type 1, n_2 identical objects of type 2, . . . , and n_t identical objects of type t. Then the number of orderings of S is*

$$\frac{n!}{n_1!n_2! \cdots n_t!}.$$

PROOF. We assign positions to each of the n items to create an ordering of S. We may assign positions to the n_1 items of type 1 in $C(n, n_1)$ ways. Having made these assignments, we may assign positions to the n_2 items of type 2 in $C(n - n_1, n_2)$ ways, and so on. By the First Counting Principle, the number of orderings is

$$C(n, n_1)C(n - n_1, n_2)C(n - n_1 - n_2, n_3)$$

$$\cdots C(n - n_1 - \cdots - n_{t-1}, n_t)$$

$$= \frac{n!}{n_1!(n - n_1)!} \frac{(n - n_1)!}{n_2!(n - n_1 - n_2)!}$$

$$\cdots \frac{(n - n_1 - \cdots - n_{t-1})!}{n_t!0!}$$

$$= \frac{n!}{n_1!n_2! \cdots n_t!}. \qquad ■$$

Example 6.4.12. In how many ways can eight distinct books be divided among three students if Bill gets four books and Shizuo and Marian each get two books?

Put the books in some fixed order. Now consider orderings of four B's, two S's, and two M's. An example is

BBBSMBMS.

Each such ordering determines a distribution of books. For the example ordering, Bill gets books 1, 2, 3, and 6, Marian gets books 5 and 7, and Shizuo gets books 4 and 8. Thus the number of ways of ordering *BBBBSSMM* is the number of ways to distribute the books.

By Theorem 6.4.11, this number is

$$\frac{8!}{4!2!2!} = 420.$$

EXERCISES

In Exercises 1–3, let $X = \{a, b, c, d\}$.

1H. Compute the number of 3-combinations of X.

2H. List the 3-combinations of X.

3. Show the relationship between the 3-permutations and the 3-combinations of X by drawing a picture like that in Figure 6.4.1.

4. In how many ways can we select a committee of three from a group of 11 persons?

5H. In how many ways can we select a committee of four from a group of 12 persons?

In Exercises 6–11, find the number of strings that can be formed by ordering the letters given.

6H. *DECIMAL*   **7.** *MESSAGE*   **8.** *LEADING*

9H. *APPLIED*   **10.** *BOOKKEEPER*   **11.** *TOOT*

12H. At one point in the Illinois state lottery Lotto game, a person was required to choose six numbers (in any order) from among 44 numbers. In how many ways can this be done? The state was considering changing the game so that a person would be required to choose six numbers from among 48 numbers. In how many ways can this be done?

13. Find a formula for the number of edges in K_n, the complete graph on n vertices.

14. How many simple paths (of positive length) are there in a tree with n vertices?

Exercises 15–20 refer to a club consisting of six distinct men and seven distinct women.

15H. In how many ways can we select a committee of five persons?

16. In how many ways can we select a committee of three men and four women?

17. In how many ways can we select a committee of four persons that has at least one woman?

18H. In how many ways can we select a committee of four persons that has at most one man?

19. In how many ways can we select a committee of four persons that has persons of both sexes?

20. In how many ways can we select a committee of four persons so that Mabel and Ralph do not serve together?

21H. In how many ways can we select a committee of four Republicans, three Democrats, and two Independents from a group of 10 distinct Republicans, 12 distinct Democrats, and four distinct Independents?

22. How many eight-bit strings contain exactly one 1?

23. How many eight-bit strings contain exactly two 1's?

24H. How many eight-bit strings contain at least one 1?

25. How many eight-bit strings contain exactly three 0's?

26. How many eight-bit strings contain three 0's in a row and five 1's?

*27H. How many eight-bit strings contain at least two 0's in a row?

In Exercises 28–36, find the number of (unordered) five-card poker hands, selected from an ordinary 52-card deck, having the properties indicated.

28H. Containing four aces

29. Containing four of a kind, that is, four cards of the same denomination

30. Containing all spades

31H. Containing cards of exactly two suits

32. Containing cards of all suits

33. Of the form A2345 of the same suit

34H. Consecutive and of the same suit. (Assume that the ace is the lowest denomination.)

35. Consecutive. (Assume that the ace is the lowest denomination.)

36. Containing two of one denomination, two of another denomination, and one of a third denomination

37H. Find the number of (unordered) 13-card bridge hands selected from an ordinary 52-card deck.

38. How many bridge hands are all of the same suit?

39. How many bridge hands contain exactly two suits?

40H. How many bridge hands contain all four aces?

41. How many bridge hands contain five spades, four hearts, three clubs, and one diamond?

42. How many bridge hands contain five of one suit, four of another suit, three of another suit, and one of another suit?

43H. How many bridge hands contain four cards of three suits and one card of the fourth suit?

44. How many bridge hands contain no face cards? (A face card is one of 10, J, Q, K, A.)

In Exercises 45–49, a coin is flipped 10 times.

45H. How many outcomes are possible? (An *outcome* is a list of five H's and five T's that gives the result of each of 10 tosses. For example, the outcome

$$HHTHTHHHTH$$

represents 10 tosses, where a head was obtained on the first two tosses, a tail was obtained on the third toss, a head was obtained on the fourth toss, etc.)

46. How many outcomes have exactly three heads?

47. How many outcomes have at most three heads?

48H. How many outcomes have a head on the fifth toss?

49. How many outcomes have as many heads as tails?

Exercises 50–53 refer to a shipment of 50 microprocessors of which four are defective.

50H. In how many ways can we select a set of four microprocessors?

51. In how many ways can we select a set of four nondefective microprocessors?

52. In how many ways can we select a set of four microprocessors containing exactly two defective microprocessors?

53H. In how many ways can we select a set of four microprocessors containing at least one defective microprocessor?

54. In how many ways can 10 distinct books be divided among three students if the first student gets five books, the second three books, and the third two books?

55. In how many ways can 12 distinct books be divided among three students if the first student gets five books, the second four books, and the third three books?

56. What is wrong with the following argument, which purports to show that $4C(39, 13)$ bridge hands contain three or fewer suits?

There are $C(39, 13)$ hands that contain only clubs, diamonds, and spades. In fact, for any three suits, there are $C(39, 13)$ hands that contain only those three suits. Since there are four 3-combinations of the suits, the answer is $4C(39, 13)$.

57. What is wrong with the following argument, which purports to show that there are $13^4 \cdot 48$ (unordered) five-card poker hands containing cards of all suits?

Pick one card of each suit. This can be done in $13 \cdot 13 \cdot 13 \cdot 13 = 13^4$ ways. Since the fifth card can be chosen in 48 ways, the answer is $13^4 \cdot 48$.

58H. What is wrong with following argument, which supposedly counts the number of partitions of a 10-element set into eight (nonempty) subsets?

List the elements of the set with blanks between them:

$$x_1 \underline{\hspace{0.5em}} x_2 \underline{\hspace{0.5em}} x_3 \underline{\hspace{0.5em}} x_4 \underline{\hspace{0.5em}} x_5 \underline{\hspace{0.5em}} x_6 \underline{\hspace{0.5em}} x_7 \underline{\hspace{0.5em}} x_8 \underline{\hspace{0.5em}} x_9 \underline{\hspace{0.5em}} x_{10}.$$

Every time we fill seven of the nine blanks with seven vertical bars, we obtain a partition of $\{x_1, \ldots , x_{10}\}$ into eight subsets. For example, the partition $\{x_1\}, \{x_2\}, \{x_3, x_4\}, \{x_5\}, \{x_6\}, \{x_7, x_8\}, \{x_9\}, \{x_{10}\}$ would be represented as

$$x_1 \mid x_2 \mid x_3 \, x_4 \mid x_5 \mid x_6 \mid x_7 \, x_8 \mid x_9 \mid x_{10}.$$

Thus the solution to the problem is $C(9, 7)$.

59. Find the number of partitions of a 10-element set into eight subsets.

60. By interpreting $C(i, j)$ as the number of j-element subsets of an i-element set, show that

$$C(n, r) = C(n, n - r).$$

*61H. How many routes are there from A to B in the following figure if we are restricted to traveling only to the right or upward? (One such route is shown.)

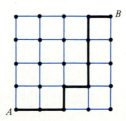

*62. Show that the number of bit strings of length $n \geq 4$ that contain exactly two occurrences of 10 is $C(n + 1, 5)$.

*63. Show that the number of n-bit strings having exactly k 0's, with no two 0's consecutive, is $C(n - k + 1, k)$.

*64H. Show that the product of any positive integer and its $k - 1$ successors is divisible by $k!$.

*65. Show that there are $(2n - 1)(2n - 3) \cdot \ldots \cdot 3 \cdot 1$ ways to pick n pairs from $2n$ distinct items.

66. Suppose that we have n objects, r distinct and $n - r$ identical. Give another derivation of the formula

$$P(n, r) = r!C(n, r)$$

by counting the number of orderings of the n objects in two ways:

(a) Count the orderings by first choosing positions for the r distinct objects.

(b) Count the orderings by first choosing positions for the $n - r$ identical objects.

6.5 The Binomial Theorem

At first glance the expression $(a + b)^n$ does not have much to do with combinations; but as we will see in this section, we can obtain the formula for the expansion of $(a + b)^n$ by using the formula for the number of r-combinations of n objects. Frequently, we can relate an algebraic expression to some counting process. Several advanced counting techniques use such methods (see [Riordan, Tucker]).

The **Binomial Theorem** gives a formula for the coefficients in the expansion of $(a + b)^n$. Since

$$(a + b)^n = \underbrace{(a + b)(a + b) \cdots (a + b)}_{n \text{ factors}},$$

$$(6.5.1)$$

the expansion results from selecting either a or b from each of the n factors, multiplying the selections together, and then summing all such products obtained. For example, in the expansion of $(a + b)^3$, we select either a or b from the first factor $(a + b)$; either a or b from the second factor $(a + b)$; and either a or b from the third factor $(a + b)$; multiply

Table 6.5.1

Selection From First Factor $(a + b)$	Selection From Second Factor $(a + b)$	Selection From Third Factor $(a + b)$	Product of Selections
a	a	a	$aaa = a^3$
a	a	b	$aab = a^2b$
a	b	a	$aba = a^2b$
a	b	b	$abb = ab^2$
b	a	a	$baa = a^2b$
b	a	b	$bab = ab^2$
b	b	a	$bba = ab^2$
b	b	b	$bbb = b^3$

the selections together; and then sum the products obtained. If we select a from all factors and multiply, we obtain the term aaa. If we select a from the first factor, b from the second factor, and a from the third factor and multiply, we obtain the term aba. Table 6.5.1 shows all the possibilities. If we sum the products of all the selections, we obtain

$$(a + b)^3 = (a + b)(a + b)(a + b)$$

$$= aaa + aab + aba + abb + baa + bab + bba + bbb$$

$$= a^3 + a^2b + a^2b + ab^2 + a^2b + ab^2 + ab^2 + b^3$$

$$= a^3 + 3a^2b + 3ab^2 + b^3.$$

In (6.5.1), a term of the form $a^{n-k}b^k$ arises from choosing b from k factors and a from the other $n - k$ factors. But this can be done in $C(n, k)$ ways, since $C(n, k)$ counts the number of ways of selecting k things from n items. Thus $a^{n-k}b^k$ appears $C(n, k)$ times. It follows that

$$(a + b)^n = C(n, 0)a^nb^0 + C(n, 1)a^{n-1}b^1 + C(n, 2)a^{n-2}b^2 \quad (6.5.2)$$
$$+ \cdots + C(n, n - 1)a^1b^{n-1} + C(n, n)a^0b^n.$$

This result is known as the Binomial Theorem.

Theorem 6.5.1 Binomial Theorem. *If a and b are real numbers and n is a positive integer, then*

$$(a + b)^n = \sum_{k=0}^{n} C(n, k)a^{n-k}b^k.$$

PROOF. The proof precedes the statement of the theorem. ■

The Binomial Theorem can also be proved using induction on n (see Exercise 25).

The numbers $C(n, r)$ are known as **binomial coefficients** because they appear in the expansion (6.5.2) of the binomial $a + b$ raised to a power.

Example 6.5.2.   Taking $n = 3$ in Theorem 6.5.1, we obtain

$$(a + b)^3 = C(3, 0)a^3b^0 + C(3, 1)a^2b^1 + C(3, 2)a^1b^2$$
$$+ C(3, 3)a^0b^3$$

$$= a^3 + 3a^2b + 3ab^2 + b^3.$$

Example 6.5.3.   Expand $(3x - 2y)^4$ using the Binomial Theorem.

If we take $a = 3x$, $b = -2y$, and $n = 4$ in Theorem 6.5.1, we obtain

$$(3x - 2y)^4 = (a + b)^4$$

$$= C(4, 0)a^4b^0 + C(4, 1)a^3b^1 + C(4, 2)a^2b^2$$
$$+ C(4, 3)a^1b^3 + C(4, 4)a^0b^4$$

$$= C(4, 0)(3x)^4(-2y)^0 + C(4, 1)(3x)^3 (-2y)^1$$
$$+ C(4, 2)(3x)^2(-2y)^2 + C(4, 3)(3x)^1(-2y)^3$$
$$+ C(4, 4)(3x)^0(-2y)^4$$

$$= 3^4x^4 + 4 \cdot 3^3x^3(-2y) + 6 \cdot 3^2x^2(-2)^2y^2$$
$$+ 4(3x)(-2)^3y^3 + (-2)^4y^4$$

$$= 81x^4 - 216x^3y + 216x^2y^2 - 96xy^3 + 16y^4.$$

Example 6.5.4.   Find the fifth term in the expansion of $(a + b)^9$.

The first term is $C(9, 0)a^9b^0$; the second term is $C(9, 1)a^8b^1$; the third term is $C(9, 2)a^7b^2$; and so on. Thus the fifth term is

$$C(9, 4)a^5b^4 = 126a^5b^4.$$

Example 6.5.5.   The seventh term in the expansion of $(4u - 2v)^{10}$ is

$$C(10, 6)(4u)^4(-2v)^6 = 210 \cdot 4^4(-2)^6u^4v^6 = 3{,}440{,}640u^4v^6.$$

The method of proving the Binomial Theorem can be applied just as well to a trinomial $(x + y + z)^n$ [and, in general, to a multinomial $(x_1 + x_2 + \cdots + x_m)^n$].

Example 6.5.6. Find the coefficient of $x^2 y^3 z^4$ in the expansion of $(x + y + z)^9$.
Since

$$(x + y + z)^9 = (x + y + z)(x + y + z)$$
$$\cdots (x + y + z) \quad \text{(nine terms)},$$

we obtain $x^2 y^3 z^4$ each time we multiply together x chosen from two of the nine terms, y chosen from three of the nine terms, and z chosen from four of the nine terms. We can choose two terms for the x's in $C(9, 2)$ ways. Having made this selection, we can choose three terms for the y's in $C(7, 3)$ ways. This leaves the remaining four terms for the z's. Thus the coefficient of $x^2 y^3 z^4$ in the expansion of $(x + y + z)^9$ is

$$C(9, 2)C(7, 3) = \frac{9!}{2!7!} \frac{7!}{3!4!} = \frac{9!}{2!3!4!} = 1260.$$

We can write the binomial coefficients in a triangular form known as **Pascal's triangle** (see Figure 6.5.1). The border consists of 1's and any interior value is the sum of the two numbers above it. This relationship is stated formally in the next theorem.

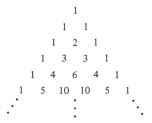

Figure 6.5.1

Theorem 6.5.7

$$C(n + 1, k) = C(n, k - 1) + C(n, k)$$

for $1 \leq k \leq n$.

PROOF. We will give a combinatorial argument.[†]
Let X be a set with n elements. Choose $a \notin X$. Then $C(n + 1, k)$ is the number of k-element subsets of $Y = X \cup \{a\}$. Now the k-element

---

†An identity that results from some counting process is called a combinatorial identity and the argument that leads to its formulation is called a combinatorial argument.

Blaise Pascal (*1623–1662*)

Blaise Pascal was a French mathematician and physicist. He is one of the founders of the theory of probability. During his life, he invented several calculating machines. The Pascal programming language was named after him in honor of these achievements. In his later years, Pascal spent a great deal of time in religious contemplation. In 1658, a tooth ceased to ache at the same time some results in geometry occurred to him. Pascal regarded this as a divine signal to apply himself to geometry. During the next days he developed a theory of the cycloid curve. [*Photo courtesy of the Louvre Museum, Paris*]

subsets of Y can be divided into two disjoint classes:

1. Subsets of Y not containing a.
2. Subsets of Y containing a.

The subsets of class 1 are just k-element subsets of X and there are $C(n, k)$ of these. The subsets of class 2 consist of a $(k - 1)$-element subset of X together with a and there are $C(n, k - 1)$ of these. Therefore,

$$C(n + 1, k) = C(n, k - 1) + C(n, k).$$ ∎

Theorem 6.5.7 can also be proved using Theorem 6.4.4 (Exercise 26). If we set $a = b = 1$ in Theorem 6.5.1, we obtain the identity

$$2^n = (1 + 1)^n = \sum_{k=0}^{n} C(n, k)1^{n-k}1^k = \sum_{k=0}^{n} C(n, k).   (6.5.3)$$

This equation can also be proved by giving a combinatorial argument. Given an n-element set X, $C(n, k)$ counts the number of k-element subsets. Thus the right side of equation (6.5.3) counts the total number of subsets of X. But in Theorem 2.4.5 we showed that the number of subsets of X is 2^n; we have reproved equation (6.5.3).

EXERCISES

Expand each expression in Exercises 1–10 using the Binomial Theorem.

1H. $(x + y)^4$ **2.** $(x + y)^5$ **3.** $(u - v)^3$

4H. $(c + 3d)^3$ **5.** $(c - 3d)^3$ **6.** $(3c + 2d)^4$

7H. $(2c - 3d)^3$ **8.** $(2c - 3d)^5$ **9.** $(u + 2v)^6$

10H. $(3u - 2v)^6$

In Exercises 11–20, find the coefficient of each term when the given expression is expanded.

11H. Third term; $(x + y)^5$ **12.** Fourth term; $(x - 2y)^6$

13. Eighth term; $(x + y)^{11}$ **14H.** Seventh term; $(2s - t)^{12}$

15. $x^2y^3z^5$; $(x + y + z)^{10}$ **16.** $x^3y^2z^6$; $(x + y + z)^{11}$

17H. $x^2y^3z^5$; $(2x + y + z)^{10}$ **18.** $x^3y^2z^6$; $(2x - 3y + z)^{11}$

19. $w^2x^3y^2z^5$; $(w + x + y + z)^{12}$

20H. $w^2x^3y^2z^5$; $(2w + x + 3y - z)^{12}$

21H. Find the next row of Pascal's triangle given the row

$$1 \quad 7 \quad 21 \quad 35 \quad 35 \quad 21 \quad 7 \quad 1.$$

22. Find the next row of Pascal's triangle given the row

$$1 \quad 9 \quad 36 \quad 84 \quad 126 \quad 126 \quad 84 \quad 36 \quad 9 \quad 1.$$

23. (a) Show that $C(n, k) < C(n, k + 1)$ if and only if $k < (n - 1)/2$.
   (b) Use part (a) to deduce that the maximum of $C(n, k)$ for $k = 0, 1, \ldots, n$ is $C(n, \lfloor n/2 \rfloor)$.

24H. Use the Binomial Theorem to show that

$$0 = \sum_{k=0}^{n} (-1)^k C(n, k).$$

25. Use induction on n to prove the Binomial Theorem.

26. Prove Theorem 6.5.7 by using Theorem 6.4.4.

27H. Use the Binomial Theorem to show that

$$\sum_{k=0}^{n} 2^k C(n, k) = 3^n.$$

28. Suppose that n is even. Use the Binomial Theorem to show that

$$\sum_{k=0}^{n/2} C(n, 2k) = 2^{n-1} = \sum_{k=0}^{n/2} C(n, 2k - 1).$$

*29. Give a combinatorial argument to prove that

$$\sum_{k=0}^{n} C(n, k)^2 = C(2n, n).$$

*30H. Use induction on n to prove that

$$n2^{n-1} = \sum_{k=1}^{n} kC(n, k).$$

†6.6   Algorithms for Generating Permutations and Combinations

Solving a problem by enumerating all possibilities sometimes involves generating permutations and combinations. To mechanize the process, algorithms are required. For example, the version of the traveling sales-person problem in Example 3.1.5 asks for a minimum length path from vertex a to vertex e that passes through each of the vertices a, b, c, d, e exactly one time. We solved this problem by listing all possible paths

$$(a, b, c, d, e)$$
$$(a, b, d, c, e)$$
$$(a, c, b, d, e)$$
$$(a, c, d, b, e)$$
$$(a, d, b, c, e)$$
$$(a, d, c, b, e)$$

and choosing one of minimum length. Notice that the list of paths is obtained by writing all six permutations of $\{b, c, d\}$ and adding a at the beginning and e at the end.

In this section, we develop algorithms to list all permutations of $\{1, 2, \ldots, n\}$ and all r-combinations of $\{1, 2, \ldots, n\}$. These algorithms will list permutations and combinations in **lexicographic order**. Lexicographic order generalizes ordinary dictionary order.

Given two distinct words, to determine whether one precedes the other in the dictionary we compare the letters in the words. There are two possibilities:

1. Each letter in the shorter word is identical to the corresponding letter in the longer word.
2. At some position, the letters in the words differ.   (6.6.1)

If 1 holds, the shorter word precedes the longer. (For example, "dog" precedes "doghouse" in the dictionary.) If 2 holds, we locate the left-most position p at which the letters differ. The order of the words is

---

$\dagger$This section can be omitted without loss of continuity.

determined by the order of the letters at position p. (For example, "gladiator" precedes "gladiolus" in the dictionary. At the leftmost position at which the letters differ we find "a" in "gladiator" and "o" in "gladiolus" and "a" precedes "o" in the alphabet.)

Lexicographic order generalizes ordinary dictionary order by replacing the alphabet by any set of symbols on which an order has been defined. We will be concerned with strings of integers.

Definition 6.6.1.   Let $\alpha = s_1 s_2 \cdots s_p$ and $\beta = t_1 t_2 \cdots t_q$ be strings over $\{1, 2, \ldots, n\}$. We say that α is *lexicographically less than* β and write $\alpha < \beta$ if either

     (a) $p < q$ and $s_i = t_i$ for $i = 1, \ldots, p$;

or

     (b) for some i, $s_i \neq t_i$, and for the smallest such i, we have $s_i < t_i$.

In Definition 6.6.1, case (a) corresponds to possibility 1 of (6.6.1) and case (b) corresponds to possibility 2 of (6.6.1).

Example 6.6.2.   Let $\alpha = 132$ and $\beta = 1324$ be strings over $\{1, 2, 3, 4\}$. In the notation of Definition 6.6.1, $p = 3$, $q = 4$, $s_1 = 1$, $s_2 = 3$, $s_3 = 2$, $t_1 = 1$, $t_2 = 3$, $t_3 = 2$, and $t_4 = 4$. Since $p = 3 < 4 = q$ and $s_i = t_i$ for $i = 1, 2, 3$, condition (a) of Definition 6.6.1 is satisfied. Therefore, $\alpha < \beta$.

Example 6.6.3.   Let $\alpha = 13246$ and $\beta = 1342$ be strings over $\{1, 2, 3, 4, 5, 6\}$. In the notation of Definition 6.6.1, $p = 5$, $q = 4$, $s_1 = 1$, $s_2 = 3$, $s_3 = 2$, $s_4 = 4$, $s_5 = 6$, $t_1 = 1$, $t_2 = 3$, $t_3 = 4$, and $t_4 = 2$. The smallest i for which $s_i \neq t_i$ is $i = 3$. Since $s_3 < t_3$, by condition (b) of Definition 6.6.1, $\alpha < \beta$.

Example 6.6.4.   Let $\alpha = 1324$ and $\beta = 1342$ be strings over $\{1, 2, 3, 4\}$. In the notation of Definition 6.6.1, $p = q = 4$, $s_1 = 1$, $s_2 = 3$, $s_3 = 2$, $s_4 = 4$, $t_1 = 1$, $t_2 = 3$, $t_3 = 4$, and $t_4 = 2$. The smallest i for which $s_i \neq t_i$ is $i = 3$. Since $s_3 < t_3$, by condition (b) of Definition 6.6.1, $\alpha < \beta$.

Example 6.6.5. Let $\alpha = 13542$ and $\beta = 21354$ be strings over $\{1, 2, 3, 4, 5\}$. In the notation of Definition 6.6.1, $s_1 = 1$, $s_2 = 3$, $s_3 = 5$, $s_4 = 4$, $s_5 = 2$, $t_1 = 2$, $t_2 = 1$, $t_3 = 3$, $t_4 = 5$, and $t_5 = 4$. The smallest i for which $s_i \neq t_i$ is $i = 1$. Since $s_1 < t_1$, by condition (b) of Definition 6.1.1, $\alpha < \beta$.

As illustrated in Examples 6.6.4 and 6.6.5, for strings of the same length over $\{1, 2, \ldots, 9\}$, lexicographic order is the same as the order on the positive integers if we interpret the strings as decimal numbers. (For strings of unequal length, lexicographic order may be different from numerical order; e.g., 911, 1642.) Throughout the remainder of this section, *order* will refer to lexicographic order.

First we consider the problem of listing all r-combinations of $\{1, 2, \ldots, n\}$. In our algorithm, we will list the r-combination $\{x_1, \ldots, x_r\}$ as the string $s_1 \cdots s_r$ where $s_1 < s_2 < \cdots < s_r$ and $\{x_1, \ldots, x_r\} = \{s_1, \ldots, s_r\}$. For example, the 3-combination $\{6, 2, 4\}$ will be listed 246.

We will list the r-combinations of $\{1, 2, \ldots, n\}$ in lexicographic order. Thus the first listed string will be $12 \cdots r$ and the last listed string will be $(n - r + 1) \cdots n$.

Example 6.6.6. Consider the order in which the 5-combinations of $\{1, 2, 3, 4, 5, 6, 7\}$ will be listed. The first string is 12345, which is followed by 12346 and 12347. The next string is 12356 followed by 12457. The last string will be 34567.

Example 6.6.7. Find the string that follows 13467 when we list the 5-combinations of $X = \{1, 2, 3, 4, 5, 6, 7\}$.

No string that begins 134 and represents a 5-combination of X exceeds 13467. Thus the string that follows 13467 must begin 135. Since 13567 is the smallest string that begins 135 and represents a 5-combination of X, the answer is 13567.

Example 6.6.8. Find the string that follows 2367 when we list the 4-combinations of $X = \{1, 2, 3, 4, 5, 6, 7\}$.

No string that begins 23 and represents a 4-combination of X exceeds 2367. Thus the string that follows 2367 must begin 24. Since

2456 is the smallest string that begins 24 and represents a 4-combination of X, the answer is 2456.

A pattern is developing. Given a string $\alpha = s_1 \cdots s_r$ that represents the r-combination $\{s_1, \ldots, s_r\}$, to find the next string $\beta = t_1 \cdots t_r$, we find the rightmost element s_m that is not at its maximum value. (s_r may have the maximum value n, s_{r-1} may have the maximum value $n - 1$, etc.) Then

$$t_i = s_i \qquad \text{for } i = 1, \ldots, m - 1.$$

The element t_m is equal to $s_m + 1$. For the remainder of the string β we have

$$t_{m+1} \cdots t_r = (s_m + 2)(s_m + 3) \cdots .$$

The algorithm follows.

Algorithm 6.6.9 Generating Combinations.   This algorithm lists all r-combinations of $\{1, 2, \ldots, n\}$ in increasing lexicographic order.

Input: r, n
Output: A list of all r-combinations of $\{1, 2, \ldots, n\}$ in increasing lexicographic order.

1. [Initialize string.] For $i := 1$ to r, execute line 2.
2. $s_i := i$.
3. [Output first r-combination.] Output s.
4. [Loop to compute and output r-combinations.] For $i := 2$ to $C(n, r)$, execute lines 5–9.
5.   Find the rightmost s_m not at its maximum value. (The maximum value of s_k is defined to be $n - r + k$.)
6. $s_m := s_m + 1$.
7.   For $j := m + 1$ to r, execute line 8.
8. $s_j := s_{j-1} + 1$.
9.   Output s.

Example 6.6.10.   We will show how Algorithm 6.6.9 generates the 5-combination of $\{1, 2, 3, 4, 5, 6, 7\}$ following 23467. We are supposing that

$$s_1 = 2, \, s_2 = 3, \, s_3 = 4, \, s_4 = 6, \, s_5 = 7.$$

At line 5, we find that s_3 is the rightmost element not at its maximum value. At line 6, s_3 is set to 5. At lines 7 and 8, s_4 is set to 6 and s_5 is set to 7. At this point

$$s_1 = 2, \quad s_2 = 3, \quad s_3 = 5, \quad s_4 = 6, \quad s_5 = 7.$$

We have generated the 5-combination 23567 that follows 23467.

Example 6.6.11.   The 4-combinations of $\{1, 2, 3, 4, 5, 6\}$ as listed by Algorithm 6.6.9 are

$$1234, \ 1235, \ 1236, \ 1245, \ 1246, \ 1256, \ 1345, \ 1346,$$
$$1356, \ 1456, \ 2345, \ 2346, \ 2356, \ 2456, \ 3456.$$

Like the algorithm for generating r-combinations, the algorithm to generate permutations will list the permutations of $\{1, 2, \ldots, n\}$ in lexicographic order. (Exercise 28 asks for an algorithm that generates all r-permutations of an n-element set.)

Example 6.6.12.   To construct the permutation of $\{1, 2, 3, 4, 5, 6\}$ following 163542, we should keep as many digits as possible at the left the same.

Can the permutation following the given permutation have the form 1635_ _? Since the only permutation of the form 1635_ _ distinct from the given permutation is 163524 and 163524 is smaller than 163542, the permutation following the given permutation is not of the form 1635_ _.

Can the permutation following the given permutation have the form 163_ _ _? The last three digits must be a permutation of $\{2, 4, 5\}$. Since 542 is the largest permutation of $\{2, 4, 5\}$, any permutation that begins 163 is smaller than the given permutation. Thus the permutation following the given permutation is not of the form 163_ _ _.

The reason that the permutation following the given permutation cannot begin 1635 or 163 is that in either case the remaining digits in the given permutation (42 and 542, respectively) *decrease*. Therefore working from the right, we must find the first digit d whose right neighbor r satisfies $d < r$. In our case, the third digit, 3, has this property. Thus the permutation following the given permutation will begin 16.

The digit following 16 must exceed 3. Since we want the next smallest permutation, the next digit is 4, the smallest available digit.

Thus the desired permutation begins 164. The remaining digits 235 must be in increasing order to achieve the minimum value. Therefore, the permutation following the given permutation is 164235.

We see that to generate all of the permutations of $\{1, 2, \ldots, n\}$, we can begin with the permutation $12 \cdots n$ and then repeatedly use the method of Example 6.6.12 to generate the next permutation. We will end when the permutation $n(n - 1) \cdots 21$ is generated.

Example 6.6.13.   Using the method of Example 6.6.12, we can list the permutations of $\{1, 2, 3, 4\}$ in lexicographic order as

$$1234, 1243, 1324, 1342, 1423, 1432, 2134, 2143,$$
$$2314, 2341, 2413, 2431, 3124, 3142, 3214, 3241,$$
$$3412, 3421, 4123, 4132, 4213, 4231, 4312, 4321.$$

The algorithm follows.

Algorithm 6.6.14 Generating Permutations.   This algorithm lists all permutations of $\{1, 2, \ldots, n\}$ in increasing lexicographic order.

Input: n
Output: A list of all permutations of $\{1, 2, \ldots, n\}$ in increasing lexicographic order.

1. [Initialize string.] For $i := 1$ to n, execute line 2.
2. $s_i := i$.
3. [Output first permutation.] Output s.
4. [Loop to compute and output permutations.] For $i := 2$ to $n!$, execute lines 5–9.
5.   Find the largest index m satisfying $s_m < s_{m+1}$.
6.   Find the largest index k satisfying $s_k > s_m$.
7.   Swap s_m and s_k. (That is, execute $temp := s_m$, $s_m := s_k$, $s_k := temp$.)
8.   Reverse the order of the elements $s_{m+1}, \ldots, s_n$.
9.   Output s.

Example 6.6.15.   We will show how Algorithm 6.6.14 generates the permutation following 163542. Suppose that

$$s_1 = 1, \quad s_2 = 6, \quad s_3 = 3, \quad s_4 = 5, \quad s_5 = 4, \quad s_6 = 2$$

and that we are at line 5. The largest index m satisfying $s_m < s_{m+1}$ is 3. At line 6, we find that the largest index k satisfying $s_k > s_m$ is 5. At line 7, we swap s_m and s_k. At this point, we have $s = 164532$. At line 8, we reverse the order of the elements $s_4 s_5 s_6 = 532$. We obtain the desired permutation 164235.

---

†COMPUTER NOTES FOR SECTION 6.6

We give pseudocode for the algorithms to generate r-combinations and permutations.

Algorithm 6.6.16 Generating Combinations.

The algorithm lists all r-combinations of $\{1, 2, \ldots, n\}$ in increasing lexicographic order.

```
        procedure combination(r, n)
          for i := 1 to r do
            s(i) := i
          print s(1), . . . , s(r) {print the first r-combination}
          for i := 2 to C(n, r) do
            begin
1.            m := r
              max_val := n
2.            while s(m) = max_val do
                {find the rightmost element not at its maximum value}
                begin
                m := m − 1
                max_val := max_val − 1
                end
3.            s(m) := s(m) + 1
              {the rightmost element is incremented}
4.            for j := m + 1 to r do
              {the rest of the elements are the successors of s(m)}
                s(j) := s(j − 1) + 1
              print s(1), . . . , s(r) {print the ith combination}
            end
        end combination
```

†This section can be omitted without loss of continuity.

Example 6.6.17.   We will show how Algorithm 6.6.16 generates the 5-combination of $\{1, 2, 3, 4, 5, 6, 7\}$ following 23467. We are supposing that

$$s(1) = 2, \ s(2) = 3, \ s(3) = 4, \ s(4) = 6, \ s(5) = 7$$

and that we are at line 1. We set m to 5 and max_val to 7. At the first iteration of the while loop 2, the condition $s(5) = 7$ is true, so m is set to 4 and max_val to 6. The condition at line 2 $s(4) = 6$ is again true, so m is set to 3 and max_val to 5. This time the condition $s(3) = 5$ is false. At line 3, $s(3)$ is set to 5. In the for loop 4, $s(4)$ is set to 6 and $s(5)$ is set to 7. At this point

$$s(1) = 2, \quad s(2) = 3, \quad s(3) = 5, \quad s(4) = 6, \quad s(5) = 7.$$

We have generated the 5-combination 23567 that follows 23467.

Algorithm 6.6.18 Generating Permutations.   This algorithm lists all permutations of $\{1, 2, \ldots, n\}$ in increasing lexicographic order.

```
        procedure permutation(n)
            for i := 1 to n do
              s(i) := i
            print s(1), . . . , s(n) {print the first permutation}
            for i := 2 to n! do
              begin
1.              m := n − 1
2.              while s(m) > s(m + 1) do
                   {find the first decrease working from the right}
                   m := m − 1
                k := n
3.               while s(m) > s(k) do
                   {find the rightmost element s(k) with s(m) < s(k)}
                   k := k − 1
                temp := s(m) {swap s(m) and s(k)}
                s(m) := s(k)
                s(k) := temp
                p := m + 1
                q := n
4.              while p < q do
                   {swap s(m + 1) and s(n), swap s(m + 2) and s(n − 1),
                   and so on}
```

```
      begin
        temp := s(p)
        s(p) := s(q)
        s(q) := temp
        p := p + 1
        q := q - 1
      end
    print s(1), . . . , s(n) {print the ith permutation}
    end
  end permutation
```

Example 6.6.19.   We will show how Algorithm 6.6.18 generates the permutation following 163542. We are supposing that

$$s(1) = 1, \quad s(2) = 6, \quad s(3) = 3,$$
$$s(4) = 5, \quad s(5) = 4, \quad s(6) = 2$$

and that we are at line 1. We set m to 5. After while loop 2 finishes, m will be 3. We have found the first decrease working from the right. Next we set k to n. After while loop 3 finishes, k will be 5. We have found the smallest digit $s(k) = 4$ to the right of $s(m) = 3$ that exceeds $s(m)$. We now swap $s(m)$ and $s(k)$. At this point, the permutation is 164532. The digits following $s(m)$ are in decreasing order. Thus by reversing $s(m + 1), \ldots, s(n)$ (in while loop 4) we obtain the desired permutation 164235.

EXERCISES

In Exercises 1–9, find the r-combination that will be generated by Algorithm 6.6.9 after the r-combination given for the value of n specified.

1H. 1356, $n = 7$ **2.** 1356, $n = 6$ **3.** 23467, $n = 7$

4H. 13578, $n = 8$ **5.** 34678, $n = 8$ **6.** 1247, $n = 7$

7H. 35689, $n = 9$ **8.** 3456789, $n = 9$ **9.** 13456789, $n = 9$

In Exercises 10–18, find the permutation that will be generated by Algorithm 6.6.14 after the permutation given.

10H. 12354   **11.** 12345   **12.** 14235

13H. 625431   **14.** 132456   **15.** 123456

16H. 1523467   **17.** 5731246   **18.** 81234567

19H. For each string in Exercises 1–9, explain (as in Example 6.6.10) exactly how Algorithm 6.6.9 generates the next r-combination.

20. For each string in Exercises 10–18, explain (as in Example 6.6.15) exactly how Algorithm 6.6.14 generates the next permutation.

21H. Show the output from Algorithm 6.6.9 when $n = 6$ and $r = 3$.

22. Show the output from Algorithm 6.6.9 when $n = 6$ and $r = 2$.

23. Show the output from Algorithm 6.6.9 when $n = 7$ and $r = 5$.

24H. Show the output from Algorithm 6.6.14 when $n = 2$.

25. Show the output from Algorithm 6.6.14 when $n = 3$.

26H. Rewrite Algorithm 6.6.16 changing the loop

$$\textbf{for } i := 2 \textbf{ to } C(n, r) \textbf{ do}$$

to

$$\textbf{while true do}.$$

Base the terminating condition on the fact that the last r-combination has every element $s(i)$ equal to its maximum value.

27. Rewrite Algorithm 6.6.18 changing the loop

$$\textbf{for } i := 2 \textbf{ to } n! \textbf{ do}$$

to

$$\textbf{while true do}.$$

Base the terminating condition on the fact that the last permutation has the elements $s(i)$ in decreasing order.

28. Write an algorithm that generates all r-permutations of an n-element set.

†6.7 The Pigeonhole Principle

So far, all of the sections of this chapter have dealt with the question: How many items having a given property are there? Sometimes, instead of counting items having a given property, we simply want to know if there are any. The **Pigeonhole Principle** (also known as the *Dirichlet*

†This section can be omitted without loss of continuity.

Drawer Principle or the *Shoe Box Principle*) is sometimes useful in answering the question: Is there an item having a given property? When the Pigeonhole Principle is successfully applied, the Principle tells us only that the object exists; the Principle will not tell us how to find the object, nor how many there are.

The first version of the Pigeonhole Principle that we will discusss asserts that if n pigeons fly into k pigeonholes and $k < n$, some pigeonhole contains at least two pigeons (see Figure 6.7.1). The reason this statement is true can be seen by arguing by contradiction. If the conclusion is false, each pigeonhole contains at most one pigeon and, in this case, we can account for at most k pigeons. Since there are n pigeons and $n > k$, we have a contradiction.

Figure 6.7.1

Pigeonhole Principle (First Form). *If n pigeons fly into k pigeonholes and $k < n$, some pigeonhole contains at least two pigeons.*

We note that the Pigeonhole Principle tells us nothing about how to locate the pigeonhole that contains two or more pigeons. It only asserts the *existence* of a pigeonhole containing two or more pigeons.

To apply the Pigeonhole Principle, we must decide which objects will play the roles of the pigeons and which objects will play the roles of the pigeonholes. Our first example illustrates one possibility.

Example 6.7.1. Ten persons have first names Alice, Bernard, and Charles and last names Lee, McDuff, and Ng. Show that at least two persons have the same first and last names.

There are nine possible names for the 10 persons. If we think of the persons as pigeons and the names as pigeonholes, we can consider the assignment of names to people to be that of assigning pigeonholes

Peter Gustav Lejeune Dirichlet (1805–1859)

Peter Gustav Lejeune Dirichlet worked in both France and Germany in number theory, analysis, and applied mathematics. He used the Pigeonhole Principle in some of his work in number theory. He proved what we now call Dirichlet's Theorem: If gcd $(a, b) = 1$, the set

$$\{an + b \mid n = 0, 1, \ldots\}$$

contains infinitely many primes. He was interested in mathematics at an early age. Before he was 12, he was spending all of his extra money on mathematics books. [*Photo courtesy of Bildarchiv Preussischer Kulturbesitz, Berlin*]

to the pigeons. By the Pigeonhole Principle some name (pigeonhole) is assigned to at least two persons (pigeons).

We next restate the Pigeonhole Principle in an alternative form.

Pigeonhole Principle (Second Form). *If f is a function from a finite set X to a finite set Y and $|X| > |Y|$, then $f(x_1) = f(x_2)$ for some $x_1, x_2 \in X$, $x_1 \neq x_2$.*

The second form of the Pigeonhole Principle can be reduced to the first form by letting X be the set of pigeons and Y be the set of pigeonholes. We assign pigeon x to pigeonhole $f(x)$. By the first form of the Pigeonhole Principle, at least two pigeons, $x_1, x_2 \in X$, are assigned to the same pigeonhole; that is $f(x_1) = f(x_2)$.

Our next examples illustrate the use of the second form of the Pigeonhole Principle.

Example 6.7.2.   If 20 processors are interconnected, show that at least two processors are directly connected to the same number of processors.

Designate the processors $1, 2, \ldots, 20$. Let a_i be the number of processors to which processor i is directly connected. We are to show that $a_i = a_j$, for some $i \neq j$. The domain of the function a is $X = \{1, 2, \ldots, 20\}$ and the range Y is some subset of $\{0, 1, \ldots, 19\}$. Unfortunately, $|X| = |\{0, 1, \ldots, 19\}|$ and we cannot immediately use the second form of the Pigeonhole Principle.

Let us examine the situation more closely. Notice that we cannot have $a_i = 0$, for some i, *and* $a_j = 19$, for some j, for then we would have one processor (the ith processor) not connected to any other processor while, at the same time, some other processor (the jth processor) is connected to all the other processors (including the ith processor). Thus the range Y is a subset of either $\{0, 1, \ldots, 18\}$ or $\{1, 2, \ldots, 19\}$. In either case, $|Y| < 20 = |X|$. By the second form of the Pigeonhole Principle, $a_i = a_j$, for some $i \neq j$, as desired.

Example 6.7.3.   Show that if we select 151 distinct computer science courses numbered between 1 and 300 inclusive, at least two are consecutively numbered.

Let the selected course numbers be

$$c_1, c_2, \ldots, c_{151}. \qquad (6.7.1)$$

The 302 numbers consisting of (6.7.1) together with

$$c_1 + 1, c_2 + 1, \ldots, c_{151} + 1 \qquad (6.7.2)$$

range in value between 1 and 301. By the second form of the Pigeonhole Principle, at least two of these values coincide. The numbers (6.7.1) are all distinct and hence the numbers (6.7.2) are also distinct. It must then be that one of (6.7.1) and one of (6.7.2) are equal. Thus we have

$$c_i = c_j + 1$$

and course c_i follows course c_j.

Example 6.7.4.   An inventory consists of a list of 80 items, each marked available or unavailable. There are 45 available items. Show that there are at least two available items in the list exactly nine items apart. (For example, available items at positions 13 and 22 or positions 69 and 78 satisfy the condition.)

Let a_i denote the position of the ith available item. We must show that $a_i - a_j = 9$ for some i and j. Consider the numbers

$$a_1, a_2, \ldots, a_{45} \qquad (6.7.3)$$

and

$$a_1 + 9, a_2 + 9, \ldots, a_{45} + 9. \qquad (6.7.4)$$

The 90 numbers in (6.7.3) and (6.7.4) have possible values only from 1 to 89. By the second form of the Pigeonhole Principle, two of the numbers must coincide. We cannot have two of (6.7.3) or two of (6.7.4) identical, thus some number in (6.7.3) is equal to some number in (6.7.4). Therefore, $a_i - a_j = 9$ for some i and j, as desired.

We next state yet another form of the Pigeonhole Principle.

Pigeonhole Principle (Third Form).   *Let f be a function from a finite set X into a finite set Y. Suppose that $|X| = n$ and $|Y| = m$. Let $k = \lceil n/m \rceil$. Then there are at least k values $a_1, \ldots, a_k \in X$ such that*

$$f(a_1) = f(a_2) = \cdots = f(a_k).$$

To verify the truth of the third form of the Pigeonhole Principle, we argue by contradiction. Let $Y = \{y_1, \ldots, y_m\}$. Suppose that the conclusion is false. Then there are at most $k - 1$ values $x \in X$ with $f(x) =$

y_1; there are at most $k - 1$ values $x \in X$ with $f(x) = y_2$; . . . ; there are at most $k - 1$ values $x \in X$ with $f(x) = y_m$. Thus there are at most $m(k - 1)$ members in the domain of f. But

$$m(k - 1) < m \, \frac{n}{m} = n,$$

which is a contradiction. Therefore, there are at least k values, $a_1, \ldots, a_k \in X$ such that

$$f(a_1) = f(a_2) = \cdots = f(a_k).$$

Our last example illustrates the use of the third form of the Pigeonhole Principle.

Example 6.7.5.   A useful feature of black-and-white pictures is the average brightness of the picture. Let us say that two pictures are similar if their average brightness differs by no more than some fixed value. Show that among six pictures, there are either three that are mutually similar or there are three that are mutually dissimilar.

Denote the pictures $P_1, P_2, \ldots, P_6$. Each of the five pairs

$$(P_1, P_2), (P_1, P_3), (P_1, P_4), (P_1, P_5), (P_1, P_6),$$

has the value "similar" or "dissimilar." By the third form of the Pigeonhole Principle, there are at least $\lceil 5/2 \rceil = 3$ pairs with the same value; that is, there are three pairs

$$(P_1, P_i), (P_1, P_j), (P_1, P_k)$$

all similar or all dissimilar. Suppose that each pair is similar. (The case that each pair is dissimilar is Exercise 8.) If any pair

$$(P_i, P_j), (P_i, P_k), (P_j, P_k) \tag{6.7.5}$$

is similar, then these two pictures together with P_1 are mutually similar and we have found three mutually similar pictures. Otherwise, each of the pairs (6.7.5) is dissimilar and we have found three mutually dissimilar pictures.

EXERCISES

1H.   Thirteen persons have first names Dennis, Evita, and Ferdinand and last names Oh, Pietro, Quine, and Rostenkowski. Show that at least two persons have the same first and last names.

2. Eighteen persons have first names Alfie, Ben, and Cissi and last names Dumont and Elm. Show that at least three persons have the same first and last names.

3. Show that every set of 11 shoes chosen from among 10 pairs of shoes contains at least one matched pair.

4H. Is it possible to interconnect five processors so that exactly two processors are directly connected to an identical number of processors? Explain.

5. An inventory consists of a list of 115 items, each marked available or unavailable. There are 60 available items. Show that there are at least two available items in the list exactly four items apart.

6. An inventory consists of a list of 100 items, each marked available or unavailable. There are 55 available items. Show that there are at least two available items in the list exactly nine items apart.

*7H. An inventory consists of a list of 80 items, each marked available or unavailable. There are 50 available items. Show that there are at least two unavailable items in the list either three or six items apart.

8. Complete Example 6.7.5 by showing that if the pairs (P_1, P_i), (P_1, P_j), (P_1, P_k) are dissimilar, there are three pictures that are mutually similar or mutually dissimilar.

9. Does the conclusion to Example 6.7.5 necessarily follow if there are fewer than six pictures? Explain.

10. Does the conclusion to Example 6.7.5 necessarily follow if there are more than six pictures? Explain.

Answer Exercises 11–14 to give an argument that shows that if X is any $(n + 2)$-element subset of $\{1, 2, \ldots , 2n + 1\}$ and m is the greatest element in X, there exist distinct i and j in X with $m = i + j$.

For each element $k \in X - \{m\}$, let

$$
a_k = \begin{cases} k & \text{if } k \leq \dfrac{m}{2} \\[2ex] m - k & \text{if } k > \dfrac{m}{2}. \end{cases}
$$

11H. How many elements are in the domain of a?

12H. Show that the range of a is contained in $\{1, 2, \ldots, n\}$.

13H. Explain why Exercises 11 and 12 imply that $a_i = a_j$ for some $i \neq j$.

14H. Explain why Exercise 13 implies that there exist distinct i and j in X with $m = i + j$.

15. Give an example of an $(n + 1)$-element subset X of $\{1, 2, \ldots, 2n + 1\}$ having the property: For no distinct $i, j \in X$ do we have $i + j \in X$.

An **increasing sequence** is a sequence satisfying

$$x_1 < x_2 < \cdots .$$

A **decreasing sequence** is a sequence satisfying

$$x_1 > x_2 > \cdots .$$

A **subsequence of length** k of the sequence $x_1, x_2, \ldots$ is a sequence of the form

$$x_{f(1)}, \ldots, x_{f(k)},$$

where f is a function from $\{1, \ldots, k\}$ into the positive integers satisfying

$$f(1) < f(2) < \cdots < f(k).$$

EXAMPLE: The sequence

$$2, 5, 9, 8, 3, 1, 4, 10, 7, 6$$

contains the decreasing subsequence

$$9, 8, 7, 6$$

of length 4.

Answer Exercises 16–19 to give an argument that proves the following result.

A sequence $a_1, a_2, \ldots, a_{n^2+1}$ of $n^2 + 1$ distinct numbers contains either an increasing subsequence of length $n + 1$ or a decreasing subsequence of length $n + 1$.

Suppose by way of contradiction that every increasing or decreasing subsequence has length n or less. Let b_i be the length of a longest increasing subsequence starting at a_i and let c_i be the length of a longest decreasing subsequence starting at a_i.

16.   Show that the ordered pairs (b_i, c_i), $i = 1, \ldots, n^2 + 1$, are distinct.

17.   How many ordered pairs (b_i, c_i) are there?

18.   Explain why $1 \le b_i \le n$ and $1 \le c_i \le n$.

19.   What is the contradiction?

Answer Exercises 20–23 to give an argument that shows that in a group of 10 persons there are at least two such that either the difference or sum of their ages is divisible by 16. Assume that the ages are given as whole numbers.

Let $a_1, \ldots, a_{10}$ denote the ages. Let $r_i = a_i \bmod 16$ and let

$$s_i = \begin{cases} r_i & \text{if } r_i \le 8 \\ 16 - r_i & \text{if } r_i > 8. \end{cases}$$

20.   Show that $s_1, \ldots, s_{10}$ range in value from 0 to 8.

21.   Explain why $s_j = s_k$ for some $j \ne k$.

22.   Explain why if $s_j = r_j$ and $s_k = r_k$ or $s_j = 16 - r_j$ and $s_k = 16 - r_k$, then 16 divides $a_j - a_k$.

23.   Show that if the conditions in Exercise 22 fail, then 16 divides $a_j + a_k$.

24H.   Show that in the decimal expansion of the quotient of two integers, eventually some block of digits repeats.

EXAMPLES: $1/6 = 0.1\underline{6}66\ldots$, $217/660 = 0.32\underline{87}8787\ldots$

*25.   Sixteen basketball players, whose uniforms are numbered 1 through 12, stand around the center ring on the court in an arbitrary arrangement. Show that some three consecutive players have the sum of their numbers at least 26.

*26.   Let f be a one-to-one function from $X = \{1, 2, \ldots, n\}$ onto X. Let $f^k = f \circ f \circ \cdots \circ f$ denote the n-fold composition of f with itself. Show that there are distinct positive integers i and j such that $f^i(x) = f^j(x)$ for all $x \in X$. Show that for some positive integer k, $f^k(x) = x$ for all $x \in X$.

6.8 Notes

An elementary book on counting methods is [Niven]. References on combinatorics are [Bogart; Brualdi; Even, 1973; Liu, 1968; Riordan; and Roberts]. [Vilenkin] contains many worked-out combinatorial examples. The general discrete mathematical references [Johnsonbaugh; Liu, 1985; and Tucker] devote several sections to the topics of Chapter 6. [Even, 1973; Hu, 1982; and Reingold] treat combinatorial algorithms. Algorithms for generating combinations and permutations can be found in [Even, 1973; and Reingold].

Computer Exercises

1. Write a program that generates all r-combinations of the elements $\{1, \ldots, n\}$.

2. Write a program that generates all permutations of the elements $\{1, \ldots, n\}$.

3. Write a program that generates all r-permutations of the elements $\{1, \ldots, n\}$.

4. Write a program that lists all permutations of *ABCDEF* in which *A* appears before *D*.

5. Write a program that lists all permutations of *ABCDEF* in which *C* and *E* are side by side in either order.

6. Write a program that generates Pascal's triangle to level n, for arbitrary n.

7. Write a program that finds an increasing or decreasing subsequence of length $n + 1$ of a sequence of $n^2 + 1$ distinct numbers.

8. [*Project*]   Report on other algorithms for generating combinations and permutations (see, e.g., [Even, 1973; Reingold]). Implement some of these algorithms as programs.

Chapter Review

SECTION 6.1
First Counting Principle

SECTION 6.2
Second Counting Principle

SECTION 6.3
Permutation of $x_1, \ldots, x_n$: ordering of $x_1, \ldots, x_n$
$n!$ = Number of permutations of an n-element set
r-permutation of $x_1, \ldots, x_n$: ordering of r elements of $x_1, \ldots, x_n$
$P(n, r)$: number of r-permutations of an n-element set; $P(n, r) =$
$\quad n(n - 1) \cdots (n - r + 1)$

SECTION 6.4
r-combination of $\{x_1, \ldots, x_n\}$: (unordered) subset of $\{x_1, \ldots, x_n\}$
$\quad$ containing r elements
$C(n, r)$: number of r-combinations of an n-element set; $C(n, r) =$
$\quad P(n, r)/r! = n!/[(n - r)!r!]$
Number of orderings of n items of t types with n_i identical objects of
$\quad$ type $i = n!/[n_1! \cdots n_t!]$

SECTION 6.5
Binomial Theorem
Binomial coefficient: $C(n, r)$
kth term in $(a + b)^n = C(n, k - 1)a^{n-k+1}b^{k-1}$
Pascal's triangle
$C(n + 1, k) = C(n, k - 1) + C(n, k)$

SECTION 6.6
Lexicographic order
Algorithm for generating r-combinations: Algorithms 6.6.9 and 6.6.16
Algorithm for generating permutations: Algorithms 6.6.14 and 6.6.18

SECTION 6.7
Pigeonhole Principle (Three forms)

Chapter Self-Test

SECTION 6.1

1H. A woman has seven sweaters, four skirts, five pairs of shoes, and three pairs of stockings. How many different outfits are possible?

2H. How many eight-bit strings begin with 0 and end with 101?

3H. Strings of length 4 are to be formed using the letters *ABCDEF*. How many strings contain *C* and end with *F* if we do not allow repetitions?

4H. A seven-person committee composed of Greg, Hwang, Isaac, Jasmine, Kirk, Lynn, and Manuel is to select a chairperson, vice-chairperson, social events chairperson, secretary, and treasurer. In how many ways can this be done if Hwang is not an officer and Isaac is an officer?

SECTION 6.2

5H. Two dice are rolled, one blue and one red. How many outcomes give either a sum of 6 or a sum of 8?

6H. How many eight-bit strings begin either 1 or 011?

7H. A seven-person committee composed of Greg, Hwang, Isaac, Jasmine, Kirk, Lynn, and Manuel is to select a chairperson, vice-chairperson, social events chairperson, secretary, and treasurer. How many ways can the officers be chosen if either Greg is secretary or he is not an officer?

8H. How many ways can we select three books each from a different subject from a set of six distinct history books, nine distinct classics books, seven distinct law books, and four distinct education books?

SECTION 6.3

9H. How many permutations are there of eight objects?

10H. How many strings that contain the substring *FED* can be formed by ordering the letters *ABCDEF*?

11H. How many strings that contain the letters *BEAD* together in any order can be formed by ordering the letters *ABCDEFG*?

12H. In how many ways can six distinct Martians and nine distinct Jovians wait in line if no two Martians stand together?

SECTION 6.4

13H. How many 3-combinations are there of six objects?

14H. In how many ways can a club consisting of eight women and six men select a committee consisting of three women and three men?

15H. How many five-card poker hands contain four cards of one suit and one card of another suit?

16H. How many strings can be formed by ordering the letters ILLINOIS?

SECTION 6.5

17H. Expand the expression $(s - r)^4$ using the Binomial Theorem.

18H. Expand the expression $(2c + d)^5$ using the Binomial Theorem.

19H. Find the ninth term in the expansion of $(x - 2y)^{12}$.

20H. Rotate Pascal's triangle counterclockwise so that the top row consists of 1's. Explain why the second row lists the positive integers in order 1, 2, ...

SECTION 6.6

21H. Find the 5-combination that will be generated by Algorithm 6.6.9 after 12467 if $n = 7$.

22H. Find the 6-combination that will be generated by Algorithm 6.6.9 after 145678 if $n = 8$.

23H. Find the permutation that will be generated by Algorithm 6.6.14 after 6427135.

24H. Find the permutation that will be generated by Algorithm 6.6.14 after 625431.

SECTION 6.7

25H. Show that every set of 15 socks chosen from among 14 pairs of socks contains at least one matched pair.

26H. Seventeen persons have first names Zeke, Wally, and Linda; middle names Lee and David; and last names Yu, Zamora, and Smith. Show that at least two persons have the same first, middle, and last names.

27H. An inventory consists of a list of 200 items, each marked available or unavailable. There are 110 available items. Show that there are at least two available items in the list exactly 19 items apart.

28H. Each edge of K_6, the complete graph on six vertices, is colored either red or blue. Show that there is a cycle of length three with all edges the same color.

7

You want to tell me now?

Tell you what?

What it is you're trying to find out. You know, it's a funny thing. You're trying to find out what your father hired me to find out and I'm trying to find out why you want to find out.

You could go on forever, couldn't you?

—from *The Big Sleep*

Recurrence Relations

This chapter offers an introduction to recurrence relations. Recurrence relations are useful in certain counting problems. A recurrence relation relates the nth element of a sequence to its predecessors. Because recurrence relations are closely related to recursive algorithms (algorithms that call themselves), recurrence relations arise naturally in the analysis of recursive algorithms.

7.1 Introduction

Consider the following instructions for generating a sequence:

1. Start with 5.
2. Given any term, add 3 to get the next term.

If we list the terms of the sequence, we obtain

$$5, 8, 11, 14, 17, \ldots . \tag{7.1.1}$$

The first term is 5 because of instruction 1. The second term is 8 because instruction 2 says to add 3 to 5 to get the next term 8. The third term is 11 because instruction 2 says to add 3 to 8 to get the next term 11. By following instructions 1 and 2, we can compute any term in the sequence. Instructions 1 and 2 do not give an explicit formula for the nth term of the sequence in the sense of providing a formula that we can "plug n into" to obtain the value of the nth term, but by computing term by term we can eventually compute any term of the sequence.

If we denote the sequence (7.1.1) as $\{a_n\}_{n=1}^{\infty}$, we may rephrase instruction 1 as

$$a_1 = 5 \tag{7.1.2}$$

and we may rephrase instruction 2 as

$$a_n = a_{n-1} + 3, \qquad n \geq 2. \tag{7.1.3}$$

Taking $n = 2$ in (7.1.3), we obtain

$$a_2 = a_1 + 3.$$

By (7.1.2), $a_1 = 5$; thus

$$a_2 = a_1 + 3 = 5 + 3 = 8.$$

Taking $n = 3$ in (7.1.3), we obtain

$$a_3 = a_2 + 3.$$

Since $a_2 = 8$,

$$a_3 = a_2 + 3 = 8 + 3 = 11.$$

By using (7.1.2) and (7.1.3), we can compute any term in the sequence just as we did using instructions 1 and 2. We see that (7.1.2) and (7.1.3) are equivalent to instructions 1 and 2.

Equation (7.1.3) furnishes an example of a **recurrence relation**. A recurrence relation defines a sequence by giving the nth value in terms of certain of its predecessors. In (7.1.3), the nth value is given in terms of the immediately preceding value. In order for a recurrence relation such as (7.1.3) to define a sequence, a ''start up'' value or values, such as (7.1.2), must be given. These start up values are called **initial conditions**. The formal definitions follow.

Definition 7.1.1.   A *recurrence relation* for the sequence $a_0, a_1, \ldots$ is an equation that relates a_n to certain of its predecessors $a_0, a_1, \ldots, a_{n-1}$.

Initial conditions for the sequence $a_0, a_1, \ldots$ are explicitly given values for a finite number of the terms of the sequence.

We have seen that it is possible to define a sequence by a recurrence relation together with certain initial conditions. We give several examples of recurrence relations.

Example 7.1.2 Lucas Sequence.   The sequence

$$1, 3, 4, 7, 11, 18, \ldots$$

is known as the **Lucas sequence**. Find a recurrence relation and initial conditions that define the Lucas sequence.

Evidently, each term in the sequence is the sum of the two preceding

François-Édouard-Anatole Lucas (1842–1891)

The French mathematician François-Édouard-Anatole Lucas was interested in number theory and recreational mathematics. He was a co-inventor of the Tower of Hanoi puzzle and edited a classic four-volume treatise on recreational mathematics. He was the first person to call the sequence 1, 1, 2, 3, 5, . . . the Fibonacci sequence. He died from complications resulting from a freak accident when his cheek was struck by a fragment of a plate that had been dropped. [*Photo courtesy of the New York Public Library*]

terms. Thus we have the recurrence relation

$$L_n = L_{n-1} + L_{n-2}, \qquad n \geq 2.$$

To apply this recurrence relation for $n = 2$, we need to know the values of L_0 and L_1. Thus we must provide the initial conditions

$$L_0 = 1, \qquad L_1 = 3.$$

We previously defined the Fibonacci sequence (see Example 2.3.10) by the recurrence relation

$$f_n = f_{n-1} + f_{n-2}, \qquad n \geq 2$$

and initial conditions

$$f_0 = 1, \qquad f_1 = 1.$$

We note that the Lucas sequence and the Fibonacci sequence satisfy the same recurrence relation; only the initial conditions are different. It is not surprising that the Lucas sequence and Fibonacci sequence are related. One formula that relates the two sequences is

$$L_n = f_{n-1} + f_{n+1}.$$

Example 7.1.3. A person invests \$1000 at 12 percent compounded annually. If A_n represents the amount at the end of n years, find a recurrence relation and initial conditions that define the sequence $\{A_n\}$.

At the end of $n - 1$ years, the amount is A_{n-1}. After one more year, we will have the amount A_{n-1} plus the interest. Thus

$$A_n = A_{n-1} + (0.12)A_{n-1}$$

$$= (1.12)A_{n-1}, \qquad n \geq 1.$$

To apply this recurrence relation for $n = 1$, we need to know the value of A_0. Since A_0 is the beginning amount, we have the initial condition

$$A_0 = 1000.$$

Example 7.1.4. Find A_3 where $\{A_n\}$ is the sequence defined in Example 7.1.3. We have

$$A_3 = (1.12)A_2$$

$$= (1.12)(1.12)A_1$$

$$\text{(7.1.4)}$$

$$= (1.12)(1.12)(1.12)A_0$$

$$= (1.12)^3(1000) = 1404.93.$$

Thus, at the end of the third year, the amount is $1404.93.

The computation (7.1.4) can be carried out for an arbitrary value of n to obtain

$$A_n = (1.12)A_{n-1}$$

$$.$$
$$.$$
$$.$$

$$= (1.12)^n(1000).$$

We see that sometimes an explicit formula can be derived from a recurrence relation and initial conditions. Finding explicit formulas from recurrence relations is the topic of Sections 7.2 and 7.3.

Although it is easy to obtain an explicit formula from the recurrence relation and initial condition for the sequence of Example 7.1.3, it is not immediately apparent how to obtain explicit formulas for the Lucas and Fibonacci sequences. In the next section we give a method that yields explicit formulas for the Fibonacci and Lucas sequences. Of course, the lack of an explicit formula is no deterrent to writing an algorithm to compute the Fibonacci or Lucas sequence.

Example 7.1.5.   Let S_n denote the number of subsets of an n-element set. Since going from an $(n - 1)$-element set to an n-element set doubles the number of subsets (see Theorem 2.4.5), we obtain the recurrence relation

$$S_n = 2S_{n-1}.$$

The initial condition is

$$S_0 = 1.$$

One of the main reasons for using recurrence relations is that sometimes it is easier to determine the nth term of a sequence in terms of its predecessors than it is to find an explicit formula for the nth term in terms of n. The remaining examples in this section are intended to illustrate this thesis.

Example 7.1.6 Tower of Hanoi.   The Tower of Hanoi is a puzzle consisting of three pegs mounted on a board and n disks of various sizes with holes in their centers (see Figure 7.1.1). It is assumed that if a disk is on a peg, only a disk of smaller diameter can be placed on top of the first disk. Given all the disks stacked on one peg as in Figure 7.1.1, the problem is to transfer the disks to another peg by moving one disk at a time.

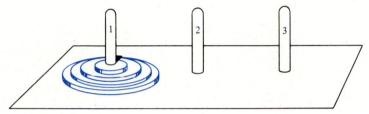

Figure 7.1.1

We will discuss a solution and find a recurrence relation and an initial condition for the sequence $c_1, c_2, \ldots$ where c_n denotes the number of moves required to solve the n-disk puzzle.

Suppose that we have n disks on peg 1 as in Figure 7.1.1. Then, in c_{n-1} moves, we can move the top $n-1$ disks to peg 2 (see Figure 7.1.2). During these moves, the bottom disk on peg 1 stays fixed. Next, we move the remaining disk on peg 1 to peg 3. Finally, in c_{n-1} moves, we can move the $n-1$ disks on peg 2 to peg 3. Therefore,

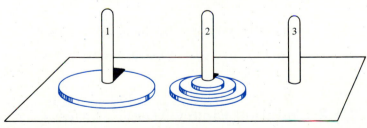

Figure 7.1.2

the recurrence relation is

$$c_n = 2c_{n-1} + 1. \qquad (7.1.5)$$

The initial condition is

$$c_1 = 1.$$

Example 7.1.7.   Let S_n denote the number of n-bit strings that do not contain the pattern 111. Develop a recurrence relation for $S_0, S_1, \ldots$. Also, give initial conditions that define the sequence S.

We will count the number of n-bit strings that do not contain the pattern 111

(a) that begin with 0;
(b) that begin with 10;
(c) that begin with 11.

Since the sets of strings of types (a), (b), and (c) are disjoint, by the Second Counting Principle S_n will equal the sum of the numbers of strings of types (a), (b), and (c). Suppose that an n-bit string begins with 0 and does not contain the pattern 111. Then the $(n - 1)$-bit string following the initial 0 does not contain the pattern 111. Since any $(n - 1)$-bit string not containing 111 can follow the initial 0, there are S_{n-1} strings of type (a). If an n-bit string begins with 10 and does not contain the pattern 111, then the $(n - 2)$-bit string following the initial 10 cannot contain the pattern 111; therefore, there are S_{n-2} strings of type (b). If an n-bit string begins with 11 and does not contain the pattern 111, then the third bit must be 0. The $(n - 3)$-bit string following the initial 110 cannot contain the pattern 111; therefore, there are S_{n-3} strings of type (c). Thus

$$S_n = S_{n-1} + S_{n-2} + S_{n-3}, \qquad n \geq 4.$$

By inspection, we find the initial conditions

$$S_1 = 2, \quad S_2 = 4, \quad S_3 = 7.$$

Example 7.1.8 Counting Binary Trees.   We will develop a recurrence relation for the sequence $\{a_n\}$, where a_n is the number of binary trees having n vertices. For example, $a_1 = 1$ since a binary tree with one vertex consists of that vertex and no edges. Also, $a_0 = 1$ since there is one binary tree having no edges and no vertices. There are two binary trees with two vertices and five binary trees with three vertices (see Figure 7.1.3); thus $a_2 = 2$ and $a_3 = 5$.

Suppose that we want to construct an n-vertex binary tree and that we know how to construct all binary trees with fewer than n vertices. One vertex must be the root. Since there are $n - 1$ vertices remaining, if the left subtree has k vertices, the right subtree must have $n - k - 1$ vertices. We will construct an n-vertex binary tree whose left subtree has k vertices and whose right subtree has $n - k - 1$ vertices by a two-step process: Construct the left subtree, construct the right subtree. (Figure 7.1.4 shows this construction for $n = 6$ and $k = 2$.) By the First Counting Principle, this construction can be carried out in $a_k a_{n-k-1}$ ways. Different values of k give distinct n-vertex binary trees, so by the Second Counting Principle, the total number of binary trees is

$$\sum_{k=0}^{n-1} a_k a_{n-k-1}.$$

We obtain the recurrence relation

$$a_n = \sum_{k=0}^{n-1} a_k a_{n-k-1}, \qquad n \geq 1. \tag{7.1.6}$$

For example, with $n = 1$, (7.1.6) becomes

$$a_1 = a_0 a_0 = 1 \cdot 1 = 1.$$

With $n = 2$, (7.1.6) becomes

$$a_2 = a_0 a_1 + a_1 a_0 = 1 \cdot 1 + 1 \cdot 1 = 2.$$

With $n = 3$, (7.1.6) becomes

$$a_3 = a_0 a_2 + a_1 a_1 + a_2 a_0 = 1 \cdot 2 + 1 \cdot 1 + 2 \cdot 1 = 5.$$

It can be shown (see [Standish] or [Kruse]) that the solution of (7.1.6), subject to the initial condition $a_0 = 1$, is

$$a_n = \frac{C(2n, n)}{n + 1}.$$

The numbers a_n occur so frequently that they are known by a special name; a_n is called the nth **Catalan number**.

Eugène-Charles Catalan   *(1814–1894)*

The Catalan numbers are named after the Belgian mathematician Eugène-Charles Catalan. He gave an elementary proof that the nth Catalan number is equal to the number of ways of partitioning a convex polygon with $n + 2$ sides into triangles using $n - 1$ nonintersecting diagonals. After the 1851 coup d'état in France, Catalan refused to swear allegiance to the new government. As a result, he lost his official appointments in mathematics. Until he became professor of mathematics at the University of Liége in Belgium in 1865, he was in charge of preparatory education at the École Polytechnique. Catalan published numerous papers in analysis, combinatorics, algebra, geometry, probability, and number theory. [*Photo courtesy of the Royal Academy of Sciences, Letters, and Fine Arts of Belgium and the Prime Minister's Office, Brussels*]

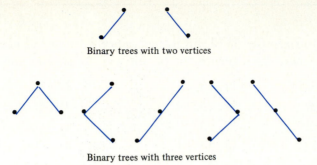

Binary trees with two vertices

Binary trees with three vertices

Figure 7.1.3

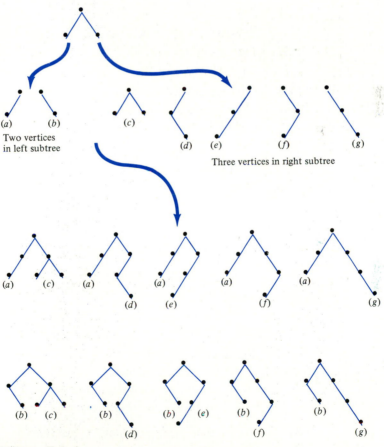

Figure 7.1.4

EXERCISES

In Exercises 1–10, find a recurrence relation and initial conditions that generate a sequence that begins with the given terms.

1H. $3, 7, 11, 15, \ldots$ **2.** $15, 13, 11, 9, \ldots$

3. $2, 8, 17, 29, 44, \ldots$ **4H.** $3, 6, 9, 15, 24, 39, \ldots$

5. $3, 4, 1, -3, -4, -1, 3, \ldots$ **6.** $1, 2, 5, 12, 29, 70, \ldots$

7H. $2, 6, 18, 54, \ldots$ **8.** $1, 3, 7, 15, 31, \ldots$

9. $2, 3, 6, 18, 108, \ldots$

10H. $1, 1, 2, 4, 16, 128, 4096, \ldots$

In Exercises 11–15, assume that a person invests \$2000 at 14 percent compounded annually. Let A_n represent the amount at the end of n years.

11H.   Find a recurrence relation for the sequence $A_0, A_1, \ldots$.

12H.   Find an initial condition for the sequence $A_0, A_1, \ldots$.

13H.   Find $A_1, A_2,$ and A_3.

14H.   Find an explicit formula for A_n.

15H.   How long will it take for a person to double the initial investment?

In Exercises 16–20, assume that a person invests \$3000 at 16 percent compounded annually. Let A_n represent the amount at the end of n years.

16.   Find a recurrence relation for the sequence $A_0, A_1, \ldots$.

17.   Find an initial condition for the sequence $A_0, A_1, \ldots$.

18.   Find $A_1, A_2,$ and A_3.

19.   Find an explicit formula for A_n.

20.   How long will it take for a person to double the initial investment?

If a person invests in a tax-sheltered annuity, the money invested, as well as the interest earned, is not subject to taxation until withdrawn from the account. In Exercises 21–24, assume that a person invests \$2000 *each* year in a tax-sheltered annuity at 10 percent compounded annually. Let A_n represent the amount at the end of n years.

21.   Find a recurrence relation for the sequence $A_0, A_1, \ldots$.

22.   Find an initial condition for the sequence $A_0, A_1, \ldots$.

23.   Find A_1, A_2, and A_3.

24.   Find an explicit formula for A_n.

In Exercises 25–29, assume that a person invests \$3000 at 12 percent annual interest compounded quarterly. Let A_n represent the amount at the end of n years.

25H. Find a recurrence relation for the sequence A_0, A_1, ...

26H. Find an initial condition for the sequence A_0, A_1, ...

27H. Find A_1, A_2, and A_3.

28H. Find an explicit formula for A_n.

29H. How long will it take for a person to double the initial investment?

In Exercises 30–34, assume that a person invests \$4000 at 16.5 percent annual interest compounded quarterly. Let A_n represent the amount at the end of n years.

30.   Find a recurrence relation for the sequence A_0, A_1, ...

31.   Find an initial condition for the sequence A_0, A_1, ...

32.   Find A_1, A_2, and A_3.

33.   Find an explicit formula for A_n.

34.   How long will it take for a person to double the initial investment?

35H. Use the recurrence relation (7.1.6) to compute the number of binary trees having four vertices. Draw all binary trees having four vertices.

36.   By drawing trees as in Figure 7.1.4, illustrate the construction given in Example 7.1.8 for $n = 5$ and $k = 2$.

37.   By drawing trees as in Figure 7.1.4, illustrate the construction given in Example 7.1.8 for $n = 7$ and $k = 3$.

38H. Compute a_5, a_6, and a_7, where a_n denotes the nth Catalan number.

39.   Write explicit solutions for the cases $n = 3, 4, 5$ of the Tower of Hanoi puzzle.

40.   Find a solution to the Tower of Hanoi puzzle in which there is an extra peg (otherwise, all the rules are the same). Your solution should require fewer moves than the solution given in Example 7.1.6 for the three-peg version. Develop a recurrence relation for the number of moves required by your solution.

41H. A network consists of n nodes. Each node has communications facilities and local storage. Periodically, all files must be shared. A *link* consists of two nodes sharing files. Specifically, when nodes A and B are linked, A transmits all its files to B and B transmits all its files to A. Only one link exists at a time and after a link is established and the files are shared the link is deleted. Let a_n be the minimum number of links required by n nodes so that all files are known to all nodes.
(a) Show that $a_2 = 1$, $a_3 \leq 3$, $a_4 \leq 4$.
(b) Show that $a_n \leq a_{n-1} + 2$, $n \geq 3$.

42. If P_n denotes the number of permutations of n distinct objects, find a recurrence relation and an initial condition for the sequence $P_1, P_2, \ldots$.

43. Suppose that we have n dollars and that each day we buy either orange juice ($1), milk ($2), or beer ($2). If C_n is the number of ways of spending all the money, show that

$$C_n = C_{n-1} + 2C_{n-2}.$$

Order is taken into account. For example, there are 11 ways to spend four dollars: *MB*, *BM*, *OOM*, *OOB*, *OMO*, *OBO*, *BOO*, *MOO*, *OOOO*, *MM*, *BB*.

44H. Suppose that we have n dollars and that each day we buy either tape ($1), paper ($1), pens ($2), pencils ($2), or binders ($3). If C_n is the number of ways of spending all the money, derive a recurrence relation for the sequence $C_1, C_2, \ldots$.

45. Let C_n denote the number of regions into which the plane is divided by n lines. Assume that each pair of lines meets in a point, but that no three lines meet in a point. Derive a recurrence relation for the sequence $C_1, C_2, \ldots$.

Exercises 46 and 47 refer to the sequence S_n defined by

$$S_1 = 0, \qquad S_2 = 1$$

$$S_n = \frac{S_{n-1} + S_{n-2}}{2}, \qquad n = 3, 4, \ldots .$$

46H. Compute S_3 and S_4.

*47. Guess a formula for S_n and show that it is correct by using induction.

48H. A robot can move forward in steps of size 1 meter or 2 meters. Let C_n denote the number of ways the robot can walk n meters.

Find a recurrence relation and initial conditions for the sequence $\{C_n\}$. Show that $C_n = f_n$, $n = 1, 2, \ldots$, where f denotes the Fibonacci sequence.

49. Let S_n denote the number of n-bit strings that do not contain the pattern 000. Find a recurrence relation and initial conditions for the sequence $\{S_n\}$.

Exercises 50–52 refer to the sequence S where S_n denotes the number of n-bit strings that do not contain the pattern 00.

50H. Find a recurrence relation and initial conditions for the sequence $\{S_n\}$.

51. Show that $S_n = f_{n+1}$, $n = 1, 2, \ldots$, where f denotes the Fibonacci sequence.

52. By considering the number of n-bit strings with exactly i 0's and Exercise 51, show that

$$f_{n+1} = \sum_{i=0}^{\lfloor (n+1)/2 \rfloor} C(n + 1 - i, i), \qquad n = 1, 2, \ldots,$$

where f denotes the Fibonacci sequence.

Exercises 53–55 refer to the sequence S where S_n denotes the number of n-bit strings that do not contain the pattern 010.

53H. Compute S_1, S_2, S_3, and S_4.

54. By considering the number of n-bit strings that do not contain the pattern 010 that have no leading 0's (i.e., that begin with 1); that have one leading 0 (i.e., that begin 01); that have two leading 0's; and so on; derive the recurrence relation

$$S_n = S_{n-1} + S_{n-3} + S_{n-4} + S_{n-5} \qquad (7.1.7)$$
$$+ \cdots + S_1 + 3.$$

55. By replacing n by $n - 1$ in (7.1.7), write a formula for S_{n-1}. Subtract the formula for S_{n-1} from the formula for S_n and use the result to derive the recurrence relation

$$S_n = 2S_{n-1} - S_{n-2} + S_{n-3}.$$

*56H. Let P_n be the number of partitions of an n-element set. Show that the sequence $P_0, P_1, \ldots$ satisfies the recurrence relation

$$P_n = \sum_{k=0}^{n-1} C(n - 1, k)P_k.$$

Wilhelm Ackermann *(1896–1962)*

Wilhelm Ackermann was a German mathematician and logician. He was a student of David Hilbert, one of the greatest mathematicians of all time. He was the co-author, with Hilbert, of *Principles of Mathematical Logic* (English translation of *Grundzüge der theoretischen Logik*). According to [Reid, 1970], Hilbert withdrew his support of Ackermann after Ackermann married and had a child. Hilbert considered such activity crazy. As a result, Ackermann was unable to obtain a university position for some time. [*Photo courtesy of the German Information Center and the Estate of Wilhelm Ackermann*]

*57. Let F_n denote the number of functions f from $X = \{1, \ldots, n\}$ into X having the property that if i is in the range of f, then $1, 2, \ldots, i$ are also in the range of f. (Set $F_0 = 1$.) Show that the sequence $F_0, F_1, \ldots$ satisfies the recurrence relation

$$F_n = \sum_{j=0}^{n-1} C(n, j)F_j.$$

It is possible to extend the definition of recurrence relation to include functions indexed over n-tuples of positive integers. Ackermann's function can be defined by the recurrence relations

$$A(m, 0) = A(m - 1, 1), \qquad m = 1, 2, \ldots,$$
$$A(m, n) = A(m - 1, A(m, n - 1)) \qquad m = 1, 2, \ldots,$$
$$n = 1, 2, \ldots,$$

and the initial conditions

$$A(0, n) = n + 1, \qquad n = 0, 1, \ldots.$$

Ackermann's function is of theoretical importance because of its rapid rate of growth. Functions related to Ackermann's function appear in the time complexity of certain algorithms such as the time to execute union/find algorithms (see [Horowitz, 1978]). Exercises 58–61 refer to Ackermann's function.

58H. Compute $A(1, 1)$, $A(1, 2)$, $A(2, 2)$, and $A(2, 3)$.

59H. Use induction to show that

$$A(1, n) = n + 2, \qquad n = 0, 1, \ldots.$$

60. Use induction to show that

$$A(2, n) = 3 + 2n, \qquad n = 0, 1, \ldots.$$

61. Guess a formula for $A(3, n)$ and prove it by using induction.

7.2 Solving Recurrence Relations

To solve a recurrence relation involving the sequence $a_0, a_1, \ldots$ is to find an explicit formula for the general term a_n. In this section we discuss two methods of solving recurrence relations: **iteration** and a special method that applies to **linear homogeneous recurrence relations with constant**

coefficients. For more powerful methods, such as methods which make use of generating functions, consult [Brualdi].

To solve a recurrence relation involving the sequence $a_0, a_1, \ldots$ by iteration, we use the recurrence relation to write the nth term a_n in terms of certain of its predecessors $a_{n-1}, \ldots, a_0$. We then successively use the recurrence relation to replace each of $a_{n-1}, \ldots$ by certain of their predecessors. We continue until an explicit formula is obtained. The iterative method was used to solve the recurrence relation of Example 7.1.4.

Example 7.2.1.   We can solve the recurrence relation

$$a_n = a_{n-1} + 3, \qquad\qquad (7.2.1)$$

subject to the initial condition

$$a_1 = 2,$$

by iteration. Replacing n by $n - 1$ in (7.2.1), we obtain

$$a_{n-1} = a_{n-2} + 3.$$

If we substitute this expression for a_{n-1} into (7.2.1), we obtain

$$a_n = a_{n-1} + 3 = (a_{n-2} + 3) + 3 = a_{n-2} + 2 \cdot 3. \quad (7.2.2)$$

Replacing n by $n - 2$ in (7.2.1), we obtain

$$a_{n-2} = a_{n-3} + 3.$$

If we substitute this expression for a_{n-2} into (7.2.2), we obtain

$$a_n = a_{n-2} + 2 \cdot 3 = (a_{n-3} + 3) + 2 \cdot 3 = a_{n-3} + 3 \cdot 3.$$

In general, we have

$$a_n = a_{n-k} + k \cdot 3.$$

If we set $k = n - 1$ in this last expression, we have

$$a_n = a_1 + (n - 1) \cdot 3.$$

Since $a_1 = 2$, we obtain the explicit formula

$$a_n = 2 + 3(n - 1)$$

for the sequence a.

Example 7.2.2. We can solve the recurrence relation

$$S_n = 2S_{n-1}$$

of Example 7.1.5, subject to the initial condition

$$S_0 = 1,$$

by iteration:

$$S_n = 2S_{n-1}$$
$$= 2(2S_{n-2})$$
$$\vdots$$
$$= 2^n S_0$$
$$= 2^n.$$

Example 7.2.3. Find an explicit formula for c_n, the number of moves in which the n-disk Tower of Hanoi puzzle can be solved (see Example 7.1.6).

In Example 7.1.6 we obtained the recurrence relation

$$c_n = 2c_{n-1} + 1 \qquad (7.2.3)$$

and initial condition

$$c_1 = 1.$$

Applying the iterative method to (7.2.3), we obtain

$$c_n = 2c_{n-1} + 1$$
$$= 2(2c_{n-2} + 1) + 1$$
$$= 2^2 c_{n-2} + 2 + 1$$
$$= 2^2(2c_{n-3} + 1) + 2 + 1$$
$$= 2^3 c_{n-3} + 2^2 + 2 + 1$$
$$\vdots$$
$$= 2^{n-1} c_1 + 2^{n-2} + 2^{n-3} + \cdots + 2 + 1$$
$$= 2^{n-1} + 2^{n-2} + 2^{n-3} + \cdots + 2 + 1$$
$$= 2^n - 1.$$

The last step results from the formula for the geometric sum (see Example 2.4.3).

We turn next to a special class of recurrence relations.

Definition 7.2.4.   A *linear homogeneous recurrence relation of order k with constant coefficients* is a recurrence relation of the form

$$a_n = c_1 a_{n-1} + c_2 a_{n-2} + \cdots + c_k a_{n-k}. \qquad (7.2.4)$$

Notice that a linear homogeneous recurrence relation of order k with constant coefficients (7.2.4), together with the k initial conditions

$$a_0 = C_0, a_1 = C_1, \ldots, a_{k-1} = C_{k-1},$$

uniquely defines a sequence $a_0, a_1, \ldots$.

Example 7.2.5.   The recurrence relations

$$S_n = 2S_{n-1} \qquad (7.2.5)$$

of Example 7.2.2 and

$$f_n = f_{n-1} + f_{n-2}, \qquad (7.2.6)$$

which defines the Fibonacci sequence, are both linear homogeneous recurrence relations with constant coefficients. The recurrence relation (7.2.5) is of order 1 and (7.2.6) is of order 2.

Example 7.2.6.   The recurrence relation

$$a_n = 3a_{n-1}a_{n-2} \qquad (7.2.7)$$

is not a linear homogeneous recurrence relation with constant coefficients. In a linear homogeneous recurrence relation with constant coefficients, each term is of the form ca_k. Terms such as $a_{n-1}a_{n-2}$ are not permitted. Recurrence relations such as (7.2.7) are said to be *nonlinear*.

Example 7.2.7.   The recurrence relation

$$a_n - a_{n-1} = 2n$$

is not a linear homogeneous recurrence relation with constant coefficients because the expression on the right side of the equation is not zero. (Such an equation is said to be *inhomogeneous*. Linear inhomogeneous recurrence relations with constant coefficients are discussed in Exercises 29–35.)

Example 7.2.8. The recurrence relation

$$a_n = 3na_{n-1}$$

is not a linear homogeneous recurrence relation with constant coefficients because the coefficient $3n$ is not constant. It is a linear homogeneous recurrence relation with nonconstant coefficients.

We will illustrate the general method of solving linear homogeneous recurrence relations with constant coefficients by finding an explicit formula for the sequence defined by the recurrence relation

$$a_n = 5a_{n-1} - 6a_{n-2} \tag{7.2.8}$$

and initial conditions

$$a_0 = 7, \ a_1 = 16. \tag{7.2.9}$$

Often in mathematics, when trying to solve a more difficult instance of some problem, we begin with an expression that solved a simpler version. For the first-order recurrence relation (7.2.5), we found in Example 7.2.2 that the solution was of the form

$$S_n = t^n;$$

thus for our first attempt at finding a solution of the second-order recurrence relation (7.2.8), we will search for a solution of the form $V_n = t^n$.

If $V_n = t^n$ is to solve (7.2.8), we must have

$$V_n = 5V_{n-1} - 6V_{n-2}$$

or

$$t^n = 5t^{n-1} - 6t^{n-2}$$

or

$$t^n - 5t^{n-1} + 6t^{n-2} = 0.$$

Dividing by t^{n-2}, we obtain the equivalent equation

$$t^2 - 5t + 6 = 0. \qquad (7.2.10)$$

Solving (7.2.10), we find the solutions

$$t = 2, \qquad t = 3.$$

At this point, we have two solutions S and T of (7.2.8), given by

$$S_n = 2^n, \qquad T_n = 3^n. \qquad (7.2.11)$$

We can verify (see Theorem 7.2.11) that if S and T are solutions of (7.2.8), then $bS + dT$, where b and d are any numbers whatever, is also a solution of (7.2.8). In our case, if we define the sequence U by the equation

$$U_n = bS_n + dT_n$$

$$= b2^n + d3^n,$$

U is a solution of (7.2.8).

To satisfy the initial conditions (7.2.9), we must have

$$7 = U_0 = b2^0 + d3^0 = b + d$$

$$16 = U_1 = b2^1 + d3^1 = 2b + 3d.$$

Solving these equations for b and d, we obtain

$$b = 5, \qquad d = 2.$$

Therefore, the sequence U defined by

$$U_n = 5 \cdot 2^n + 2 \cdot 3^n$$

satisfies the recurrence relation (7.2.8) and the initial conditions (7.2.9). We conclude that

$$a_n = U_n = 5 \cdot 2^n + 2 \cdot 3^n, \qquad \text{for } n = 0, 1, \ldots .$$

We may summarize the preceding method as follows. To solve the recurrence relation

$$a_n - c_1 a_{n-1} - c_2 a_{n-2} = 0 \qquad (7.2.12)$$

with initial conditions

$$a_0 = C_0, \qquad a_1 = C_1,$$

first solve the quadratic equation

$$t^2 - c_1 t - c_2 = 0$$

to obtain roots r_1 and r_2. Assume that $r_1 \neq r_2$. (The case that $r_1 = r_2$ will be considered shortly.) The solution of (7.2.12) is

$$a_n = b r_1^n + d r_2^n$$

for some constants b and d. The values of these constants can be determined from the initial conditions by solving

$$b + d = C_0, \qquad b r_1 + d r_2 = C_1.$$

Example 7.2.9. Solve the recurrence relation

$$a_n + a_{n-1} - 2a_{n-2} = 0$$

with initial conditions

$$a_0 = -1, \qquad a_1 = 11.$$

We first solve the quadratic equation

$$t^2 + t - 2 = 0$$

to obtain roots 1 and -2. The sequence a is of the form

$$a_n = b \cdot 1^n + d(-2)^n = b + d(-2)^n.$$

To meet the initial conditions, we must have

$$-1 = a_0 = b + d$$

$$11 = a_1 = b - 2d.$$

Solving for b and d, we find that $b = 3$ and $d = -4$. Thus a_n is given by

$$a_n = 3 - 4(-2)^n.$$

Example 7.2.10. Find an explicit solution for the Fibonacci sequence.

The Fibonacci sequence is defined by the linear, homogeneous, second order recurrence relation

$$f_n - f_{n-1} - f_{n-2} = 0.$$

We begin by using the quadratic formula to solve

$$t^2 - t - 1 = 0.$$

The solutions are

$$t = \frac{1 \pm \sqrt{5}}{2}.$$

Thus the solution is of the form

$$f_n = b\left(\frac{1 + \sqrt{5}}{2}\right)^n + d\left(\frac{1 - \sqrt{5}}{2}\right)^n.$$

To satisfy the initial conditions

$$f_0 = 1 = f_1,$$

we must have

$$b + d = 1$$

$$b\left(\frac{1 + \sqrt{5}}{2}\right) + d\left(\frac{1 - \sqrt{5}}{2}\right) = 1.$$

Solving these equations for b and d, we obtain

$$b = \frac{1}{\sqrt{5}}\left(\frac{1 + \sqrt{5}}{2}\right), \qquad d = -\frac{1}{\sqrt{5}}\left(\frac{1 - \sqrt{5}}{2}\right).$$

Therefore, an explicit formula for the Fibonacci sequence is

$$f_n = \frac{1}{\sqrt{5}}\left(\frac{1 + \sqrt{5}}{2}\right)^{n+1} - \frac{1}{\sqrt{5}}\left(\frac{1 - \sqrt{5}}{2}\right)^{n+1}.$$

Surprisingly, even though f_n is an integer, the preceding formula involves the irrational number $\sqrt{5}$.

At this point, we will summarize and justify the techniques we have been using.

Theorem 7.2.11.   *Let*

$$a_n - c_1 a_{n-1} - c_2 a_{n-2} = 0 \tag{7.2.13}$$

be a second-order linear homogeneous recurrence relation with constant coefficients.

If S and T are solutions of (7.2.13), then $U = bS + dT$ is also a solution of (7.2.13).

If r is a root of

$$t^2 - c_1 t - c_2 = 0, \qquad (7.2.14)$$

then the sequence S defined by $S_n = r^n$ is a solution of (7.2.13).

If a is the sequence defined by (7.2.13),

$$a_0 = C_0, a_1 = C_1, \qquad (7.2.15)$$

and r_1 and r_2 are roots of (7.2.14) with $r_1 \neq r_2$, then there exist constants b and d such that

$$a_n = b r_1^n + d r_2^n, \qquad n = 0, 1, \ldots .$$

PROOF. Since S and T are solutions of (7.2.13),

$$S_n = c_1 S_{n-1} + c_2 S_{n-2}$$

$$T_n = c_1 T_{n-1} + c_2 T_{n-2}.$$

If we multiply the first equation by b and the second by d and add, we obtain

$$U_n = b S_n + d T_n = c_1(b S_{n-1} + d T_{n-1}) + c_2(b S_{n-2} + d T_{n-2})$$

$$= c_1 U_{n-1} + c_2 U_{n-2}.$$

Therefore, U is a solution of (7.2.13).

Since r is a root of (7.2.14),

$$r^2 = c_1 r + c_2.$$

Now

$$c_1 r^{n-1} + c_2 r^{n-2} = r^{n-2}(c_1 r + c_2)$$

$$= r^{n-2} r^2 = r^n;$$

thus the sequence S defined by $S_n = r^n$ is a solution of (7.2.13).

If we set $U_n = b r_1^n + d r_2^n$, then U is a solution of (7.2.13). To meet the initial conditions (7.2.15), we must have

$$U_0 = b + d = C_0$$

$$U_1 = b r_1 + d r_2 = C_1.$$

If we multiply the first equation by r_1 and subtract, we obtain

$$d(r_1 - r_2) = r_1 C_0 - C_1.$$

Since $r_1 - r_2 \neq 0$, we can solve for d. Similarly, we can solve for b. With these choices for b and d, we have

$$U_0 = C_0, \qquad U_1 = C_1.$$

Let a be the sequence defined by (7.2.13) and (7.2.15). Since U also satisfies (7.2.13) and (7.2.15), it follows that $U_n = a_n$, $n = 0, 1, \ldots$. ∎

Theorem 7.2.11 states that any solution of (7.2.13) may be given in terms of two basic solutions r_1^n and r_2^n. However, in case (7.2.14) has two equal roots r, we obtain only one basic solution r^n. Theorem 7.2.13 shows that in this case, nr^n furnishes the other basic solution. Before proving the theorem, we illustrate this case with an example.

Example 7.2.12.   Solve the recurrence relation

$$a_n - 4a_{n-1} - 4a_{n-2} = 0 \tag{7.2.16}$$

subject to the initial conditions

$$a_0 = 1 = a_1.$$

According to Theorem 7.2.11, the sequence S defined by $S_n = r^n$ is a solution of (7.2.16) where r is a solution of

$$t^2 - 4t - 4 = 0. \tag{7.2.17}$$

Thus we obtain the solution S of (7.2.16) defined by

$$S_n = 2^n.$$

Since 2 is the only solution of (7.2.17), the sequence T defined by

$$T_n = n2^n$$

is also a solution of (7.2.16). Thus the solution of (7.2.16) is of the form

$$a_n = bS_n + dT_n = b2^n + dn2^n.$$

We must have

$$a_0 = 1 = a_1.$$

These last equations become

$$b + 0d = 1$$

$$2b + 2d = 1.$$

Solving for b and d, we obtain

$$b = 1, \qquad d = -\tfrac{1}{2}.$$

Therefore, the sequence a is defined by

$$a_n = 2^n - n2^{n-1}.$$

Theorem 7.2.13. *Let*

$$a_n - c_1 a_{n-1} - c_2 a_{n-2} = 0 \qquad (7.2.18)$$

be a second-order linear homogeneous recurrence relation with constant coefficients.

Let a be the sequence satisfying (7.2.18) and

$$a_0 = C_0, \qquad a_1 = C_1.$$

If both roots of

$$t^2 - c_1 t - c_2 = 0 \qquad (7.2.19)$$

are equal to r, then there exist constants b and d such that

$$a_n = br^n + dnr^n, \qquad n = 0, 1, \ldots .$$

PROOF. The proof of Theorem 7.2.11 shows that the sequence S defined by $S_n = r^n$ is a solution of (7.2.18). We show that the sequence T defined by $T_n = nr^n$ is also a solution of (7.2.18).

Since r is the only solution of (7.2.19), we must have

$$t^2 - c_1 t - c_2 = (t - r)^2.$$

It follows that

$$c_1 = 2r, c_2 = -r^2.$$

Now

$$c_1[(n-1)r^{n-1}] + c_2[(n-2)r^{n-2}] = 2r(n-1)r^{n-1}$$
$$- r^2(n-2)r^{n-2}$$
$$= r^n[2(n-1) - (n-2)]$$
$$= nr^n.$$

Therefore, T is a solution of (7.2.18).

By Theorem 7.2.11, the sequence U defined by $U_n = br^n + dnr^n$ is a solution of (7.2.18).

The proof that there are constants b and d such that $U_0 = C_0$ and $U_1 = C_1$ is similar to the argument given in Theorem 7.2.11 and is left as an exercise (Exercise 37). It follows that $U_n = a_n$, $n = 0, 1, \ldots$. ■

EXERCISES

State whether or not each recurrence relation in Exercises 1–10 is a linear homogeneous recurrence relation with constant coefficients. Give the order of each linear homogeneous recurrence relation with constant coefficients.

1H. $a_n = -3a_{n-1}$

2. $a_n = 2na_{n-1}$

3. $a_n = 2na_{n-2} - a_{n-1}$

4H. $a_n = a_{n-1} + n$

5. $a_n = 7a_{n-2} - 6a_{n-3}$

6. $a_n = a_{n-1} + 1 + 2^{n-1}$

7H. $a_n = (\lg 2n)a_{n-1} - [\lg (n - 1)]a_{n-2}$

8. $a_n = 6a_{n-1} - 9a_{n-2}$

9. $a_n = -a_{n-1} - a_{n-2}$

10H. $a_n = -a_{n-1} + 5a_{n-2} - 3a_{n-3}$

In Exercises 11–25, solve the given recurrence relation for the initial conditions given.

11H. Exercise 1; $a_0 = 2$

12. Exercise 2; $a_0 = 1$

13. Exercise 4; $a_0 = 0$

14H. $a_n = 6a_{n-1} - 8a_{n-2}$; $a_0 = 1$, $a_1 = 0$

15. $a_n = 7a_{n-1} - 10a_{n-2}$; $a_0 = 5$, $a_1 = 16$

16. $a_n = 2a_{n-1} + 8a_{n-2}$; $a_0 = 4$, $a_1 = 10$

17H. $2a_n = 7a_{n-1} - 3a_{n-2}$; $a_0 = a_1 = 1$

18. Exercise 6; $a_0 = 0$

19. Exercise 8; $a_0 = a_1 = 1$

20H. $a_n = -8a_{n-1} - 16a_{n-2}$; $a_0 = 2$, $a_1 = -20$

21. $9a_n = 6a_{n-1} - a_{n-2}$; $a_0 = 6$, $a_1 = 5$

22. The recurrence relation that defines the Lucas sequence (Example 7.1.2)

23H. Exercise 43, Section 7.1

24. Exercise 45, Section 7.1

25. The recurrence relation preceding Exercise 46, Section 7.1

Sometimes a nonlinear recurrence relation can be transformed into a linear recurrence relation by making a substitution. In Exercises 26–28, make the given substitution and solve the resulting recurrence relation. Then find the solution to the original recurrence relation.

26H. Solve the recurrence relation

$$\sqrt{a_n} = \sqrt{a_{n-1}} + 2\sqrt{a_{n-2}}$$

with initial conditions $a_0 = a_1 = 1$ by making the substitution $b_n = \sqrt{a_n}$.

27. Solve the recurrence relation

$$a_n = \sqrt{\frac{a_{n-2}}{a_{n-1}}}$$

with initial conditions $a_0 = 8$, $a_1 = 1/2\sqrt{2}$ by taking the logarithm of both sides and making the substitution $b_n = \lg a_n$.

28. Solve the recurrence relation

$$a_n = -2na_{n-1} + 3n(n-1)a_{n-2}$$

with the initial conditions $a_0 = 1$, $a_1 = 2$ by dividing both sides by $n!$ and making the substitution $b_n = a_n/n!$.

*29H. The equation

$$a_n = c_1 a_{n-1} + c_2 a_{n-2} + f(n) \qquad (7.2.20)$$

is called a **second-order linear inhomogeneous recurrence relation with constant coefficients.**

Let $g(n)$ be a solution of (7.2.20). Show that any solution U of (7.2.20) is of the form

$$U_n = V_n + g(n), \qquad (7.2.21)$$

where V is a solution of the homogeneous equation (7.2.12).

If $f(n) = C$ in (7.2.20), it can be shown that $g(n) = C'$ in (7.2.21). Also, if $f(n) = Cn$, $g(n) = C_1'n + C_0'$; if $f(n) = Cn^2$, $g(n) = C_2'n^2 + C_1'n + C_0'$; and if $f(n) = C^n$, $g(n) = C'C^n$. Use these facts together with Exercise 29 to find the general solution of the recurrence relations of Exercises 30–35.

30H. $a_n = 6a_{n-1} - 8a_{n-2} + 3$

31. $a_n = 7a_{n-1} - 10a_{n-2} + 16n$

32. $a_n = 2a_{n-1} + 8a_{n-2} + 81n^2$

33H. $2a_n = 7a_{n-1} - 3a_{n-2} + 2^n$

34. $a_n = -8a_{n-1} - 16a_{n-2} + 3n$

35. $9a_n = 6a_{n-1} - a_{n-2} + 5n^2$

36H. The equation

$$a_n = f(n)a_{n-1} + g(n)a_{n-2} \qquad (7.2.22)$$

is called a **second-order linear homogeneous recurrence relation**. The coefficients $f(n)$ and $g(n)$ are not necessarily constant. Show that if S and T are solutions of (7.2.22), then $bS + dT$ is also a solution of (7.2.22).

37. Suppose that both roots of

$$t^2 - c_1 t - c_2 = 0$$

are equal to r and suppose that a_n satisfies

$$a_n = c_1 a_{n-1} + c_2 a_{n-2},$$
$$a_0 = C_0, \qquad a_1 = C_1.$$

Show that there exist constants b and d such that

$$a_n = br^n + dnr^n, \qquad n = 0, 1, \ldots,$$

thus completing the proof of Theorem 7.2.13.

*38. Let a_n be the common value of the off-diagonal elements of A^n where A is the adjacency matrix of K_5, the complete graph on five vertices. Show that

$$a_n = 3a_{n-1} + 4a_{n-2}$$

and $a_1 = 1$ and $a_2 = 3$.

39H. Solve the recurrence relation of Exercise 38.

40. Let a_n be the minimum number of links required to solve the n-node communication problem (see Exercise 41, Section 7.1). Use iteration to show that $a_n \leq 2n - 4$, $n \geq 4$.

7.3 Recursive Algorithms and Recurrence Relations

A **recursive algorithm** is an algorithm that calls itself. Many higher-level programming languages, such as Pascal, Ada, ALGOL, PL/1, and C, directly support recursion by allowing routines to call themselves. Other languages, such as FORTRAN, do not allow routines to call themselves. As we will see, recursive algorithms and recurrence relations go hand in hand.

Algorithmn 7.3.1 computes the value of A_n, where $\{A_n\}$ is the sequence defined by the recurrence relation

$$A_n = (1.12)A_{n-1} \tag{7.3.1}$$

and the initial condition

$$A_0 = 1000. \tag{7.3.2}$$

A_n is the amount of money at the end of n years assuming an initial amount of \$1000 and an interest rate of 12 percent compounded annually (see Example 7.1.3).

Algorithm 7.3.1 Computing Compound Interest (Recursive Version).

This recursive algorithm computes the amount of money at the end of n years assuming an initial amount of \$1000 and an interest rate of 12 percent compounded annually.

Input: n
Output: *amount*

1. [Case: $n = 0$.] If $n = 0$, execute *amount* $:= 1000$ and stop.
2. [Recursive call.] Use this algorithm to compute $amount_{n-1}$, the amount at the end of $n - 1$ years.
3. [Case n.] *amount* $:= 1.12amount_{n-1}$.

Example 7.3.2. We show how Algorithm 7.3.1 computes the amount for $n = 0$, 1, 2.

In case $n = 0$, at line 1 we simply set *amount* to 1000 (the correct value) and stop.

In case $n = 1$, we skip immediately to line 2. We use Algorithm 7.3.1 to compute the amount at the end of 0 years. In the preceding paragraph, we found that for $n = 0$, Algorithm 7.3.1 computes the value 1000. At line 3, we set *amount* to $(1.12)1000 = 1120$. This is the correct value.

In case $n = 2$, we skip immediately to line 2. We use Algorithm 7.3.1 to compute the amount at the end of 1 year. In the preceding paragraph, we found that for $n = 1$, Algorithm 7.3.1 computes the value 1120. At line 3, we set *amount* to $(1.12)1120 = 1254.40$. This is the correct value.

A recursive algorithm always computes one or more cases explicitly and returns without calling itself. In Algorithm 7.3.1, the explicit case is $n = 0$ and the value 1000 is computed without calling itself. The explicit cases correspond to the initial conditions of a recurrence relation. In Algorithm 7.3.1, the explicit case at line 1 corresponds to the initial condition (7.3.2). In case $n > 0$, in Algorithm 7.3.1 we first compute the amount at the end of $n - 1$ years (line 2). This amount is obtained by recursively calling Algorithm 7.3.1. At line 3, we compute the amount at the end of n years. The recursive call and the calculation of the desired quantity correspond to the recurrence relation (7.3.1).

It is often easy to prove a recursive algorithm correct. We can give a quick proof using mathematical induction that Algorithm 7.3.1 computes the sequence defined by equations (7.3.1) and (7.3.2).

BASIS STEP   ($n = 0$). If $n = 0$, Algorithm 7.3.1 computes the value 1000. Since $A_0 = 1000$, Algorithm 7.3.1 is correct in case $n = 0$.

INDUCTIVE STEP. Suppose that Algorithm 7.3.1 is correct for the case $n - 1$. Consider the case $n > 0$. Since $n > 0$, at step 1 the condition $n = 0$ fails and we proceed to step 2. By the inductive assumption, Algorithm 7.3.1 is correct for $n - 1$; thus after we execute line 2, the variable $amount_{n-1}$ is equal to A_{n-1}. Because $A_n = (1.12)A_{n-1}$, after line 3, *amount* will be equal to A_n. Thus Algorithm 7.3.1 is correct for the case n.

Since the basis step and the inductive step have been verified, by the Principle of Mathematical Induction, Algorithm 7.3.1 is correct for all values of n.

The following nonrecursive algorithm (Algorithm 7.3.3), which also computes the sequence $\{A_n\}$ given by equations (7.3.1) and (7.3.2), is a more straightforward algorithm than Algorithm 7.3.1 and when implemented as a computer program, is generally more efficient than a program based on Algorithm 7.3.1. This does not mean that recursive algorithms are useless. Some problems can only be solved by inherently recursive methods. As we have seen, it is often easy to prove that a recursive algorithm is correct. Having proved that a recursive algorithm is correct, we can then produce a nonrecursive version, if desired, either by hand or automatically and be confident that the nonrecursive version is also correct.

Algorithm 7.3.3 Computing Compound Interest (Nonrecursive Version).

This nonrecursive algorithm computes the amount of money at the end of n years assuming an initial amount of \$1000 and an interest rate of 12 percent compounded annually.

Input: n
Output: *amount*

1. [Initialization.] *amount* := 1000.
2. [Compute the amount after n years.] For $i := 1$ to n, execute line 3.
3.   *amount* := 1.12*amount*

Algorithm 7.3.4, based on the recurrence relation and initial conditions that define the Fibonacci sequence (see Example 2.3.10), computes the Fibonacci sequence.

Algorithm 7.3.4 Computing the Fibonacci Sequence.

This algorithm computes the nth Fibonacci number.

Input: n
Output: f_n

1. [Case: $n = 0$.] If $n = 0$, execute $f_0 := 1$ and stop.
2. [Case: $n = 1$.] If $n = 1$, execute $f_1 := 1$ and stop.
3. [Recursive call.] Use this algorithm to compute f_{n-2} and f_{n-1}.
4. [Case n.] $f_n := f_{n-2} + f_{n-1}$.

We close this section by writing a recursive algorithm to solve the Tower of Hanoi puzzle. This algorithm is a direct translation of the method of Example 7.1.6.

Algorithm 7.3.5 Solving the Tower of Hanoi Puzzle.   This algorithm prints the moves required to solve the Tower of Hanoi puzzle (see Example 7.1.6). The input to the algorithm consists of the names a, b, and c of the poles and the number n of disks to be moved. The disks are moved from the first referenced pole to the second referenced pole using the third referenced pole as the odd pole.

Input: a, b, c, n
Output: Moves to solve the n-disk Tower of Hanoi puzzle

1. [Case: $n = 1$.] If $n = 1$, output: "Move disk from pole" a "to pole" b and stop.
2. [First recursive call.] Call this algorithm with input: a, c, b, $n - 1$.
3. [Move one disk.] Output: "Move disk from pole" a "to pole" b.
4. [Second recursive call.] Call this algorithm with input: c, b, a, $n - 1$.

Example 7.3.6.   We show how Algorithm 7.3.5 executes in case $n = 3$. Since $n \neq 1$, we move immediately to line 2. When we call Algorithm 7.3.5 with input a, c, b, 2, we solve the two-disk problem of moving two disks from pole a to pole c using pole b as the odd pole. The output will be

Move disk from pole a to pole b.
Move disk from pole a to pole c.
Move disk from pole b to pole c.

At line 3, we output:

Move disk from pole a to pole b.

When we call Algorithm 7.3.5 with input c, b, a, 2, we solve the two-disk problem of moving two disks from pole c to pole b using pole a as the odd pole. The output will be:

Move disk from pole c to pole a.
Move disk from pole c to pole b.
Move disk from pole a to pole b.

EXERCISES

In Exercises 1–9, write a recursive algorithm to compute the sequence.

1H. $a_n = a_{n-1} + 3, a_1 = 2$

2. $c_n = 2c_{n-1} + 1, c_1 = 1$

3. $a_n = 3a_{n-1}a_{n-2}, a_0 = 1, a_1 = 2$

4H. $a_n = 3a_{n-1} - 2a_{n-2} + 5a_{n-3}, a_0 = 1, a_1 = 2, a_2 = 3$

5. $a_n = 5a_{n-1} + 2a_{n-2}, a_0 = 2, a_1 = 1$

6. $a_n = a_{n-1} + 6a_{n-2} + 2n - 1, a_0 = a_1 = 1$

7H. Lucas sequence (Example 7.1.2)

8. $C_1, C_2, \ldots$, where C_n is the nth Catalan number (see Example 7.1.8)

9. Ackermann's function (see Exercises 58–61, Section 7.1)

10H. Suppose that we have n dollars and that each day we buy orange juice (1\$), milk (\$2), or beer (\$2). Write a recursive algorithm that lists all the ways of spending all the money, order being taken into account.

11. Write a recursive algorithm that prints the moves to solve the four-peg Tower of Hanoi puzzle (see Exercise 40, Section 7.1). Your solution should require fewer moves than the solution presented to the three-peg Tower of Hanoi puzzle.

12. Write a recursive algorithm that prints a solution to the n-node communication problem (see Exercise 41, Section 7.1).

13H. Write a recursive algorithm that generates all r-combinations of an n-element set.

14. Write a recursive algorithm that generates all permutations of an n-element set.

15. Write a recursive algorithm based on Theorem 5.3.8 that computes the greatest common divisor of two nonnegative integers a and b, not both zero.

16H. Write a recursive version of the binary search algorithm (Algorithm 5.4.15).

†7.4 Applications to the Analysis of Algorithms

In Section 7.3 we began with a recurrence relation and initial conditions for some sequence and then developed an algorithm to compute the sequence. In this section the process is reversed. We begin with an algorithm and develop a recurrence relation and initial conditions that define a sequence $\{a_n\}$, where a_n is the worst-case or best-case time required for the algorithm to execute an input of size n. By solving the recurrence relation, we can determine the time needed by the algorithm.

Our first algorithm is a version of the selection sorting algorithm. This algorithm is based on the idea of selecting the largest item, placing it last; then recursively repeating this process.

Algorithm 7.4.1 Selection Sort.   This algorithm sorts the array

$$s(1), \ldots, s(n)$$

in increasing order by first selecting the largest item and placing it last and then recursively sorting the remaining elements.

```
procedure selection_sort (s, n)
    {sort the elements s(1), . . . , s(n) is increasing order}
    {if n = 1 there is nothing to do, so simply return}
    if n = 1 then
        return
    {nontrivial case; first find the largest element}
    max := s(1)
    max_index := 1
    for i := 2 to n do
1.      if s(i) > max then
            {found a larger element, so update max and max_index}
            begin
            max := s(i)
```

†This section can be omitted without loss of continuity. A prerequisite for this section is Section 5.2.

$$max\_index := i$$
end
{move the largest element to the end}
$$temp := s(n)$$
$$s(n) := s(max\_index)$$
$$s(max\_index) := temp$$
{recursively sort $s(1), \ldots, s(n-1)$}
2.   **call** $selection\_sort\ (s, n-1)$
end $selection\_sort$

As a measure of the time required by this algorithm, we will count the number of comparisons b_n at line 1 required to sort n items. We see that

$$b_1 = 0.$$

If $n > 1$, $n - 1$ comparisons are required at line 1. By the definition of b_{n-1}, an additional b_{n-1} comparisons are required at line 2. Therefore,

$$b_n = b_{n-1} + n - 1.$$

This recurrence relation can be solved by iteration:

$$b_n = b_{n-1} + n - 1$$

$$= (b_{n-2} + n - 2) + (n - 1)$$

$$= (b_{n-3} + n - 3) + (n - 2) + (n - 1)$$

$$\vdots$$

$$= b_1 + 1 + 2 + \cdots + (n - 2) + (n - 1)$$

$$= 0 + 1 + 2 + \cdots + (n - 1) = \frac{(n-1)n}{2} = O(n^2).$$

Thus the time required by Algorithm 7.4.1 is $O(n^2)$.

Example 7.4.2.   Find a recurrence relation for the time b_n required by the binary search algorithm (Algorithm 5.4.15) in the worst case for input of size n. Define b_n to be the number of times the test in the while loop is true. In case $n = 1$, the while loop test is true only the first time. Thus

$$b_1 = 1. \tag{7.4.1}$$

Assume that $n > 1$. The first time we enter the while loop, the test is true. Later in the loop, either *left* or *right* are reset so that the next time we enter the while loop, we are effectively looking at an array of size $\lfloor n/2 \rfloor$ in the worst case. By assumption, an array of size $\lfloor n/2 \rfloor$ requires $b_{\lfloor n/2 \rfloor}$ iterations of the while loop in the worst case. Therefore,

$$b_n = 1 + b_{\lfloor n/2 \rfloor}. \qquad (7.4.2)$$

In case n is a power of 2, we can solve recurrence relation (7.4.2) explicitly by iteration. To simplify the notation, set $n = 2^k$ and $a_k = b_{2^k}$. The recurrence relation (7.4.2) becomes

$$a_k = 1 + a_{k-1} \qquad (7.4.3)$$

and the initial condition (7.4.1) becomes

$$a_0 = 1.$$

Solving (7.4.3) by iteration, we obtain

$$a_k = 1 + a_{k-1}$$
$$= 1 + (1 + a_{k-2})$$
$$= 2 + a_{k-2}$$
$$= 2 + (1 + a_{k-3})$$
$$= 3 + a_{k-3}$$
$$\vdots$$
$$= k + a_{k-k} = k + a_0 = k + 1.$$

Since $n = 2^k$, $k = \lg n$, and we have

$$b_n = b_{2^k} = a_k = 1 + k = 1 + \lg n.$$

We conclude that if n is a power of 2, in the worst case the binary search algorithm is $O(\lg n)$. By using induction, it can be proved that $b_n = 1 + \lfloor \lg n \rfloor$ for any n (not necessarily a power of 2). Thus, in the worst case, the binary search algorithm is $O(\lg n)$.

EXERCISES

Exercises 1–4 refer to the following algorithm.

Algorithm 7.4.3 Computing an Exponential. This algorithm computes a^n recursively.

```
procedure exp1(a, n)
    if n = 1 then
        return(a)
    m := ⌊n/2⌋
    temp1 := exp1(a, m)
    temp2 := exp1(a, n − m)
1.   return(temp1*temp2)
    end exp1
```

Let b_n be the number of multiplications (line 1) required to compute a^n.

1H. Explain how Algorithm 7.4.3 computes a^n.

2H. Find a recurrence relation and initial conditions for the sequence $\{b_n\}$.

3H. Compute b_2, b_3, and b_4.

4H. Solve the recurrence relation of Exercise 2 in case n is a power of 2.

Exercises 5–8 refer to the following algorithm.

Algorithm 7.4.4 Computing an Exponential. This algorithm computes a^n recursively.

```
procedure exp2(a, n)
    if n = 1 then
        return(a)
    m := ⌊n/2⌋
    x := exp2(a, m)
1.   x := x*x
    if n is even then
        return(x)
2.   return(a*x)
    end exp2
```

Let b_n be the number of multiplications (lines 1 and 2) required to compute a^n.

5. Explain how Algorithm 7.4.4 computes a^n.

6. Show that

$$b_n = \begin{cases} b_{(n-1)/2} + 2 & \text{if } n \text{ is odd;} \\ b_{n/2} + 1 & \text{if } n \text{ is even.} \end{cases}$$

7. Find b_1, b_2, b_3, and b_4.

8. Solve the recurrence relation of Exercise 6 in case n is a power of 2.

Exercises 9–12 refer to the following algorithm.

Algorithm 7.4.5 Finding the Largest and Smallest Elements in an Array. This recursive algorithm finds the largest and smallest elements in the array

$$s(i), \ldots, s(j).$$

The largest element is returned in *large* and the smallest in *small*.

```
     procedure large_small(s, i, j, large, small)
        if i = j then
           begin
           large := s(i)
           small := s(i)
           return
1.         end
2.      m := ⌊(i + j)/2⌋
        call large_small(s, i, m, large_left, small_left)
        call large_small(s, m + 1, j, large_right, small_right)
3.      if large_left > large_right then
           large := large_left
        else
           large := large_right
4.      if small_left > small_right then
           small := small_right
        else
           small := small_left
     end large_small
```

Let b_n be the number of comparisons (lines 3 and 4) required for an input of size n.

9. Show that $b_1 = 0$ and $b_2 = 2$.

10. Find b_3.

11. Establish the recurrence relation

$$b_{\lfloor n/2 \rfloor} + b_{\lfloor (n+1)/2 \rfloor} + 2 = b_n \qquad (7.4.4)$$

for $n > 1$.

12. Solve the recurrence relation (7.4.4) in case n is a power of 2 to obtain

$$b_n = 2n - 2, \qquad n = 1, 2, 4, \ldots .$$

Exercises 13–16 refer to Algorithm 7.4.5 with the following inserted after line 1.

```
        if j = i + 1 then
            begin
1a.         if s(i) > s(j) then
                begin
                large := s(i)
                small := s(j)
                end
            else
                begin
                small := s(i)
                large := s(j)
                end
            return
            end
```

Let b_n be the number of comparisons (lines 1a, 3, and 4) for an input of size n.

13H. Show that $b_1 = 0$ and $b_2 = 1$.

14H. Compute b_3 and b_4.

15H. Show that the recurrence relation (7.4.4) holds for $n > 2$.

16H. Solve the recurrence relation (7.4.4) in case n is a power of 2 to obtain

$$b_n = \tfrac{3}{2}n - 2, \qquad n = 2, 4, 8, \ldots .$$

Exercises 17–21 refer to the following algorithm.

Algorithm 7.4.6 Insertion Sort.   This algorithm sorts the array

$$s(1), \ldots, s(n)$$

in increasing order by recursively sorting the first $n - 1$ elements and then inserting $s(n)$ in the correct position.

```
procedure insert_sort(s, n)
   if n = 1 then
      return
   call insert_sort(s, n - 1)
   i := n - 1
   temp := s(n)
   while i ≥ 1 and s(i) > temp do
      begin
         s(i + 1) := s(i)
         i := i - 1
      end
   s(i + 1) := temp
end insert_sort
```

Let b_n be the number of times the test $s(i) > temp$ is made in the worst case.

17. Explain how *insert_sort* sorts.
18. Which input produces the worst case behavior for *insert_sort*?
19. Find b_1, b_2, and b_3.
20. Find a recurrence relation for the sequence $\{b_n\}$.
21. Solve the recurrence relation of Exercise 20.

Exercises 22–27 refer to an algorithm that accepts as input the array

$$s(i), \ldots, s(j).$$

If $j > i$, the subproblems

$$s(i), \ldots, s\left(\left\lfloor \frac{i + j}{2} \right\rfloor\right) \text{ and } s\left(\left\lfloor \frac{i + j}{2} \right\rfloor + 1\right), \ldots, s(j)$$

are solved recursively. Solutions to subproblems of sizes m and k can be combined in time $c_{m,k}$ to solve the original problem. Let b_n be the time required by the algorithm for an input of size n.

22H. Write a recurrence relation for b_n assuming that $c_{m,k} = 3$.

23. Write a recurrence relation for b_n assuming that $c_{m,k} = m + k$.

24. Solve the recurrence relation of Exercise 22 assuming that $b_1 = 0$ in case n is a power of 2.

25H. Solve the recurrence relation of Exercise 22 assuming that $b_1 = 1$ in case n is a power of 2.

26. Solve the recurrence relation of Exercise 23 assuming that $b_1 = 0$ in case n is a power of 2.

27. Solve the recurrence relation of Exercise 23 assuming that $b_1 = 1$ in case n is a power of 2.

An alternative form of mathematical induction, known as the **Strong Form of Mathematical Induction**, can be used to prove the statements of Exercises 28–34. The difference between the Strong Form of Mathematical Induction and the form of mathematical induction discussed in Section 2.4 is that when using the strong form, at the inductive step we can assume not only that the preceding case is true, but that *all* preceding cases are true.

Strong Form of Mathematical Induction.   *Suppose that for each positive integer k we have a statement S(k) which is either true or false. Suppose that*

> *S(1) is true;*
> *if S(k) is true for all k < n, then S(n) is true.*

Then S(k) is true for every positive integer k.

28H. $b_n = n - 1$, $n \geq 1$; b_n as in Exercises 1–4

*29. $b_n \leq 2 \lg n$, $n \geq 1$; b_n as in Exercises 5–8

30. $b_n = 2n - 2$, $n \geq 1$; b_n as in Exercises 9–12

31H. $b_n \leq 1 + \lg n$, $n \geq 1$; b_n as in Example 7.4.2

*32. $b_n = \lfloor 1 + \lg n \rfloor$, $n \geq 1$; b_n as in Example 7.4.2

*33. Modify Algorithm 7.4.5 by inserting the lines preceding Exercise 13 after line 1 and replacing line 2 with the following lines.

> **if** $j - i$ is odd **and** $(1 + j - i)/2$ is odd **then**
> $\quad m := \lfloor (i + j)/2 \rfloor - 1$
> **else**
> $\quad m := \lfloor (i + j)/2 \rfloor$

Show that in the worst case, this modified algorithm requires at most $\lceil (3n/2) - 2 \rceil$ comparisons to find the largest and smallest elements in an array of size n.

*34H. Use the notation of Exercises 22–27. Assume that if $m_1 \geq m_2$ and $k_1 \geq k_2$, then $c_{m_1,k_1} \geq c_{m_2,k_2}$. Show that the sequence $b_1, b_2, \ldots$ is increasing.

*35. Use the notation of Exercises 22–27. Assuming that $c_{m,k} = m + k$ and $b_1 = 0$, show that $b_n \leq 4n \lg n$.

7.5 Notes

Recurrence relations are treated more fully in [Bogart; Liu, 1985; Roberts; and Tucker]. Several applications to the analysis of algorithms are presented in [Horowitz, 1978].

[Cull] gives algorithms for solving certain Tower of Hanoi problems with minimum space and time complexity.

[Standish] gives a proof using generating functions that the number of binary trees having n vertices is equal to the nth Catalan number. An elementary proof (i.e., a proof not using generating functions) of this result appears in [Kruse].

Recurrence relations are also called **difference equations**. [Goldberg] contains a discussion of difference equations and applications.

Computer Exercises

1. Write a program that prints the amount accumulated yearly if a person invests n dollars at p percent compounded annually.
2. Write a program that prints the amount accumulated yearly if a person invests n dollars at p percent annual interest compounded m times yearly.
3. Write a program to compute the Lucas sequence.
4. Write a program that solves the three-peg Tower of Hanoi puzzle.
5. Write a program that solves the four-peg Tower of Hanoi puzzle.

6. Write a program that describes all binary trees having n vertices.

7. Write a program that prints a solution to the n-node communication problem (see Exercise 41, Section 7.1).

8. Write a program that prints all the ways to spend n dollars under the conditions of Exercise 43, Section 7.1.

9. Write a program that prints all n-bit strings that do not contain the pattern 010.

10. Implement Algorithm 7.4.1 and other sorting algorithms as programs and compare the times needed to sort n items.

11. Implement a computation of a^n nonrecursively that uses repeated multiplication and Algorithms 7.4.3 and 7.4.4 as computer programs, and compare the times needed to execute each.

12. Implement methods of computing the largest and smallest elements in an array (see Exercises 9–16 and 33, Section 7.4) and compare the times needed to execute each.

Chapter Review

SECTION 7.1
Recurrence relation
Initial condition
Lucas sequence
Compound interest
Tower of Hanoi
Recursive construction of binary trees
nth Catalan number $= C(2n, n)/(n + 1) =$ number of binary trees having n vertices

SECTION 7.2
Solving a recurrence relation by iteration
nth-order linear homogeneous recurrence relation with constant coefficients and how to solve a second-order relation

SECTION 7.3
Recursive algorithm
Relationship among a recursive algorithm, induction, and a recurrence relation

SECTION 7.4
How to find a recurrence relation that describes the time required by a recursive algorithm

Chapter Self-Test

SECTION 7.1

1H. Answer parts (a)–(c) for the sequence defined by the rules:

1. The first term is 3.
2. The nth term is n plus the previous term.

   (a) Write the first four terms of the sequence.
   (b) Find an initial condition for the sequence.
   (c) Find a recurrence relation for the sequence.

2H. Assume that a person invests \$4000 at 17 percent compounded annually. Let A_n represent the amount at the end of n years. Find a recurrence relation and an initial condition for the sequence $A_0, A_1, \ldots$.

3H. How many binary trees have six vertices?

4H. Suppose that we have a $2 \times n$ rectangular board divided into $2n$ squares. Let a_n denote the number of ways to exactly cover this board by 1×2 dominoes. Show that the sequence $\{a_n\}$ satisfies the recurrence relation

$$a_n = a_{n-1} + a_{n-2}.$$

Show that $a_n = f_n$, where $\{f_n\}$ is the Fibonacci sequence.

SECTION 7.2

5H. Is the recurrence relation

$$a_n = a_{n-1} + a_{n-3}$$

a linear homogeneous recurrence relation with constant coefficients?

In Exercises 6–8, solve the recurrence relation subject to the initial conditions.

6H. $a_n = 2a_{n-1} + 3; a_0 = 1$

7H. $a_n = -4a_{n-1} - 4a_{n-2}; a_0 = 2, a_1 = 4$

8H. $a_n = 3a_{n-1} + 10a_{n-2}; a_0 = 4, a_1 = 13$

SECTION 7.3

9H. Write a recursive algorithm to compute the sequence of Exercise 7.

10H. Write a recursive algorithm to compute the sequence defined by

$$a_n = 6a_{n-1} - 4a_{n-3} + a_{n-4},$$
$$a_0 = a_1 = 1, \qquad a_2 = a_3 = 0.$$

11H. Suppose that we have n dollars and that each day we buy a disk ($1), a printer cover ($3), or computer paper ($5). Write a recursive algorithm that lists all the ways of spending all the money, order being taken into account.

12H. Write a recursive algorithm to compute the number of binary trees having n vertices.

SECTION 7.4

Exercises 13–16 refer to the following algorithm.

Algorithm: Polynomial Evaluation. This algorithm evaluates the polynomial

$$p(x) = \sum_{k=0}^{n} c(k)x^{n-k}$$

at the point t.

procedure *poly_eval*(c, n, t)
   if $n = 0$ **then**
      return($c(0)$)
   $y = $ *poly_eval*(c, $n - 1$, t)
   $y = y*t + c(n)$
   return(y)
end *poly_eval*

Let b_n be the number of multiplications required to compute $p(t)$.

13H. Find a recurrence relation and an initial condition for the sequence $\{b_n\}$.

14H. Compute b_1, b_2, and b_3.

15H. Solve the recurrence relation of Exercise 13.

16H. Suppose that we compute $p(t)$ by a straightforward technique that requires $n - k$ multiplications to compute $c(k)t^{n-k}$. How many multiplications would be required to compute $p(t)$? Would you prefer this method or the algorithm above? Explain.

8

That's not fair, Vinnie. When I talk about *my* relatives, I *criticize* them.

—from *Life With Father*

Relations

A relation is a generalization of a function. In Section 2.5 we found that a function associates a *unique* element of its range with each element of its domain. A relation may associate *several* elements with a single element of its domain. Section 8.1 gives the fundamental terminology of relations. In Section 8.2 we examine equivalence relations that form an important subclass of relations. Matrices are useful for representing relations and for working with relations as we will see in Section 8.3. The mathematical concept of relation is the theoretical basis of the relational data base that we discuss briefly in Section 8.4.

8.1 Introduction

A **relation** can be thought of as a table that lists the relationship of elements to other elements (see Table 8.1.1). Table 8.1.1 tells which students are taking which courses. For example, Bill is taking Computer Science and Art, and Mary is taking Mathematics. In the terminology of relations, we would say that Bill is related to Computer Science and Art, and that Mary is related to Mathematics.

Of course, Table 8.1.1 is really just a set of ordered pairs. Abstractly, we *define* a relation to be a set of ordered pairs. In this setting we consider the first element of the ordered pair to be related to the second element of the ordered pair.

Table 8.1.1

Student	*Course*
Bill	CompSci
Mary	Math
Bill	Art
Beth	History
Beth	CompSci
Dave	Math

Definition 8.1.1. A *(binary) relation R from a set X to a set Y* is a set of ordered pairs (x, y) where $x \in X$ and $y \in Y$. If $(x, y) \in R$, we write xRy and say that *x is related to y.* In case $X = Y$, we call R a *(binary) relation on X.*

The set

$$\{x \in X \mid (x, y) \in R \text{ for some } y \in Y\}$$

is called the *domain* of R. The set

$$\{y \in Y \mid (x, y) \in R \text{ for some } x \in X\}$$

is called the *range* of R.

If a relation is given as a table, the domain consists of the members of the first column and the range consists of the members of the second column.

Example 8.1.2. If we let

$$X = \{\text{Bill, Mary, Beth, Dave}\}$$

and

$$Y = \{\text{CompSci, Math, Art, History}\},$$

our relation R of Table 8.1.1 can be written

$$R = \{(\text{Bill, CompSci}), (\text{Mary, Math}), (\text{Bill, Art}),$$
$$(\text{Beth, History}), (\text{Beth, CompSci}), (\text{Dave, Math})\}.$$

Since (Beth, History) $\in R$, we may write Beth R History. The domain (first column) of R is the set X and the range (second column) of R is the set Y.

Example 8.1.2 shows that a relation can be given simply by specifying which ordered pairs belong to the relation. A function f from X to Y (see Section 2.5) is a relation from X to Y having the properties:

(a) The domain of f is equal to X.
(b) For each $x \in X$, there is exactly one $y \in Y$ such that $(x, y) \in f$.

Our next example shows that sometimes it is possible to define a relation by giving a rule for membership in the relation.

Example 8.1.3.   Let

$$X = \{2, 3, 4\} \quad \text{and} \quad Y = \{3, 4, 5, 6, 7\}.$$

If we define a relation R from X to Y by

$$(x, y) \in R \text{ if } x \text{ divides } y,$$

we obtain

$$R = \{(2, 4), (2, 6), (3, 3), (3, 6), (4, 4)\}.$$

If we rewrite R as a table, we obtain

X	Y
2	4
2	6
3	3
3	6
4	4

The domain of R is the set $\{2, 3, 4\}$ and the range of R is the set $\{3, 4, 6\}$.

Example 8.1.4.   Let R be the relation on $X = \{1, 2, 3, 4\}$ defined by $(x, y) \in R$ if $x \le y$, $x, y \in X$. Then

$$R = \{(1, 1), (1, 2), (1, 3), (1, 4), (2, 2),$$
$$(2, 3), (2, 4), (3, 3), (3, 4), (4, 4)\}.$$

The domain and range of R are both equal to X.

An informative way to picture a relation on a set is to draw its digraph. To draw the digraph of a relation on a set X, we first draw vertices to represent the elements of X. In Figure 8.1.1, we have drawn four vertices to represent the elements of the set X of Example 8.1.4. Next, if the element (x, y) is in the relation, we draw a directed edge from x to y. In Figure 8.1.1, we have drawn directed edges to represent the members of the relation R of Example 8.1.4. Notice that an element of the form (x, x) in a relation corresponds to a loop. There is a loop at every vertex in Figure 8.1.1.

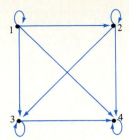

Figure 8.1.1

Example 8.1.5.   The relation R on $X = \{a, b, c, d\}$ given by the digraph of Figure 8.1.2 is

$$R = \{(a, a), (b, c), (c, b), (d, d)\}.$$

We next define several useful properties that some relations have.

Figure 8.1.2

Definition 8.1.6.   A relation R on a set X is called *reflexive* if $(x, x) \in R$ for every $x \in X$.

Example 8.1.7.   The relation R on $X = \{1, 2, 3, 4\}$ of Example 8.1.4 is reflexive because for each element $x \in X$, $(x, x) \in R$; specifically, $(1, 1)$, $(2, 2)$, $(3, 3)$, and $(4, 4)$ are each in R. The digraph of a reflexive relation has a loop at every vertex. Notice that the digraph of the reflexive relation of Example 8.1.4 (see Figure 8.1.1) has a loop at every vertex.

Example 8.1.8.   The relation R on $X = \{a, b, c, d\}$ of Example 8.1.5 is not reflexive. For example $b \in X$, but $(b, b) \notin R$. That this relation is not reflexive can also be seen by looking at its digraph (see Figure 8.1.2); vertex b does not have a loop.

Definition 8.1.9.   A relation R on a set A is called *symmetric* if for all $(x, y) \in R$, we have $(y, x) \in R$.

Example 8.1.10.   The relation R of Example 8.1.5 is symmetric because for all $(x, y) \in R$, we have $(y, x) \in R$. For example, (b, c) is in R and (c, b) is also in R. The digraph of a symmetric relation has the property that whenever there is a directed edge from v to w, there is also a directed edge from w to v. Notice that the digraph of the relation of Example 8.1.5 (see Figure 8.1.2) has the property that for every directed edge from v to w, there is also a directed edge from w to v.

Example 8.1.11.   The relation R of Example 8.1.4 is not symmetric. For example, $(2, 3) \in R$, but $(3, 2) \notin R$. The digraph of this relation (see Figure 8.1.1) has a directed edge from 2 to 3, but there is no directed edge from 3 to 2.

Definition 8.1.12.   A relation R on a set A is called *antisymmetric* if for all $(x, y) \in R$ with $x \neq y$, we have $(y, x) \notin R$.

Example 8.1.13.   The relation R of Example 8.1.4 is antisymmetric because for all $(x, y) \in R$ with $x \neq y$, we have $(y, x) \notin R$. For example, $(1, 2) \in R$, but $(2, 1) \notin R$. The digraph of an antisymmetric relation has the property that between any two vertices there is at most one directed edge. Notice that the digraph of the relation of Example 8.1.4 (see Figure 8.1.1) has at most one directed edge between each pair of vertices.

Example 8.1.14.   The relation R of Example 8.1.5 is not antisymmetric because both (b, c) and (c, b) are in R. Notice that in the digraph of the relation of Example 8.1.5 (see Figure 8.1.2) there are two directed edges between b and c.

Example 8.1.15. If a relation R has no members of the form (x, y) with $x \neq y$, then R is antisymmetric; Definition 8.1.12 is said to be vacuously satisfied. For example, the relation

$$R = \{(a, a), (b, b), (c, c)\}$$

on $X = \{a, b, c\}$ is antisymmetric. The digraph of R shown in Figure 8.1.3 has at most one directed edge between each pair of vertices. Notice that R is also reflexive and symmetric. This example shows that "antisymmetric" is not the same as "not symmetric."

Figure 8.1.3

Definition 8.1.16. A relation R on a set X is called *transitive* if for all (x, y), $(y, z) \in R$, we have $(x, z) \in R$.

Example 8.1.17. The relation R of Example 8.1.4 is transitive because for all (x, y), $(y, z) \in R$, we have $(x, z) \in R$. To formally verify that this relation satisfies Definition 8.1.16, we would have to list all pairs of pairs of the form (x, y) and (y, z) and then verify that in every case, $(x, z) \in R$:

Pairs of Form (x, y), (y, z)		(x, z)
$(1, 1)$	$(1, 1)$	$(1, 1)$
$(1, 1)$	$(1, 2)$	$(1, 2)$
$(1, 1)$	$(1, 3)$	$(1, 3)$
$(1, 1)$	$(1, 4)$	$(1, 4)$
$(1, 2)$	$(2, 2)$	$(1, 2)$
$(1, 2)$	$(2, 3)$	$(1, 3)$
$(1, 2)$	$(2, 4)$	$(1, 4)$
$(1, 3)$	$(3, 3)$	$(1, 3)$
$(1, 3)$	$(3, 4)$	$(1, 4)$
$(1, 4)$	$(4, 4)$	$(1, 4)$
$(2, 2)$	$(2, 2)$	$(2, 2)$
$(2, 2)$	$(2, 3)$	$(2, 3)$

Pairs of Form (x, y), (y, z)		(x, z)
(2, 2)	(2, 4)	(2, 4)
(2, 3)	(3, 3)	(2, 3)
(2, 3)	(3, 4)	(2, 4)
(2, 4)	(4, 4)	(2, 4)
(3, 3)	(3, 3)	(3, 3)
(3, 3)	(3, 4)	(3, 4)
(3, 4)	(4, 4)	(3, 4)
(4, 4)	(4, 4)	(4, 4)

In determining whether a relation R is transitive directly from Definition 8.1.16, in case $x = y$ or $y = z$ we need not explicitly verify that the condition,

$$\text{if } (x, y) \text{ and } (y, z) \in R, \quad \text{then } (x, z) \in R,$$

is satisfied since it will automatically be true. Suppose, for example, that $x = y$ and (x, y) and (y, z) are in R. Since $x = y$, $(x, z) = (y, z)$ is in R and the condition is satisfied. Eliminating the cases $x = y$ and $y = z$ leaves only the following to be explicitly checked to verify that the relation of Example 8.1.4 is transitive:

Pairs of Form (x, y), (y, z)		(x, z)
(1, 2)	(2, 3)	(1, 3)
(1, 2)	(2, 4)	(1, 4)
(1, 3)	(3, 4)	(1, 4)
(2, 3)	(3, 4)	(2, 4)

The digraph of a transitive relation has the property that whenever there are directed edges from x to y and from y to z, there is also a directed edge from x to z. Notice that the digraph of the relation of Example 8.1.4 (see Figure 8.1.1) has this property.

Example 8.1.18. The relation of Example 8.1.5 is not transitive. For example, (b, c) and (c, b) are in R, but (b, b) is not in R. Notice that in the digraph of the relation of Example 8.1.5 (see Figure 8.1.2) there are directed edges from b to c and from c to b, but there is no directed edge from b to b.

Definition 8.1.19. A relation R on a set X is called a *partial order* if R is reflexive, antisymmetric, and transitive.

Example 8.1.20. The relations of Examples 8.1.4 and 8.1.15 are partial orders. The relation of Example 8.1.5 is not a partial order.

Given a relation R from X to Y we may define a relation from Y to X by reversing the order of each ordered pair in R. The formal definition follows.

Definition 8.1.21. Let R be a relation from X to Y. The *converse* of R, denoted R^{-1}, is the relation from Y to X defined by

$$R^{-1} = \{(y, x) \mid (x, y) \in R\}.$$

Example 8.1.22. The converse of the relation R of Example 8.1.3 is

$$R^{-1} = \{(4, 2), (6, 2), (3, 3), (6, 3), (4, 4)\}.$$

In words, we might describe this relation as "is divisible by."

In case the relation is a one-to-one, onto function, the converse relation is simply the inverse function.

If we have a relation R_1 from X to Y and a relation R_2 from Y to Z, we can form a relation from X to Z by applying first relation R_1 and then relation R_2. The resulting relation is denoted $R_2 \circ R_1$. Notice the order in which the relations are written. The formal definition follows.

Definition 8.1.23. Let R_1 be a relation from X to Y and R_2 be a relation from Y to Z. The *composition* of R_1 and R_2, denoted $R_2 \circ R_1$, is the relation from X to Z defined by

$$R_2 \circ R_1 = \{(x, z) \mid (x, y) \in R_1 \text{ and } (y, z) \in R_2 \text{ for some } y \in Y\}.$$

In case the relations are functions, composition of relations is simply composition of functions.

Example 8.1.24. The composition of the relations

$$R_1 = \{(1, 2), (1, 6), (2, 4), (3, 4), (3, 6), (3, 8)\}$$

and

$$R_2 = \{(2, u), (4, s), (4, t), (6, t), (8, u)\}$$

is

$$R_2 \circ R_1 = \{(1, u), (1, t), (2, s), (2, t), (3, s), (3, t), (3, u)\}.$$

EXERCISES

In Exercises 1–4, write the relation as a set of ordered pairs.

1H.

8840	Hammer
9921	Pliers
452	Paint
2207	Carpet

2.

a	3
b	1
b	4
c	1

3.

Sally	Math
Ruth	Physics
Sam	Econ

4H.

a	a
b	b

In Exercises 5–8, write the relation as a table.

5H. $R = \{(a, 6), (b, 2), (a, 1), (c, 1)\}$

6. $R = \{(\text{Roger, Music}), (\text{Pat, History}), (\text{Ben, Math}), (\text{Pat, PolySci})\}$.

7. The relation R on $\{1, 2, 3, 4\}$ defined by $(x, y) \in R$ if $x^2 \geq y$.

8H. The relation R from the set X of states whose names begin with the letter "M" to the set Y of cities defined by $(S, C) \in X \times Y$ if C is the capital of S.

In Exercises 9–12, draw the digraph of the relation.

9H. The relation of Exercise 4 on $\{a, b, c\}$

10. The relation $R = \{(1, 2), (2, 1), (3, 3), (1, 1), (2, 2)\}$ on $X = \{1, 2, 3\}$

11. The relation $R = \{(1, 2), (2, 3), (3, 4), (4, 1)\}$ on $\{1, 2, 3, 4\}$

12H. The relation of Exercise 7

In Exercises 13–16, write the relation as a set of ordered pairs.

13H.   **14.**

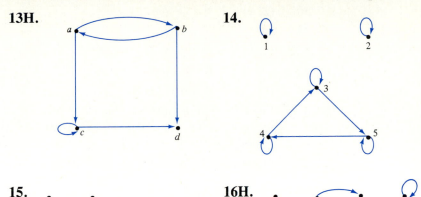

15.   **16H.**

17H. Find the domain and range of each relation in Exercises 1–16.

18.   Find the converse (as a set of ordered pairs) of each relation in Exercises 1–16.

Exercises 19–24 refer to the relation R on the set $\{1, 2, 3, 4, 5\}$ defined by the rule $(x, y) \in R$ if 3 divides $x - y$.

19H. List the elements of R.

20.   List the elements of R^{-1}.

21.   Find the domain of R.

22H. Find the range of R.

23.   Find the domain of R^{-1}.

24.   Find the range of R^{-1}.

25H. Repeat Exercises 19–24 for the relation R on the set $\{1, 2, 3, 4, 5\}$ defined by the rule $(x, y) \in R$ if $x + y \leq 6$.

26.   Repeat Exercises 19–24 for the relation R on the set $\{1, 2, 3, 4, 5\}$ defined by the rule $(x, y) \in R$ if $x = y - 1$.

27.   Is the relation of Exercise 25 reflexive; symmetric; antisymmetric; transitive; a partial order?

28H. Is the relation of Exercise 26 reflexive; symmetric; antisymmetric; transitive; a partial order?

In Exercises 29–34, determine whether each relation defined on the set of positive integers is reflexive, symmetric, antisymmetric, transitive, or a partial order.

29H. $(x, y) \in R$ if $x = y^2$

30. $(x, y) \in R$ if $x > y$

31. $(x, y) \in R$ if $x \geq y$

32H. $(x, y) \in R$ if $x = y$

33. $(x, y) \in R$ if the greatest common divisor of x and y is 1

34. $(x, y) \in R$ if 3 divides $x - y$

35H. Let X be a nonempty set. Define a relation on $\mathscr{P}(X)$, the power set of X, as $(A, B) \in R$ if $A \subseteq B$. Is this relation reflexive; symmetric; antisymmetric; transitive; a partial order?

36.   Let R_1 and R_2 be the relations on $\{1, 2, 3, 4\}$ given by

$$R_1 = \{(1, 1), (1, 2), (3, 4), (4, 2)\}$$

$$R_2 = \{(1, 1), (2, 1), (3, 1), (4, 4), (2, 2)\}.$$

List the elements of $R_1 \circ R_2$ and $R_2 \circ R_1$.

Give examples of relations on $\{1, 2, 3, 4\}$ having the properties specified in Exercises 37–41.

37H. Reflexive, symmetric, not transitive

38.   Reflexive, not symmetric, not transitive

39.   Reflexive, antisymmetric, not transitive

40H. Not reflexive, symmetric, not antisymmetric, transitive

41.   Not reflexive, not symmetric, transitive

Since relations on X are subsets of $X \times X$, it makes sense to talk about unions and intersections of relations on X. In Exercises 42–46, give examples of relations R and S on $\{1, 2, 3, 4\}$ having the properties specified.

42H. R and S transitive, $R \cup S$ not transitive

43. R and S transitive, $R \circ S$ not transitive

44. R and S symmetric, $R \circ S$ not symmetric

45H. R and S antisymmetric, $R \cup S$ not antisymmetric

46. R and S antisymmetric, $R \circ S$ not antisymmetric

47H. What is wrong with the following argument, which supposedly shows that any relation R on X that is symmetric and transitive is reflexive?

Let $x \in X$. Using symmetry, we have (x, y) and (y, x) both in R. Since (x, y), $(y, x) \in R$, by transitivity we have $(x, x) \in R$. Therefore, R is reflexive.

8.2 Equivalence Relations

Suppose that we have a set X of 10 balls, each of which is either red, blue, or green (see Figure 8.2.1). If we divide the balls into sets R, B, and G according to color, the family $\{R, B, G\}$ is a partition of X. (Recall that in Section 2.2 we defined a partition of a set X to be a pairwise disjoint collection of nonempty subsets $X_1, X_2, \ldots, X_n$ of X such that $X = X_1 \cup X_2 \cup \cdots \cup X_n$.)

A partition can be used to define a relation. If $\{X_1, X_2, \ldots, X_n\}$ is a partition of X, we may define xRy to mean that for some set X_i, both x and y belong to X_i. For the example of Figure 8.2.1, the relation obtained could be described as "is the same color as." The next theorem shows that such a relation is always reflexive, symmetric, and transitive.

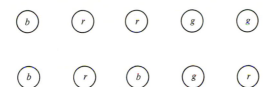

Figure 8.2.1

Theorem 8.2.1. *Let $\{X_1, X_2, \ldots, X_n\}$ be a partition of a set X. Define xRy to mean that, for some set X_i, both x and y belong to X_i. Then R is reflexive, symmetric, and transitive.*

PROOF. Let $x \in X$. Since $X = X_1 \cup X_2 \cup \cdots \cup X_n$, $x \in X_i$ for some X_i. Thus xRx and R is reflexive.

Suppose that xRy. Then both x and y belong to some set X_i. Since both y and x belong to X_i, yRx and R is symmetric.

Finally, suppose that xRy and yRz. Then both x and y belong to some set X_i and both y and z belong to some set X_j. In particular, $y \in X_i \cap X_j$. Since $\{X_1, X_2, \ldots, X_n\}$ is a pairwise disjoint family, we must have $X_i = X_j$. Thus x and z belong to X_i. Therefore, xRz and R is transitive. ∎

Example 8.2.2. Consider the partition

$$\{\{1, 3, 5\}, \{2, 6\}, \{4\}\}$$

of $X = \{1, 2, 3, 4, 5, 6\}$. The relation R on X given by Theorem 8.2.1 contains the ordered pairs $(1, 1)$, $(1, 3)$, and $(1, 5)$ because $\{1, 3, 5\}$ is a member of the partition. The complete relation is

$$R = \{(1, 1), (1, 3), (1, 5), (3, 1), (3, 3), (3, 5), (5, 1)$$
$$(5, 3), (5, 5), (2, 2), (2, 6), (6, 2), (6, 6), (4, 4)\}.$$

If $\{X_1, X_2, \ldots, X_n\}$ is a partition of X, we can regard the members of X_i as equivalent in the sense of the relation R of Theorem 8.2.1. For this reason, relations that are reflexive, symmetric, and transitive are called **equivalence relations**. In the example of Figure 8.2.1, the relation is "is the same color as"; hence "equivalent" means "is the same color as." Each set in the partition consists of all of the balls of a particular color.

Definition 8.2.3.   A relation that is reflexive, symmetric, and transitive on a set X is called an *equivalence relation on X*.

Example 8.2.4.   The relation R of Example 8.2.2 is an equivalence relation on $\{1, 2, 3, 4, 5, 6\}$ because of Theorem 8.2.1. We can also verify directly that R is reflexive, symmetric, and transitive.

The digraph of the relation R of Example 8.2.2 is shown in Figure 8.2.2. Again, we see that R is reflexive (there is a loop at every vertex), symmetric (for every directed edge from v to w, there is also a directed edge from w to v), and transitive (if there is a directed edge from x to y and a directed edge from y to z, there is a directed edge from x to z).

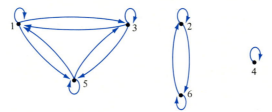

Figure 8.2.2

Example 8.2.5.   Consider the relation

$$R = \{(1, 1), (1, 3), (1, 5), (2, 2), (2, 4), (3, 1), (3, 3),$$
$$(3, 5), (4, 2), (4, 4), (5, 1), (5, 3), (5, 5)\}$$

on $\{1, 2, 3, 4, 5\}$. R is reflexive because $(1, 1), (2, 2), (3, 3), (4, 4),$ $(5, 5),$ ϵ R. R is symmetric because whenever (x, y) is in R, (y, x) is also in R. Finally, R is transitive because whenever (x, y) and (y, z) are in R, (x, z) is also in R. Since R is reflexive, symmetric, and transitive, R is an equivalence relation on $\{1, 2, 3, 4, 5\}$.

Example 8.2.6. The relation R of Example 8.1.4 is not an equivalence relation because R is not symmetric.

Example 8.2.7. The relation R of Example 8.1.5 is not an equivalence relation because R is neither reflexive nor transitive.

Example 8.2.8. The relation R of Example 8.1.15 is an equivalence relation because R is reflexive, symmetric, and transitive.

Given an equivalence relation on a set X, we can partition X by grouping related members of X together. Elements related to one another may be thought of as equivalent. The next theorem gives the details.

Theorem 8.2.9. *Let R be an equivalence relation on a set $X = \{x_1, x_2, \ldots, x_m\}$. For each $x_i \in X$, let*

$$[x_i] = \{x \in X \mid xRx_i\}.$$

Then the collection of sets

$$\{[x_1], [x_2], \ldots, [x_m]\}$$

is a partition of X.

PROOF. Two facts need to be verified in order to deduce that $\{[x_1], [x_2], \ldots, [x_m]\}$ is a partition of X:

1. $X = [x_1] \cup [x_2] \cup \cdots \cup [x_m]$.
2. $\{[x_1], [x_2], \ldots, [x_m]\}$ is a pairwise disjoint family.

Let $x_i \in X$. Since x_iRx_i, $x_i \in [x_i]$. Thus

$$X = [x_1] \cup [x_2] \cup \cdots \cup [x_m],$$

and fact 1 is established.

It remains to verify fact 2. We will first show that if aRb, then $[a] = [b]$. Suppose that aRb. Let $x \in [a]$. Then xRa. Since aRb and R is transitive, xRb. Therefore, $x \in [b]$ and $[a] \subseteq [b]$. The argument that $[b] \subseteq [a]$ is the same as that just given but with the roles of a and b interchanged. Thus $[a] = [b]$.

We must show that $\{[x_1], [x_2], \ldots, [x_m]\}$ is a pairwise disjoint family. Suppose that $[x_i] \neq [x_j]$. We must show that $[x_i] \cap [x_j] = \varnothing$. Suppose, by way of contradiction, that for some x, $x \in [x_i] \cap [x_j]$. Then xRx_i and xRx_j. Our result above shows that $[x] = [x_i]$ and $[x] = [x_j]$. Thus $[x_i] = [x_j]$, which is a contradiction. Therefore, $[x_i] \cap [x_j] = \varnothing$ and $\{[x_1], [x_2], \ldots, [x_m]\}$ is a pairwise disjoint family. ∎

Definition 8.2.10.   Let R be an equivalence relation on a set X. The sets $[x_i]$ defined in Theorem 8.2.9 are called the *equivalence classes of X given by the relation R*.

Example 8.2.11.   Consider the equivalence relation R of Example 8.2.2. The equivalence class $[1]$ containing 1 consists of all x such that $xR1$. Therefore,

$$[1] = \{1, 3, 5\}.$$

The remaining equivalence classes are found similarly:

$$[3] = [5] = \{1, 3, 5\}, \qquad [2] = [6] = \{2, 6\}, \qquad [4] = \{4\}.$$

Example 8.2.12.   The equivalence classes appear as components in the digraph of an equivalence relation. The three equivalence classes of the relation R of Example 8.2.2 appear in the digraph of R (shown in Figure 8.2.2) as the three components whose vertices are $\{1, 3, 5\}$, $\{2, 4\}$, and $\{6\}$.

Example 8.2.13.   There are two equivalence classes for the equivalence relation of Example 8.2.5, namely

$$[1] = [3] = [5] = \{1, 3, 5\}, \qquad [2] = [4] = \{2, 4\}.$$

Example 8.2.14.   The equivalence classes for the equivalence relation of Example 8.1.15 are

$$[a] = \{a\}, \qquad [b] = \{b\}, \qquad [c] = \{c\}.$$

Example 8.2.15. Let $X = \{1, 2, \ldots, 10\}$. Define xRy to mean that 3 divides $x - y$. We can readily verify that the relation R is reflexive, symmetric, and transitive. Thus R is an equivalence relation on X.

Let us determine the members of the equivalence classes. The equivalence class [1] consists of all x with $xR1$. Thus

$$[1] = \{x \in X \mid 3 \text{ divides } x - 1\}$$
$$= \{1, 4, 7, 10\}.$$

Similarly,

$$[2] = \{2, 5, 8\}$$
$$[3] = \{3, 6, 9\}.$$

These three sets partition X. Note that

$$[1] = [4] = [7] = [10]$$
$$[2] = [5] = [8]$$
$$[3] = [6] = [9].$$

For this relation, equivalence is "has the same remainder when divided by 3."

EXERCISES

In Exercises 1–8, determine whether the given relation is an equivalence relation on $\{1, 2, 3, 4, 5\}$. If the relation is an equivalence relation, list the equivalence classes.

1H. $\{(1, 1), (2, 2), (3, 3), (4, 4), (5, 5), (1, 3), (3, 1)\}$

2. $\{(1, 1), (2, 2), (3, 3), (4, 4), (5, 5), (1, 3), (3, 1), (3, 4), (4, 3)\}$

3. $\{(1, 1), (2, 2), (3, 3), (4, 4)\}$

4H. $\{(1, 1), (2, 2), (3, 3), (4, 4), (5, 5), (1, 5), (5, 1), (3, 5), (5, 3), (1, 3), (3, 1)\}$

5. $\{(x, y) \mid 1 \le x \le 5, 1 \le y \le 5\}$

6. $\{(x, y) \mid 4 \text{ divides } x - y\}$

7H. $\{(x, y) \mid 3 \text{ divides } x + y\}$

8. $\{(x, y) \mid x \text{ divides } 2 - y\}$

In Exercises 9–14, list the members of the equivalence relation on $\{1, 2, 3, 4\}$ defined (as in Theorem 8.2.1) by the given partition. Also, find the equivalence classes [1], [2], [3], and [4].

9H. {{1, 2}, {3, 4}}   **10.** {{1}, {2}, {3, 4}}
11. {{1}, {2}, {3}, {4}}   **12H.** {{1, 2, 3}, {4}}
13. {{1, 2, 3, 4}}   **14.** {{1}, {2, 4}, {3}}

Exercises 15 and 16 refer to the relation R on

$$X = \{\text{San Francisco, Pittsburgh, Chicago, San Diego,}$$
$$\text{Philadelphia, Los Angeles}\}$$

defined by xRy if x and y are in the same state.

15H. Show that R is an equivalence relation (i.e., explain why R is reflexive, symmetric, and transitive).

16H. List the equivalence classes of X.

Exercises 17–19 refer to the relation R defined on the set of eight-bit strings by $b_1 R b_2$ provided that the first four bits of b_1 and b_2 coincide.

17. Show that R is an equivalence relation.

18. How many equivalence classes are there?

19. List one member of each equivalence class.

20H. If an equivalence relation has only one equivalence class, what must the relation look like?

21. If R is an equivalence relation on a set X and $|X| = |R|$, what must the relation look like?

22. By listing ordered pairs, give an example of an equivalence relation on $\{1, 2, 3, 4, 5, 6\}$ having exactly four equivalence classes.

23H. How many equivalence relations are there on the set $\{1, 2, 3\}$?

24H. Let S be a unit square including the interior as shown. Define a relation R on S by $(x, y)R(x', y')$ if $(x = x'$ and $y = y')$ or $(y = y'$ and $x = 0$ and $x' = 1)$ or $(y = y'$ and $x = 1$ and $x' = 0)$.

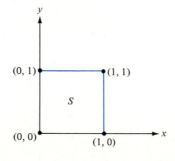

(a) Show that R is an equivalence relation on S.
(b) If points in the same equivalence class are glued together, how would you describe the figure formed?

25. Let S be a unit square including the interior (as in Exercise 24). Define a relation R' on S by $(x, y)R'(x', y')$ if $(x = x'$ and $y = y')$ or $(y = y'$ and $x = 0$ and $x' = 1)$ or $(y = y'$ and $x = 1$ and $x' = 0)$ or $(x = x'$ and $y = 0$ and $y' = 1)$ or $(x = x'$ and $y = 1$ and $y' = 0)$. Let

$$R = R' \cup \{((0, 0), (1, 1)), ((0, 1), (1, 0)),$$
$$((1, 1), (0, 0)), ((1, 0), (0, 1))\}.$$

(a) Show that R is an equivalence relation on S.
(b) If points in the same equivalence class are glued together, how would you describe the figure formed?

8.3 Matrices of Relations

A matrix is a convenient way to represent a relation R from X to Y. Such a representation can be used by a computer to analyze a relation. We label the rows with the elements of X (in some arbitrary order) and we label the columns with the elements of Y (again, in some arbitrary order). We then set the entry in row x and column y to 1 if xRy and to 0 otherwise. This matrix is called the **matrix of the relation** R (relative to the orderings of X and Y).

Example 8.3.1. Let us find the matrix of the relation

$$R = \{(1, b), (1, d), (2, c), (3, c), (3, b), (4, a)\}$$

from $X = \{1, 2, 3, 4\}$ to $Y = \{a, b, c, d\}$ relative to the orderings 1, 2, 3, 4 and a, b, c, d. We label the rows 1, 2, 3, 4 and the columns a, b, c, d:

$$\begin{array}{c} \\ 1 \\ 2 \\ 3 \\ 4 \end{array} \begin{pmatrix} \begin{array}{cccc} a & b & c & d \end{array} \\ & & & \\ & & & \\ & & & \\ & & & \end{pmatrix}.$$

We then set the entry in row 1, column b equal to 1 since $(1, b) \in R$:

$$\begin{array}{c}\\1\\2\\3\\4\end{array}\begin{pmatrix}\begin{array}{cccc}a & b & c & d\\ & 1 & & \\ & & & \\ & & & \\ & & & \end{array}\end{pmatrix}.$$

Similarly, we set the entries in row 1, column d; row 2, column c; row 3, column c; row 3, column b; and row 4, column a to 1 and the remaining entries to 0 to obtain the matrix of the relation R relative to the orderings 1, 2, 3, 4 and a, b, c, d:

$$\begin{array}{c}1\\2\\3\\4\end{array}\begin{pmatrix}\begin{array}{cccc}a & b & c & d\\0 & 1 & 0 & 1\\0 & 0 & 1 & 0\\0 & 1 & 1 & 0\\1 & 0 & 0 & 0\end{array}\end{pmatrix}.$$

Example 8.3.2. The matrix of the relation R of Example 8.3.1 relative to the orderings 2, 3, 4, 1 and d, b, a, c is

$$\begin{array}{c}2\\3\\4\\1\end{array}\begin{pmatrix}\begin{array}{cccc}d & b & a & c\\0 & 0 & 0 & 1\\0 & 1 & 0 & 1\\0 & 0 & 1 & 0\\1 & 1 & 0 & 0\end{array}\end{pmatrix}.$$

Obviously, the matrix of a relation is dependent on the orderings of the domain and range.

Example 8.3.3. The matrix of the relation R from $\{2, 3, 4\}$ to $\{5, 6, 7, 8\}$, relative to the orderings 2, 3, 4 and 5, 6, 7, 8, defined by

$$xRy \qquad \text{if } x \text{ divides } y$$

is

$$\begin{array}{c}2\\3\\4\end{array}\begin{pmatrix}\begin{array}{cccc}5 & 6 & 7 & 8\\0 & 1 & 0 & 1\\0 & 1 & 0 & 0\\0 & 0 & 0 & 1\end{array}\end{pmatrix}.$$

To compute the matrix of a relation R on a set X, we use a single ordering for both the domain and range of R.

Example 8.3.4.   The matrix of the relation

$$R = \{(a, a), (b, b), (c, c), (d, d), (b, c), (c, b)\}$$

on $\{a, b, c, d\}$, relative to the ordering a, b, c, d, is[†]

$$
\begin{array}{c c}
 & \begin{array}{cccc} a & b & c & d \end{array} \\
\begin{array}{c} a \\ b \\ c \\ d \end{array} &
\left(\begin{array}{cccc}
1 & 0 & 0 & 0 \\
0 & 1 & 1 & 0 \\
0 & 1 & 1 & 0 \\
0 & 0 & 0 & 1
\end{array}\right).
\end{array}
$$

Notice that the matrix of a relation on a set X is always a square matrix.

We can quickly determine whether a relation R on a set X is reflexive by examining the matrix A of R (relative to some ordering). The relation R is reflexive if and only if A has 1's on the main diagonal. (The main diagonal of a square matrix consists of the entries on a line from the upper left to the lower right.) The relation R is reflexive if and only if $(x, x) \in R$ for all $x \in X$. But this last condition holds precisely when the main diagonal consists of 1's. Notice that the relation R of Example 8.3.4 is reflexive and that the main diagonal of the matrix of R consists of 1's.

We can also quickly determine whether a relation R on a set X is symmetric by examining the matrix A of R (relative to some ordering). The relation R is symmetric if and only if for all i and j, the ijth entry of A is equal to the jith entry of A. (Less formally, R is symmetric if and only if A is symmetric about the main diagonal.) The reason is that R is symmetric if and only if whenever (x, y) is in R, (y, x) is also in R. But this last condition holds precisely when A is symmetric about the main diagonal. Notice that the relation R of Example 8.3.4 is symmetric and that the matrix of R is symmetric about the main diagonal.

We can also quickly determine whether a relation R is antisymmetric or a function by examining the matrix of R (relative to some ordering) (see Exercises 16 and 17). Unfortunately, there is no quick way to test whether a relation R on X is transitive by examining the matrix of R.

[†]We always use the same ordering for the rows and columns when we write the matrix of a relation on a set X.

We conclude by showing how matrix multiplication relates to composition of relations.

Example 8.3.5.   Let R_1 be the relation from $X = \{1, 2, 3\}$ to $Y = \{a, b\}$ defined by

$$R_1 = \{(1, a), (2, b), (3, a), (3, b)\}$$

and let R_2 be the relation from Y to $Z = \{x, y, z\}$ defined by

$$R_2 = \{(a, x), (a, y), (b, y), (b, z)\}.$$

The matrix of R_1 relative to the orderings 1, 2, 3 and a, b is

$$A_1 = \begin{matrix} & \begin{matrix} a & b \end{matrix} \\ \begin{matrix} 1 \\ 2 \\ 3 \end{matrix} & \begin{pmatrix} 1 & 0 \\ 0 & 1 \\ 1 & 1 \end{pmatrix} \end{matrix}$$

and the matrix of R_2 relative to the orderings a, b and x, y, z is

$$A_2 = \begin{matrix} & \begin{matrix} x & y & z \end{matrix} \\ \begin{matrix} a \\ b \end{matrix} & \begin{pmatrix} 1 & 1 & 0 \\ 0 & 1 & 1 \end{pmatrix} \end{matrix}.$$

The product of these matrices is

$$A_1 A_2 = \begin{pmatrix} 1 & 1 & 0 \\ 0 & 1 & 1 \\ 1 & 2 & 1 \end{pmatrix}.$$

Let us interpret this product.

The ikth entry in $A_1 A_2$ is computed as

$$\begin{matrix} & \begin{matrix} a & b \end{matrix} & \begin{matrix} k \end{matrix} \\ i & \begin{pmatrix} s & t \end{pmatrix} & \begin{pmatrix} u \\ v \end{pmatrix} \end{matrix} = su + tv.$$

If this value is nonzero, then either su or tv is nonzero. Suppose that $su \neq 0$. (The argument is similar if $tv \neq 0$.) Then $s \neq 0$ and $u \neq 0$. This means that $(i, a) \in R_1$ and $(a, k) \in R_2$. This implies that $(i, k) \in R_2 \circ R_1$. We have shown that if the ikth entry in $A_1 A_2$ is nonzero, then $(i, k) \in R_2 \circ R_1$. The converse is also true, as we now show.

Assume that $(i, k) \in R_2 \circ R_1$. Then, either

1. $(i, a) \in R_1$ and $(a, k) \in R_2$

or

2. $(i, b) \in R_1$ and $(b, k) \in R_2$.

If 1 holds, then $s = 1$ and $u = 1$, so $su = 1$ and $su + tv$ is nonzero. Similarly, if 2 holds, $tv = 1$ and again we have $su + tv$ nonzero. We have shown that if $(i, k) \in R_2 \circ R_1$, then the ikth entry in $A_1 A_2$ is nonzero.

We have shown that $(i, k) \in R_2 \circ R_1$ if and only if the ikth entry in $A_1 A_2$ is nonzero; thus $A_1 A_2$ is "almost" the matrix of the relation $R_2 \circ R_1$. To obtain the matrix of the relation $R_2 \circ R_1$, we need only change all nonzero entries in $A_1 A_2$ to 1. Thus the matrix of the relation $R_2 \circ R_1$, relative to the previously chosen orderings 1, 2, 3 and x, y, z, is

$$\begin{array}{c} \\ 1 \\ 2 \\ 3 \end{array} \begin{array}{ccc} x & y & z \\ \left(\begin{array}{ccc} 1 & 1 & 0 \\ 0 & 1 & 1 \\ 1 & 1 & 1 \end{array}\right). \end{array}$$

The argument given above holds for any relations. We summarize these results as Theorem 8.3.6.

Theorem 8.3.6.   *Let R_1 be a relation from X to Y and let R_2 be a relation from Y to Z. Choose orderings of X, Y, and Z. All matrices of relations are with respect to these orderings. Let A_1 be the matrix of R_1 and let A_2 be the matrix of R_2. The matrix of the relation $R_2 \circ R_1$ is obtained by replacing each nonzero term in the matrix product $A_1 A_2$ by 1.*

PROOF.   The proof precedes the statement of the theorem.   ■

EXERCISES

In Exercises 1–3, find the matrix of the relation R from X to Y relative to the orderings given.

1H. $R = \{(1, \delta), (2, \alpha), (2, \Sigma), (3, \beta), (3, \Sigma)\}$
Ordering of X: 1, 2, 3
Ordering of Y: $\alpha, \beta, \Sigma, \delta$

2. R as in Exercise 1
Ordering of X: 3, 2, 1
Ordering of Y: $\Sigma, \beta, \alpha, \delta$

3. $R = \{(x, a), (x, c), (y, a), (y, b), (z, d)\}$
Ordering of X: x, y, z
Ordering of Y: a, b, c, d

In Exercises 4–6, find the matrix of the relation R on X relative to the ordering given.

4H. $R = \{(1, 2), (2, 3), (3, 4), (4, 5)\}$
Ordering of X: 1, 2, 3, 4, 5

5. R as in Exercise 1
Ordering of X: 5, 3, 1, 2, 4

6. $R = \{(x, y) \mid x < y\}$
Ordering of X: 1, 2, 3, 4

In Exercises 7–15, write the relation R, given by the matrix, as a set of ordered pairs.

7H.
$$\begin{array}{c c} & \begin{matrix} w & x & y & z \end{matrix} \\ \begin{matrix} a \\ b \\ c \\ d \end{matrix} & \begin{pmatrix} 1 & 0 & 1 & 0 \\ 0 & 0 & 0 & 0 \\ 0 & 0 & 1 & 0 \\ 1 & 1 & 1 & 1 \end{pmatrix} \end{array}$$

8.
$$\begin{array}{c c} & \begin{matrix} f & g & h \end{matrix} \\ \begin{matrix} \alpha \\ \beta \\ \gamma \\ \delta \\ \epsilon \end{matrix} & \begin{pmatrix} 0 & 1 & 0 \\ 1 & 1 & 0 \\ 0 & 0 & 0 \\ 1 & 1 & 0 \\ 0 & 0 & 1 \end{pmatrix} \end{array}$$

9.
$$\begin{array}{c c} & \begin{matrix} 1 & 2 & 3 & 4 \end{matrix} \\ \begin{matrix} 1 \\ 2 \end{matrix} & \begin{pmatrix} 1 & 0 & 1 & 0 \\ 0 & 1 & 1 & 1 \end{pmatrix} \end{array}$$

10H.
$$\begin{array}{c c} & \begin{matrix} w & x & y & z \end{matrix} \\ \begin{matrix} w \\ x \\ y \\ z \end{matrix} & \begin{pmatrix} 1 & 0 & 1 & 0 \\ 0 & 0 & 0 & 0 \\ 1 & 0 & 1 & 0 \\ 0 & 0 & 0 & 1 \end{pmatrix} \end{array}$$

11.
$$\begin{array}{c c} & \begin{matrix} \alpha & \beta & \gamma \end{matrix} \\ \begin{matrix} \alpha \\ \beta \\ \gamma \end{matrix} & \begin{pmatrix} 1 & 1 & 0 \\ 1 & 1 & 0 \\ 0 & 0 & 1 \end{pmatrix} \end{array}$$

12.
$$\begin{array}{c c} & \begin{matrix} a & b & c & d \end{matrix} \\ \begin{matrix} a \\ b \\ c \\ d \end{matrix} & \begin{pmatrix} 1 & 0 & 1 & 0 \\ 0 & 1 & 0 & 0 \\ 1 & 0 & 1 & 0 \\ 0 & 0 & 0 & 0 \end{pmatrix} \end{array}$$

13H.
$$\begin{array}{c c} & \begin{matrix} 1 & 2 & 3 \end{matrix} \\ \begin{matrix} 1 \\ 2 \\ 3 \end{matrix} & \begin{pmatrix} 1 & 0 & 0 \\ 1 & 1 & 0 \\ 1 & 0 & 1 \end{pmatrix} \end{array}$$

14.
$$\begin{array}{c c} & \begin{matrix} w & x & y & z \end{matrix} \\ \begin{matrix} w \\ x \\ y \\ z \end{matrix} & \begin{pmatrix} 0 & 0 & 1 & 1 \\ 0 & 1 & 0 & 1 \\ 0 & 0 & 1 & 0 \\ 1 & 0 & 0 & 1 \end{pmatrix} \end{array}$$

15.
$$\begin{array}{c c} & \begin{matrix} \alpha & \beta & \gamma \end{matrix} \\ \begin{matrix} \alpha \\ \beta \\ \gamma \end{matrix} & \begin{pmatrix} 1 & 1 & 1 \\ 1 & 1 & 1 \\ 1 & 1 & 1 \end{pmatrix} \end{array}$$

16H. How can we quickly determine whether a relation R is antisymmetric by examining the matrix of R (relative to some ordering)?

17. How can we quickly determine whether a relation R is a function by examining the matrix of R (relative to some ordering)?

18. Tell whether each relation in Exercises 10–15 is reflexive, symmetric, transitive, antisymmetric, a partial order, or an equivalence relation.

19H. Given the matrix of a relation R from X to Y, how can we find the matrix of the converse relation R^{-1}?

20. Find the matrix of the converse of each of the relations of Exercises 7–9.

In Exercises 21–23, find

(a) The matrix A_1 of the relation R_1 (relative to the given orderings).

(b) The matrix A_2 of the relation R_2 (relative to the given orderings).

(c) The matrix product A_1A_2.

(d) Use the result of part (c) to find the matrix of the relation $R_2 \circ R_1$.

(e) Use the result of part (d) to find the relation $R_2 \circ R_1$ (as a set of ordered pairs).

21H. $R_1 = \{(1, x), (1, y), (2, x), (3, x)\}$
     $R_2 = \{(x, b), (y, b), (y, a), (y, c)\}$
     Orderings: 1, 2, 3; x, y; a, b, c

22. $R_1 = \{(x, y) \mid x \text{ divides } y\}$; R_1 is from X to Y
    $R_2 = \{(y, z) \mid y > z\}$; R_2 is from Y to Z
    Ordering of X and Y: 2, 3, 4, 5
    Ordering of Z: 1, 2, 3, 4

23. $R_1 = \{(x, y) \mid x + y \leq 6\}$; R_1 is from X to Y
    $R_2 = \{(y, z) \mid y = z + 1\}$; R_2 is from Y to Z
    Ordering of X, Y, and Z: 1, 2, 3, 4, 5

24H. Given the matrix of an equivalence relation R on X, how can we easily find the equivalence class containing the element $x \in X$?

25. Let A be the matrix of a function f from X to Y (relative to some orderings of X and Y). What conditions must A satisfy for f to be onto Y?

26. Let A be the matrix of a function f from X to Y (relative to some orderings of X and Y). What conditions must A satisfy for f to be one-to-one?

27H. Write an algorithm that receives as input the matrix of a relation R and tests whether R is reflexive.

28. Write an algorithm that receives as input the matrix of a relation R and tests whether R is symmetric.

29. Write an algorithm that receives as input the matrix of a relation R and tests whether R is antisymmetric.

30H. Write an algorithm that receives as input the matrix of a relation R and tests whether R is a function.

31. Write an algorithm that receives as input the matrix of a relation R and produces as output the matrix of the converse relation R^{-1}.

*32. Let R_1 be a relation from X to Y and let R_2 be a relation from Y to Z. Choose orderings of X, Y, and Z. All matrices of relations are with respect to these orderings. Let A_1 be the matrix of R_1 and let A_2 be the matrix of R_2. Show that the ikth entry in the matrix product A_1A_2 is equal to the number of elements in the set

$$\{(m) \mid (i, m) \in R_1 \text{ and } (m, k) \in R_2\}.$$

†8.4 Relational Data Bases

The "bi" in a binary relation R refers to the fact that R has two columns when we write R as a table. It is often useful to allow a table to have an arbitrary number of columns. If a table has n columns, the corresponding relation is called an **n-ary relation**.

Example 8.4.1.   Table 8.4.1 represents a 4-ary relation. This table expresses the relationship among identification numbers, names, positions, and ages.

We can also express an n-ary relation as a collection of n-tuples.

†This section can be omitted without loss of continuity.

Table 8.4.1   PLAYER

ID Number	Name	Position	Age
22012	Johnsonbaugh	c	22
93831	Glover	of	24
58199	Battey	p	18
84341	Cage	c	30
01180	Homer	1b	37
26710	Score	p	22
61049	Johnsonbaugh	of	30
39826	Singleton	2b	31

Example 8.4.2.   Table 8.4.1 can be expressed as the set

$$\{(22012, \text{Johnsonbaugh}, c, 22), (93831, \text{Glover}, of, 24),$$
$$(58199, \text{Battey}, p, 18), (84341, \text{Cage}, c, 30),$$
$$(01180, \text{Homer}, 1b, 37), (26710, \text{Score}, p, 22),$$
$$(61049, \text{Johnsonbaugh}, of, 30), (39826, \text{Singleton}, 2b, 31)\}$$

of 4-tuples.

Computer systems are capable of storing large amounts of information in **data bases**. The data are available to various applications. **Data base management systems** are programs that help users access the information in data bases. The **relational data base model**, invented by E. F. Codd in 1970, is based on the concept of an n-ary relation. We will briefly introduce some of the fundamental ideas in the theory of relational data bases. For more details on relational data bases, the reader is referred to [Codd, Date, Kroenke, and Tsichritzis]. We begin with some of the terminology.

The columns of an n-ary relation are called **attributes**. The domain of an attribute is a set to which all the elements in that attribute belong. For example, in Table 8.4.1 the attribute Age might be taken to be the set of all positive integers less than 100. The attribute Name might be taken to be all strings over the alphabet having length 30 or less.

A single attribute or a combination of attributes for a relation is a **key** if the values of the attributes uniquely define an n-tuple. For example, in Table 8.4.1, we can take the attribute ID Number as a key. (It is assumed that each person has a unique identification number.) The attribute Name is not a key because different persons can have the same

Edgar F. Codd *(1923–   )*

Edgar F. Codd was born in England and emigrated to the United States in 1948. He received bachelor's and master's degrees from Oxford University and master's and doctoral degrees from the University of Michigan. The major portion of his professional career has been spent with IBM. Most recently, he has been associated with the IBM San Jose Research Laboratory. In addition to his work on data bases, he has been involved with computer design, operating systems design, and automata. He won the prestigious Turing award from the Association for Computing Machinery in 1981. [*Photo courtesy of Dr. E. F. Codd*]

name. For the same reason, we cannot take either of the attributes, Position or Age, as keys. Name and Position, in combination, could be used as a key for Table 8.4.1, since in our example a player is uniquely defined by a name and a position.

A data base management system responds to **queries**. A query is a request for information from the data base. For example, "Find all persons who play outfield," is a meaningful query for the relation given by Table 8.4.1. We will discuss several operations on relations that are used to answer queries in the relational data base model.

Example 8.4.3 Select. The selection operator chooses certain *n*-tuples from a relation. The choices are made by giving conditions on the attributes. For example, for the relation PLAYER given by Table 8.4.1,

PLAYER [Position = c]

will select the tuples

(22012, Johnsonbaugh, c, 22), (84341, Cage, c, 30).

Example 8.4.4 Project. Whereas the selection operator chooses rows of a relation, the projection operator chooses columns. In addition, duplicates are eliminated. For example, for the relation PLAYER given by Table 8.4.1,

PLAYER [Name, Position]

will select the tuples

(Johnsonbaugh, c), (Glover, of), (Battey, p), (Cage c),
(Homer, 1b), (Score, p), (Johnsonbaugh, of), (Singleton, 2b).

Example 8.4.5 Join. The selection and projection operators manipulate a single relation; join manipulates two relations. The join operation on relations R_1 and R_2 begins by examining all pairs of tuples, one from R_1 and one from R_2. If the join condition is satisfied, the tuples are combined to form a new tuple. The join condition specifies a relationship between an attribute in R_1 and an attribute in R_2. For example, let us perform a join operation on Tables 8.4.1 and 8.4.2. As the condition we take

ID Number = PID.

Table 8.4.2
ASSIGNMENT

PID	Team
39826	Blue Sox
26710	Mutts
58199	Jackalopes
01180	Mutts

We take a row from Table 8.4.1 and a row from Table 8.4.2 and if ID Number = PID, we combine the rows. For example, the ID Number 01180 in the fifth row (01180, Homer, 1b, 37) of Table 8.4.1 matches the PID in the fourth row (01180, Mutts) of Table 8.4.2. These tuples are combined by first writing the tuple from Table 8.4.1, following it by the tuple from Table 8.4.2, and eliminating the equal entries in the specified attributes to give

(01180, Homer, 1b, 37, Mutts).

This operation is expressed as

PLAYER [ID Number = PID] ASSIGNMENT.

The relation obtained by executing this join is shown in Table 8.4.3.

Table 8.4.3   PLAYER [ID Number = PID] ASSIGNMENT

ID Number	Name	Position	Age	Team
58199	Battey	p	18	Jackalopes
01180	Homer	1b	37	Mutts
26710	Score	p	22	Mutts
39826	Singleton	2b	31	Blue Sox

Most queries to a relational data base require several operations to provide the answer.

Example 8.4.6.   Describe operations that provide the answer to the query, "Find the names of all persons who play for some team."

If we first join the relations given by Tables 8.4.1 and 8.4.2 subject to the condition ID Number = PID, we will obtain Table 8.4.3 that lists all persons who play for some team as well as other information. To obtain the names, we need only project on the attribute Name. We obtain the relation

Name
Battey
Homer
Score
Singleton

Formally, these operations would be specified as

TEMP := PLAYER [ID Number = PID] ASSIGNMENT
TEMP [Name]

Example 8.4.7. Describe operations that provide the answer to the query, "Find the names of all persons who play for the Mutts."

If we first use the selection operator to pick the rows of Table 8.4.2 that reference Mutts' players, we obtain the relation

TEMP1

PID	Team
26710	Mutts
01180	Mutts

If we now join Table 8.4.1 and the relation TEMP1 subject to ID Number = PID, we obtain the relation

TEMP2

ID Number	Name	Position	Age	Team
01180	Homer	1b	37	Mutts
26710	Score	p	22	Mutts

If we project the relation TEMP2 on the attribute Name, we obtain the relation

Name
Homer
Score

We would formally specify these operations as follows:

TEMP1 := ASSIGNMENT [Team = Mutts]
TEMP2 := PLAYER [ID Number = PID] TEMP1
TEMP2 [Name]

Notice that the operations

TEMP1 := PLAYER [ID Number = PID] ASSIGNMENT
TEMP2 := TEMP1 [Team = Mutts]
TEMP2 [Name]

would also answer the query of Example 8.4.7.

EXERCISES

1H.   Express the relation given by Table 8.4.4 as a set of *n*-tuples.

Table 8.4.4   EMPLOYEE

ID	Name	Manager
1089	Suzuki	Zamora
5620	Kaminski	Jones
9354	Jones	Yu
9551	Ryan	Washington
3600	Beaulieu	Yu
0285	Schmidt	Jones
6684	Manacotti	Jones

2.   Express the relation given by Table 8.4.5 as a set of *n*-tuples.

Table 8.4.5 DEPARTMENT

Dept	Manager
23	Jones
04	Yu
96	Zamora
66	Washington

3. Express the relation given by Table 8.4.6 as a set of *n*-tuples.

Table 8.4.6   SUPPLIER

Dept	Part No	Amount
04	335B2	220
23	2A	14
04	8C200	302
66	42C	3
04	900	7720
96	20A8	200
96	1199C	296
23	772	39

4. Express the relation given by Table 8.4.7 as a set of *n*-tuples.

Table 8.4.7   BUYER

Name	Part No
United Supplies	2A
ABC Unlimited	8C200
United Supplies	1199C
JCN Electronics	2A
United Supplies	335B2
ABC Unlimited	772
Danny's	900
United Supplies	772
Underhanded Sales	20A8
Danny's	20A8
DePaul University	42C
ABC Unlimited	20A8

In Exercises 5–20, write a sequence of operations to answer the query. Also, provide an answer to the query. Use Tables 8.4.4 to 8.4.7.

5H. Find the names of all employees. (Do not include any managers.)

6. Find the names of all managers.

7. Find all part numbers.

8H. Find the names of all buyers.

9. Find the names of all employees who are managed by Jones.

10. Find all part numbers supplied by department 96.

11H. Find all buyers of part 20A8.

12. Find all employees in department 04.

13. Find the part numbers of parts of which there are at least 100 items on hand.

14H. Find all department numbers of departments that supply parts to Danny's.

15. Find the part numbers and amounts of parts bought by United Supplies.

16. Find all managers of departments that produce parts for ABC Unlimited.

17H. Find the names of all employees who work in departments that supply parts for JCN Electronics.

18. Find all buyers who buy parts in the department managed by Jones.

19. Find all buyers who buy parts that are produced by the department for which Suzuki works.

20. Find all part numbers and amounts for Zamora's department.

21. Make up at least three *n*-ary relations with artificial data that might be used in a medical data base. Illustrate how your data base would be used by posing and answering two queries. Also, write a sequence of operations that could be used to answer the queries.

22H. Describe a *union* operation on a relational data base. Illustrate how your operator works by answering the following query using the relations of Tables 8.4.4 to 8.4.7: Find the names of all employees that work either in department 23 or department 96. Also, write a sequence of operations that could be used to answer the query.

23. Describe an *intersection* operation on a relational data base. Illustrate how your operator works by answering the following query using the relations of Tables 8.4.4 to 8.4.7: Find the names of all buyers that buy both parts 2A and 1199C. Also, write a sequence of operations that could be used to answer the query.

24. Describe a *difference* operation on a relational data base. Illustrate how your operator works by answering the following query using the relations of Tables 8.4.4 to 8.4.7: Find the names of all employees that do not work in department 04. Also, write a sequence of operations that could be used to answer the query.

8.5 Notes

Most general references on discrete mathematics contain some general discussion of relations. The relational data base model has been a favorite topic of theoretical research. Codd, the inventor of the relational data base model, has argued that this model alone is sufficiently mathematically based so as to be scientifically investigated. At first there were no commercial implementations of relational data bases. With the appearance of SQL, ORACLE, and dBASE III among others, the relational data base has passed from the purely theoretical state to operational status. [Codd; Date; Kroenke; Tsichritzis; and Ullman] are recommended references on data bases in general and the relational model in particular.

Computer Exercises

Exercises 1–8 refer to the relation

$$R = \{(a(i), b(i)) \mid i \in \{1, \ldots, n\}\},$$

where a and b are arrays indexed from 1 to n.

1. Write a program that finds the domain of R.

2. Write a program that finds the range of R.

3. Write a program to determine whether R is a reflexive relation on a set X.

4. Write a program to determine whether R is symmetric.

5. Write a program to determine whether R is antisymmetric.

6. Write a program to determine whether R is transitive.

7. Write a program that finds the converse of R.

8. Write a program to determine whether R is a function from a set X to a set Y.

9. Given arrays that represent relations R and S as described previously, write a program that finds the composition $R \circ S$.

10. [*Project*]   Prepare a report on a commercial relational data base such as SQL, ORACLE, or dBASE III. References can be found in [Kroenke].

Chapter Review

SECTION 8.1

Binary relation from X to Y: Set of ordered pairs (x, y), $x \in X$, $y \in Y$

Domain of a binary relation R: $\{x \mid (x, y) \in R\}$

Range of a binary relation R: $\{y \mid (x, y) \in R\}$

Digraph of a binary relation

Reflexive relation R on X: $(x, x) \in R$ for all $x \in X$

Symmetric relation R: $(x, y) \in R$ implies that $(y, x) \in R$

Antisymmetric relation R: $(x, y) \in R$ and $x \neq y$ implies that $(y, x) \notin R$

Transitive relation R: (x, y), $(y, z) \in R$ implies that $(x, z) \in R$

Partial order: Relation that is reflexive, antisymmetric, and transitive

Converse relation R^{-1}: $\{(y, x) \mid (x, y) \in R\}$

Composition of relations $R_2 \circ R_1$: $\{(x, z) \mid (x, y) \in R_1, (y, z) \in R_2\}$

SECTION 8.2

Partition of X: Collection of pairwise disjoint nonempty subsets of X whose union is X

Equivalence relation: Relation that is reflexive, symmetric, and transitive

Equivalence class containing a given by an equivalence relation R: $[a] = \{x \mid xRa\}$

Equivalence classes partition the set (Theorem 8.2.9)

SECTION 8.3

Matrix of a relation

R is a reflexive relation if and only if the main diagonal of the matrix of R consists of 1's

R is a symmetric relation if and only if the matrix of R is symmetric about the main diagonal

If A_1 is the matrix of the relation R_1 and A_2 is the matrix of the relation R_2, the matrix of the relation $R_2 \circ R_1$ is obtained by replacing each nonzero term in the matrix product $A_1 A_2$ by 1.

SECTION 8.4

n-ary relation: Set of n-tuples

Data base management system

Relational data base

Key

Query

Select

Project

Join

Chapter Self-Test

SECTION 8.1

Exercises 1–4 refer to the relation

$$R = \{(1, 3), (3, 1), (2, 4), (1, 1), (2, 2), (3, 3), (4, 4)\}$$

on $\{1, 2, 3, 4\}$.

1H. Draw the digraph of R.

2H. Find the domain and range of R.

3H. Find R^{-1} as a set of ordered pairs.

4H. Is R reflexive; symmetric; antisymmetric; transitive; a partial order? Explain.

SECTION 8.2

5H. Is the relation

$$\{(1, 1), (1, 2), (2, 2), (4, 4), (2, 1), (3, 3)\}$$

an equivalence relation on $\{1, 2, 3, 4\}$? Explain.

6H. Given that the relation

$$\{(1, 1), (2, 2), (3, 3), (4, 4), (1, 2), (2, 1), (3, 4), (4, 3)\}$$

is an equivalence relation on {1, 2, 3, 4}, find [3], the equivalence class containing 3. How many (distinct) equivalence classes are there?

7H. Find the equivalence relation (as a set of ordered pairs) on {a, b, c, d, e} whose equivalence classes are

$$\{a\}, \{b, d, e\}, \{c\}.$$

8H. If R is an equivalence relation on X, must the domain of R be equal to X? Explain.

SECTION 8.3

9H. Find the matrix of the relation preceding Exercise 1 relative to the ordering 1, 2, 3, 4.

10H. Determine whether the relation given by the matrix

$$\begin{pmatrix} 1 & 0 & 0 & 1 & 0 \\ 0 & 1 & 0 & 0 & 0 \\ 0 & 0 & 1 & 0 & 1 \\ 1 & 0 & 0 & 1 & 0 \\ 0 & 0 & 1 & 0 & 1 \end{pmatrix}$$

is reflexive, symmetric, transitive, antisymmetric, a partial order, an equivalence relation.

11H. Given

$$R_1 = \{(1, x), (2, x), (2, y), (3, y)\}$$
$$R_2 = \{(x, a), (x, b), (y, a), (y, c)\}$$
Orderings: 1, 2, 3; x, y; a, b, c;

find
(a) The matrix A_1 of the relation R_1 (relative to the given orderings).
(b) The matrix A_2 of the relation R_2 (relative to the given orderings).
(c) The matrix product $A_1 A_2$.
(d) Use the result of part (c) to find the matrix of the relation $R_2 \circ R_1$.
(e) Use the result of part (d) to find the relation $R_2 \circ R_1$ (as a set of ordered pairs).

12H. Write an algorithm that inputs

(a) The matrix of an equivalence relation R on

$$X = \{1, 2, \ldots, n\}$$

relative to the ordering $1, 2, \ldots, n$

and

(b) n

and outputs each element $x \in X$ followed by the members of the equivalence class containing x.

SECTION 8.4

In Exercises 13–16, write a sequence of operations to answer the query. Also, provide an answer to the query. Use Tables 8.4.1 and 8.4.2.

13H. Find all teams.

14H. Find all players' names and ages.

15H. Find the names of all teams that have a pitcher.

16H. Find the names of all teams that have players aged 30 years or older.

Appendix: More on Logic

This appendix extends the discussion of logic in Sections 1.1 (Propositions) and 1.2 (Conditional Propositions and Logical Equivalence). **Logic** studies methods of reasoning; specifically, methods to separate valid reasoning from invalid reasoning. Results in many disciplines are established by logical reasoning. We consider proofs and arguments in formal mathematical systems (Section A.1) and categorical propositions—propositions that relate members of one class to members of a second class (Section A.2).

A.1 Proofs and Arguments

A **mathematical system** consists of **axioms, definitions**, and **undefined terms**. The axioms are assumed true. Definitions are used to create new concepts in terms of existing ones. Some terms are not explicitly defined but rather are implicitly defined by the axioms.

Example A.1.1. Euclidean geometry furnishes an example of a mathematical system. Among the axioms are

1. Given two distinct points, there is exactly one line that contains them.
2. Given a line and a point not on the line, there is exactly one line parallel to the line through the point.

The terms *point* and *line* are undefined terms that are implicitly defined by the axioms that describe their properties.

Among the definitions are:

1. Two triangles are *congruent* if their vertices can be paired so that the corresponding sides and corresponding angles are equal.
2. Two angles are *supplementary* if the sum of their measures is 180°.

Example A.1.2. The real numbers furnish another example of a mathematical system. Among the axioms are

1. For all real numbers x and y, $xy = yx$.
2. There is a subset **P** of real numbers satisfying
   (a) If x and y are in **P**, then $x + y$ and xy are in **P**.
   (b) If x is a real number, then exactly one of the following statements is true

$$x \text{ is in } \mathbf{P} \qquad x = 0 \qquad -x \text{ is in } \mathbf{P}.$$

Multiplication is implicitly defined by axiom 1 (and others) which describe the properties multiplication is assumed to obey.

Among the definitions are

1. The elements in **P** (of axiom 2) are called *positive real numbers*.
2. The *absolute value* $|x|$ of a real number x is defined to be x if x is positive or 0 and $-x$ otherwise.

Within a mathematical system we can derive theorems. A **theorem** is a result that can be deduced from the axioms, definitions, and previously derived theorems. The argument establishing the truth of the theorem is called a **proof**. Special types of theorems are referred to as lemmas and corollaries. A **lemma** is a theorem that is usually not too interesting in its own right but is useful in proving another theorem. A **corollary** is a theorem that follows quickly from another theorem.

Example A.1.3.   Examples of theorems in Euclidean geometry are

1. If two sides of a triangle are equal, then the angles opposite them are equal.
2. If the diagonals of a quadrilateral bisect each other, then the quadrilateral is a parallelogram.

Example A.1.4.   An example of a corollary in Euclidean geometry is

1. If a triangle is equilateral, then it is equiangular.

This corollary follows immediately from the first theorem of Example A.1.3.

Example A.1.5.   Examples of theorems about real numbers are

1. $x \cdot 0 = 0$ for every real number x.
2. For all real numbers x, y, and z, if $x \leq y$ and $y \leq z$, then $x \leq z$.

Example A.1.6.   An example of a lemma about real numbers is

1. If n is a positive integer, then either $n - 1$ is a positive integer or $n - 1 = 0$.

Surely this result is not that interesting in its own right, but it can be used to prove other results.

Theorems are often of the form

$$\text{If } p, \text{ then } q. \tag{A.1.1}$$

A proof of (A.1.1) is an argument which shows that (A.1.1) is true. If p is false, then by Definition 1.2.4, (A.1.1) is true; thus we need only consider the case that p is true. A **direct proof** assumes that p is true and then, using p as well as other axioms, definitions, and previously derived theorems, shows directly that q is true.

Example A.1.7.   We will give a direct proof of the following proposition about real numbers d, d_1, d_2, and x:

$$\text{If } d = \min\{d_1, d_2\} \text{ and } x \le d, \text{ then } x \le d_1 \text{ and } x \le d_2.$$

PROOF.   From the definition of min, it follows that $d \le d_1$ and $d \le d_2$. From $x \le d$ and $d \le d_1$, we may derive $x \le d_1$ from a previous theorem (theorem 2 of Example A.1.5). From $x \le d$ and $d \le d_2$, we may derive $x \le d_2$ from the same previous theorem. Therefore, $x \le d_1$ and $x \le d_2$.   ■

A second technique of proof is **proof by contradiction**. A proof by contradiction establishes (A.1.1) by assuming that the hypothesis p is true and that the conclusion q is false and then, using p and $\bar{q}$ as well as other axioms, definitions, and previously derived theorems, derives a **contradiction**. A contradiction is a proposition of the form $r \wedge \bar{r}$. A proof by contradiction is sometimes called an **indirect proof** since to establish (A.1.1) using proof by contradiction, one follows an indirect route: Derive $r \wedge \bar{r}$, then conclude that (A.1.1) is true.

The only difference between the assumptions in a direct proof and a proof by contradiction is the negated conclusion. In a direct proof, the negated conclusion is not assumed, whereas in a proof by contradiction the negated conclusion is assumed.

Example A.1.8.   We will give a proof by contradiction of the following proposition about real numbers x and y:

$$\text{If } x + y \ge 2, \text{ then either } x \ge 1 \text{ or } y \ge 1.$$

PROOF.   Suppose that the conclusion is false. Then $x < 1$ and $y < 1$. (Remember that negating *or*s results in *and*s. See Example 1.2.11, De Morgan's laws for logic.) Using a previous theorem, we may add these inequalities to obtain

$$x + y < 1 + 1 = 2.$$

At this point, we have derived the contradiction $p \wedge \bar{p}$, where

$$p\colon x + y \ge 2.$$

Thus we conclude that the proposition is true.   ■

Suppose that we give a proof by contradiction of (A.1.1) in which, as in Example A.1.8, we deduce $\bar{p}$. In effect, we have proved

$$\bar{q} \to \bar{p}. \qquad (A.1.2)$$

This special case of proof by contradiction is called **proof by contrapositive**.

In constructing a proof, we must be sure that the arguments used are **valid**. In the remainder of this section we make precise the concept of a valid argument and explore this concept in some detail.

Consider the following sequence of propositions.

> The bug is either in module 17 or in module 81.
> The bug is a numerical error. $\qquad$ (A.1.3)
> Module 81 has no numerical error.

Assuming that these statements are true, it is reasonable to conclude:

> The bug is in module 17. $\qquad$ (A.1.4)

This process of drawing a conclusion from a sequence of propositions is called **deductive reasoning**. The given propositions, like (A.1.3), are called **hypotheses** or **premises** and the proposition that follows from the hypotheses, like (A.1.4), is called the **conclusion**. A (**deductive**) **argument** consists of hypotheses together with a conclusion. Many proofs in mathematics and computer science are deductive arguments.

Any argument has the form

$$\text{If } p_1 \text{ and } p_2 \text{ and } \cdots \text{ and } p_n, \text{ then } q. \qquad (A.1.5)$$

Argument (A.1.5) is said to be valid if the conclusion follows from the hypotheses; that is, if p_1 and p_2 and $\cdots$ and p_n are true, then q is also true. This discussion motivates the following definition.

Definition A.1.9. An *argument* is a sequence of propositions written

$$
\begin{array}{l}
p_1 \\
p_2 \\
\cdot \\
\cdot \\
\cdot \\
p_n \\
\hline
\therefore q
\end{array}
$$

or

$$p_1, p_2, \ldots, p_n \, / \therefore q.$$

The propositions $p_1, p_2, \ldots, p_n$ are called the *hypotheses* (or *premises*) and the proposition q is called the *conclusion*. The argument is *valid* provided that if p_1 and p_2 and $\cdots$ and p_n are all true, then q is also true; otherwise, the argument is *invalid* (or a *fallacy*).

In a valid argument, we sometimes say that the conclusion follows from the hypotheses. Notice that we are not saying that the conclusion is true; we are only saying that if you grant the hypotheses, you must also grant the conclusion. An argument is valid because of its form, not because of its content.

Example A.1.10.   Determine whether the argument

$$
\begin{array}{l}
p \to q \\
p \\
\hline
\therefore q
\end{array}
$$

is valid.

[First solution.] We construct a truth table for all the propositions involved:

p	q	$p \to q$	p	q
T	T	T	T	T
T	F	F	T	F
F	T	T	F	T
F	F	T	F	F

We observe that whenever the hypotheses $p \to q$ and p are true, the conclusion q is also true; therefore, the argument is valid.

[Second solution.] We can avoid writing the truth table by directly verifying that whenever the hypotheses are true, the conclusion is also true.

Suppose that $p \to q$ and p are true. Then q must be true for otherwise $p \to q$ would be false. Therefore, the argument is valid.

Example A.1.11. Represent the argument

$$\begin{array}{l} \text{If } 2 = 3, \text{ then I ate my hat.} \\ \underline{\text{I ate my hat.}} \\ \therefore 2 = 3 \end{array}$$

symbolically and determine whether the argument is valid.

If we let

$$p: 2 = 3, \qquad q: \text{I ate my hat.}$$

the argument may be written

$$\begin{array}{l} p \rightarrow q \\ \underline{q} \\ \therefore p \end{array}$$

If the argument is valid, then whenever $p \rightarrow q$ and q are both true, p must also be true. Suppose that $p \rightarrow q$ and q are true. This is possible if p is false and q is true. In this case, p is not true, thus the argument is invalid.

We can also determine the validity of the argument in Example A.1.11 by examining the truth table of Example A.1.10. In the third row of the table, the hypotheses are true and the conclusion is false; thus the argument is invalid.

EXERCISES

1H. Give an example (different from those of Example A.1.1) of an axiom in Euclidean geometry.

2. Give an example (different from those of Example A.1.2) of an axiom in the system of real numbers.

3. Give an example (different from those of Example A.1.1) of a definition in Euclidean geometry.

4H. Give an example (different from those of Example A.1.2) of a definition in the system of real numbers.

5. Give an example (different from those of Example A.1.3) of a theorem in Euclidean geometry.

6. Give an example (different from those of Example A.1.5) of a theorem in the system of real numbers.

7H. Justify each step of the following direct proof, which shows that if x is a real number, then $x \cdot 0 = 0$. Assume that the following are previous theorems: If a, b, and c are real numbers, then $b + 0 = b$ and $a(b + c) = ab + ac$. If $a + b = a + c$, then $b = c$.

PROOF. $x \cdot 0 + 0 = x \cdot 0 = x \cdot (0 + 0) = x \cdot 0 + x \cdot 0$, therefore $x \cdot 0 = 0$. ∎

8. Justify each step of the following proof by contradiction, which shows that if $xy = 0$, then either $x = 0$ or $y = 0$. Assume that if a, b, and c are real numbers with $ab = ac$ and $a \neq 0$, then $b = c$.

PROOF. Suppose that $xy = 0$ and $x \neq 0$ and $y \neq 0$. Since $xy = 0 = x \cdot 0$ and $x \neq 0$; $y = 0$, which is a contradiction. ∎

9. Show, by giving a proof by contradiction, that if 100 balls are placed into nine boxes, some box contains 12 or more balls.

Formulate the arguments of Exercises 10–14 symbolically and determine whether each is valid. Let

p: I study hard. q: I get A's. r: I get rich.

10H. If I study hard, then I get A's.
I study hard.
∴ I get A's.

11. If I study hard, then I get A's.
If I don't get rich, then I don't get A's.
∴ I get rich.

12. I study hard if and only if I get rich.
I get rich.
∴ I study hard.

13H. If I study hard or I get rich, then I get A's.
I get A's.
∴ If I don't study hard, then I get rich.

14. If I study hard, then I get A's or I get rich.
I don't get A's and I don't get rich.
∴ I don't study hard.

In Exercises 15–19, write the given argument in words and determine whether each argument is valid. Let

p: 64K is better than no memory at all.
q: We will buy more memory.
r: We will buy a new computer.

15H. $p \rightarrow r$
$\underline{p \rightarrow q}$
$\therefore p \rightarrow (r \wedge q)$

16. $p \rightarrow (r \vee q)$
$\underline{r \rightarrow \bar{q}}$
$\therefore p \rightarrow r$

17. $p \rightarrow r$
$\underline{r \rightarrow q}$
$\therefore q$

18H. $\bar{r} \rightarrow \bar{p}$
$\underline{r}$
$\therefore p$

19. $p \rightarrow r$
$r \rightarrow q$
$\underline{p}$
$\therefore q$

Determine whether each argument in Exercises 20–24 is valid.

20H. $p \rightarrow q$
$\underline{\bar{p}}$
$\therefore \bar{q}$

21. $p \rightarrow q$
$\underline{\bar{q}}$
$\therefore \bar{p}$

22. $\underline{p \wedge \bar{p}}$
$\therefore q$

23H. $p \rightarrow (q \rightarrow r)$
$\underline{q \rightarrow (p \rightarrow r)}$
$\therefore (p \vee q) \rightarrow r$

24. $(p \rightarrow q) \wedge (r \rightarrow s)$
$\underline{p \vee r}$
$\therefore q \vee s$

25. Show that if

$$p_1, p_2 \ / \therefore p$$

and

$$p, p_3, \ldots, p_n \ / \therefore c$$

are valid arguments, the argument

$$p_1, p_2, \ldots, p_n \ / \therefore c$$

is also valid.

26H. Comment on the following argument.

Cassette tape storage is better than nothing.
Nothing is better than a hard disk drive.
∴ Cassette tape storage is better than a hard disk drive.

A.2 Categorical Propositions

A proposition such as

$$p\text{: Some computer scientists are musicians.} \qquad (A.2.1)$$

relates members of one class to members of a second class. In (A.2.1), the first class C_1 is the class of all computer scientists and the second class C_2 is the class of all musicians. Proposition (A.2.1) states that some members of C_1 are members of C_2.

Definition A.2.1. Let C_1 and C_2 be two classes. A *categorical proposition* is a proposition of one of the forms (a)–(d).
   (a) All C_1 is C_2.
   (b) No C_1 is C_2.
   (c) Some C_1 is C_2.
   (d) Some C_1 is not C_2.

Example A.2.2. If we let C_1 denote the class of all computer scientists and C_2 denote the class of all musicians, proposition (A.2.1) is of type (c) of Definition A.2.1.

Example A.2.3. Classify each categorical proposition as of type (a)–(d) according to Definition A.2.1 and describe the classes C_1 and C_2.

   (i) All trees are connected graphs.
   (ii) Some graphs are trees.
   (iii) All squares are rectangles.
   (iv) No two distinct parallel lines intersect.
   (v) Some context-free languages are not regular languages.
   (vi) All prime numbers greater than 2 are odd.

(i) This is a type (a) categorical proposition with

$$C_1 = \text{class of all trees.}$$
$$C_2 = \text{class of all connected graphs.}$$

(ii) This is a type (c) categorical proposition with

$$C_1 = \text{class of all graphs.}$$
$$C_2 = \text{class of all trees.}$$

(iii) This is a type (a) categorical proposition with

$$C_1 = \text{class of all squares.}$$
$$C_2 = \text{class of all rectangles.}$$

(iv) This is a type (b) categorical proposition with

$$C_1 = \text{class of all distinct pairs of parallel lines.}$$
$$C_2 = \text{class of all distinct pairs of intersecting lines.}$$

(v) This is a type (d) categorical proposition with

$$C_1 = \text{class of all context-free languages.}$$
$$C_2 = \text{class of all regular languages.}$$

(vi) This is a type (a) categorical proposition with

$$C_1 = \text{class of all prime numbers greater than 2.}$$
$$C_2 = \text{class of all odd numbers.}$$

The descriptor *all* in Definition A.2.1 type (a) and the descriptor *no* in Definition A.2.1 type (b) are called **universal quantifiers** since either a type (a) or (b) categorical proposition asserts something about all elements in the *universe*. A type (a) categorical proposition states that for all elements x in the universe, if x is in class C_1, then x is also in class C_2. A type (b) categorical proposition states that for all elements x in the universe, if x is in class C_1, then x is not in class C_2. The descriptor *some* in Definition A.2.1 type (c) and (d) categorical propositions is called an **existential quantifier** since a type (c) or (d) categorical proposition asserts the *existence* of a particular element. For example, a type (d) categorical proposition asserts that there exists an element in C_1 which is not in C_2.

There are ways to state propositions that are equivalent to the categorical propositions of Definition A.2.1. It is important to recognize that these alternative forms are really categorical propositions in disguise. We regard any proposition equivalent to one of (a)–(d) of Definition A.2.1 as a categorical proposition.

Example A.2.4.   Equivalent forms of

$$\text{All } C_1 \text{ is } C_2.$$

are

    (i) If x is a member of C_1, then x is a member of C_2.
    (ii) For all x, if x is a member of C_1, then x is a member of C_2.

Example A.2.5.   Equivalent forms of

$$\text{No } C_1 \text{ is } C_2.$$

are

    (i) If x is a member of C_1, then x is not a member of C_2.
    (ii) For all x, if x is a member of C_1, then x is not a member of C_2.

The word *some* in Definition A.2.1 types (c) and (d) is interpreted as meaning at least one; thus an equivalent formulation of proposition (A.2.1) is

    At least one computer scientist is a musician.

Examples A.2.6 and A.2.7 give general equivalent formulations of Definition A.2.1 types (c) and (d).

Example A.2.6.   Equivalent forms of

$$\text{Some } C_1 \text{ is } C_2.$$

are

    (i) Some C_2 is C_1.
    (ii) There is at least one member of C_1 that is also a member of C_2.
    (iii) There exists x such that x is a member of both C_1 and C_2.

Example A.2.7.   Equivalent forms of

$$\text{Some } C_1 \text{ is not } C_2.$$

are

    (i) Some non-C_2 is C_1.
    (ii) There is at least one member of C_1 that is not a member of C_2.
    (iii) There exists x such that x is a member of C_1 and x is not a member of C_2.

Example A.2.8.   By defining classes C_1 and C_2, state each proposition as a categorical proposition in the form of Definition A.2.1.

    (i) There exists x such that $x^2 \geq 4$ and $x < 2$.
    (ii) For all x, if $x \geq 2$, then $x^2 \geq 4$.
    (iii) If x is an irrational number, then x is not representable as a repeating decimal.
    (iv) There exists an isosceles triangle that is not equilateral.

    (i) Some C_1 is C_2.

$$C_1 = \text{class of numbers } x \text{ with } x^2 \geq 4.$$
$$C_2 = \text{class of numbers } x \text{ with } x < 2.$$

    (ii) All C_1 is C_2.

$$C_1 = \text{class of numbers } x \text{ with } x \geq 2.$$
$$C_2 = \text{class of numbers } x \text{ with } x^2 \geq 4.$$

    (iii) No C_1 is C_2.

$$C_1 = \text{class of irrational numbers.}$$
$$C_2 = \text{class of numbers representable as repeating decimals.}$$

    (iv) Some C_1 is not C_2.

$$C_1 = \text{class of isosceles triangles.}$$
$$C_2 = \text{class of equilateral triangles.}$$

Categorical propositions can be pictorially represented using Venn diagrams (see Section 2.2). We first discuss a Venn diagram for a type (a)

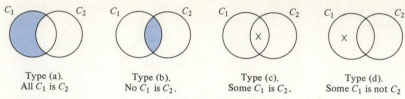

Type (a).   Type (b).   Type (c).   Type (d).
All C_1 is C_2.   No C_1 is C_2.   Some C_1 is C_2.   Some C_1 is not C_2

Figure A.2.1

categorical proposition

$$\text{All } C_1 \text{ is } C_2.$$

If every member of class C_1 is in class C_2, there can be no elements in the region representing $C_1 \cap \overline{C_2}$. We shade this region to indicate that it is empty (see Figure A.2.1a). Similarly, we obtain the Venn diagram shown in Figure A.2.1b for a type (b) categorical proposition. The Venn diagram for a type (c) categorical proposition

$$\text{Some } C_1 \text{ is } C_2.$$

shows an $\times$ in the region representing C_1 and C_2 (see Figure A.2.1c). The presence of the element $\times$ shows that at least one member of C_1 is in C_2. Similarly, we obtain the Venn diagram shown in Figure A.2.1d for a type (d) categorical proposition.

Venn diagrams are useful for testing the validity of arguments involving categorical propositions as our next examples illustrate.

Example A.2.9.   Use a Venn diagram to determine whether the following argument is valid.

$$\text{All } C_1 \text{ is } C_2.$$
$$\underline{\text{All } C_2 \text{ is } C_3.}$$
$$\therefore \text{All } C_1 \text{ is } C_3.$$

We assume that the hypotheses are true. To construct the Venn diagram, we first draw three overlapping circles to represent the classes C_1, C_2, and C_3 (see Figure A.2.2). Next, we shade the region $C_1 \cap \overline{C_2}$ (shown with vertical lines) to represent the proposition "All C_1 is C_2." Finally, we shade the region $C_2 \cap \overline{C_3}$ (shown with horizontal lines) to represent the proposition "All C_2 is C_3." We now see that all members of C_1 (if any) must be in C_3; that is, the conclusion "All C_1 is C_3" is true. The argument is valid.

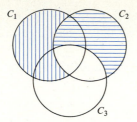

Figure A.2.2

Example A.2.10. Write the following argument using the forms of Definition A.2.1 for the categorical propositions. Use a Venn diagram to determine whether the argument is valid.

All professors are human beings.
All good teachers are human beings.
∴ All professors are good teachers.

If we let

C_1 = class of professors.
C_2 = class of human beings.
C_3 = class of good teachers.

we may rewrite the argument as

All C_1 is C_2.
All C_3 is C_2.
∴ All C_1 is C_3.

The Venn diagram representing the hypotheses is shown in Figure A.2.3. Does the conclusion follow? Consider the region with a check in it. Since this region is not shaded, it is not necessarily empty; that

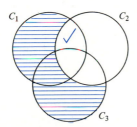

Figure A.2.3

is, there may be an element in C_1 but not in C_3. Therefore, the conclusion does not necessarily follow. The argument is invalid.

We next consider negations of categorical propositions. Consider the negation of

$$\text{All } C_1 \text{ is } C_2. \tag{A.2.2}$$

If this proposition is false, then the shaded region in Figure A.2.1a must have at least one member; that is, the proposition

$$\text{Some } C_1 \text{ is not } C_2. \tag{A.2.3}$$

is true. Similarly, if (A.2.2) is true, then (A.2.3) is false; thus, (A.2.3) is the negation of (A.2.2). By similar reasoning (see Exercise 38), we can show that the negation of

$$\text{No } C_1 \text{ is } C_2.$$

is

$$\text{Some } C_1 \text{ is } C_2.$$

Example A.2.11.

The negations of

(i) All squares are rectangles.
(ii) No two parallel lines intersect.
(iii) Some graphs are trees.
(iv) Some context-free languages are not regular languages.

are

(i) Some square is not a rectangle.
(ii) Some two parallel lines intersect.
(iii) No graph is a tree.
(iv) All context-free languages are regular languages.

Notice that the negation of a proposition involving a universal quantifier is a proposition involving an existential quantifier and that the negation of a proposition involving an existential quantifier is a proposition involving a universal quantifier.

The interpretation of certain universal quantifiers can be a bit tricky. In positive statements, "any," "all," "each," and "every" are equivalent.

$$\text{Any } C_1 \text{ is } C_2.$$
$$\text{All } C_1 \text{ is } C_2.$$
$$\text{Each } C_1 \text{ is } C_2.$$
$$\text{Every } C_1 \text{ is } C_2.$$

are considered to have the same meaning. In negative statements, the situation changes.

$$\text{Not all } C_1 \text{ is } C_2.$$
$$\text{Not each } C_1 \text{ is } C_2.$$
$$\text{Not every } C_1 \text{ is } C_2.$$

are all considered to have the same meaning as

$$\text{Some } C_1 \text{ is not } C_2.$$

whereas

$$\text{Not any } C_1 \text{ is } C_2.$$

means

$$\text{No } C_1 \text{ is } C_2.$$

We conclude this section by discussing some methods of proof used to establish theorems that can be stated as categorical propositions. To give a direct proof of a theorem of the form (A.2.2), we must examine an *arbitrary* element x of C_1 and show that x is also in C_2. To say that x is an arbitrary element in C_1, we mean that x could be *any* element in C_1. After showing that an arbitrarily chosen element in C_1 is in C_2, we can conclude that (A.2.2) is true.

Example A.2.12. The theorem

A quadratic equation has at most two solutions.

may be rephrased in the form (A.2.2) as

$$\text{All quadratic equations are equations with} \qquad \text{(A.2.4)}$$
$$\text{at most two solutions.}$$

PROOF. To prove (A.2.4), we consider an *arbitrary* quadratic equation

$$ax^2 + bx + c = 0.$$

The symbols a, b, and c denote arbitrary numbers with $a \neq 0$. The quadratic formula states that

$$x = \frac{-b \pm \sqrt{b^2 - 4ac}}{2a};$$

thus there are at most two solutions, namely

$$\frac{-b + \sqrt{b^2 - 4ac}}{2a} \quad \text{and} \quad \frac{-b - \sqrt{b^2 - 4ac}}{2a}. \qquad \blacksquare$$

In some cases, we are asked to show that (A.2.2) is *false*. This is the same as showing that the negation (A.2.3) of (A.2.2) is true. To show that (A.2.3) is true, we must exhibit an element in C_1 but not in C_2. Such an element is called a **counterexample** to (A.2.2).

Example A.2.13.   To show that

$$\text{For all integers } n, \ n^2 + n + 41 \text{ is prime.} \qquad \text{(A.2.5)}$$

is false, we must find a counterexample; that is, an integer n for which $n^2 + n + 41$ is not prime.

PROOF.   For $n = 41$,

$$n^2 + n + 41 = 41^2 + 41 + 41 = 41(41 + 1 + 1) = 41 \cdot 43$$

is not prime. Thus $n = 41$ is a counterexample to (A.2.5). $\blacksquare$

It is an interesting fact that $n^2 + n + 41$ is prime for $n = -40$, $-39, \ldots, -1, 0, 1, \ldots, 39$. It is not known whether there are any other second-degree polynomials that are prime for more than 80 consecutive values. The remarkable polynomial $n^2 + n + 41$ was discovered by Euler in 1772.

Example A.2.14 Fermat's Last Theorem.   The seventeenth-century mathematician Pierre de Fermat wrote in the margin of a book that if n is an integer greater than 2, then

$$x^n + y^n = z^n \qquad \text{(A.2.6)}$$

has no positive integer solutions. He also stated that he had a proof, but that the margin was too narrow to contain it. Fermat should have gotten some extra paper. To this day, it is unknown whether Fermat's

Last Theorem is true or false. To show that it is false, one would need a counterexample; that is, one would have to find positive integers x, y, and z and an integer $n \geq 3$, satisfying (A.2.6).

Example A.2.15. Euler conjectured that if m and n are positive integers satisfying $m < n$ and $n > 2$, then

$$z^n = x_1^n + x_2^n + \cdots + x_m^n$$

has no solution in positive integers. In words, no positive nth power is the sum of fewer than n positive nth powers. In 1966, about 200 years after Euler made his conjecture, Leon Lander and Thomas Parkin provided a counterexample:

$$27^5 + 84^5 + 110^5 + 133^5 = 144^5,$$

thus showing that Euler's conjecture is false.

To prove the proposition

$$\text{Some } C_1 \text{ is } C_2. \tag{A.2.7}$$

we must find an element in C_1 and C_2. A proof in which the element is shown to exist is called an **existence proof**.

Example A.2.16. To prove that some real numbers are irrational, it suffices to exhibit an irrational number such as $\sqrt{2}$ (and prove that $\sqrt{2}$ is irrational).

The truth of categorical propositions can sometimes be established using proof by contradiction. For example, if we want to establish (A.2.2) using proof by contradiction, we assume that (A.2.2) is false and deduce a contradiction. The negation of (A.2.2) is (A.2.3). Assuming (A.2.3), we know that there exists an element in C_1 that is not in C_2. Using this assumption as well as axioms, definitions, and previously derived theorems, we deduce a contradiction. Similarly, to establish (A.2.7) using proof by contradiction, we assume its negation:

$$\text{No } C_1 \text{ is } C_2.$$

Example A.2.17. By defining classes C_1 and C_2, state the proposition

For all x and y, if x is a rational number and y is an irrational number, then $x + y$ is an irrational number.

Pierre de Fermat (1601–1665)

Pierre de Fermat was a lawyer and a mathematician. There is a theorem in number theory named after Fermat. Fermat is one of the cofounders, with Pascal, of the theory of probability. Fermat's Last Theorem has been verified for $n < 125,000$, so if there is a counterexample, the numbers involved would be quite large. Fermat believed that

$$2^{2^n} + 1$$

is prime for $n = 0, 1, 2, \ldots$, but did not claim to have a proof. These numbers are now known as the **Fermat numbers**. In 1732, Euler showed that

$$2^{2^5} + 1 = 4,294,967,297 = 6,700,417 \cdot 641,$$

thus providing a counterexample to Fermat's claim. [*Photo courtesy of The Bettmann Archive*]

as a categorical proposition in the form of Definition A.2.1. Prove this proposition using proof by contradiction. Assume as a previous theorem: If a and b are rational numbers, then $a - b$ is a rational number.

PROOF. If we let C_1 be the class of numbers of the form $x + y$, where x is rational and y is irrational, and C_2 be the class of irrational numbers, we can rephrase the theorem as

$$\text{All } C_1 \text{ is } C_2. \tag{A.2.8}$$

Assume that the given statement is false. We are then assuming that the negation of (A.2.8),

$$\text{Some } C_1 \text{ is not } C_2.$$

is true; that is, that there exist numbers x and y with x rational and y irrational such that $z = x + y$ is rational. By the previous theorem, $y = z - x$ is rational. But now y is both rational and irrational—a contradiction. The theorem is proved. ∎

EXERCISES

1. Find an example of a categorical proposition in a newspaper.

In Exercises 2–11, classify each categorical proposition as of type (a)–(d) according to Definition A.2.1 and describe the classes C_1 and C_2. Also, give the negation of each proposition.

2H. All isosceles triangles are equilateral triangles.

3. Some similar triangles are not congruent.

4. Some expert bridge players can also play a mean game of cribbage.

5H. Some chips are not functional.

6. No one over 30 is trustworthy.

7. Some successful persons never attended college.

8H. All violent movies are R-rated movies.

9. All connected graphs are graphs with spanning trees.

10. Some linear programming algorithms are polynomial-time algorithms.

11H. Nobody is despised who can manage a crocodile.

By defining classes C_1 and C_2, state each proposition in Exercises 12–15 as a categorical proposition in the form of Definition A.2.1. Also, give the negation of each proposition.

12H. If G is a planar map, then G can be colored using at most four colors.

13. There is at least one Cubs fan who is rational.

14. There exists a continuous function that is not differentiable at any point.

15H. If software has poor documentation, then it is not worth much.

Use Venn diagrams to determine whether the arguments in Exercises 16–25 are valid.

16H. Some C_1 is C_2.
All C_2 is C_3.
∴ Some C_3 is C_1.

17. No C_1 is C_2.
Some C_2 is C_3.
∴ Some C_3 is not C_1.

18. All C_1 is C_2.
No C_3 is C_1.
∴ No C_3 is C_2.

19H. No C_1 is C_2.
Some C_3 is C_1.
∴ Some C_3 is not C_2.

20. Some C_1 is not C_2.
All C_2 is C_3.
∴ Some C_3 is not C_1.

21. Some C_1 is C_2.
All C_3 is C_1.
∴ Some C_3 is C_2.

22H. No C_1 is C_2.
All C_3 is C_1.
∴ No C_3 is C_2.

23. All C_1 is C_2.
Some C_1 is not C_3.
∴ Some C_3 is not C_2.

24. No C_1 is C_2.
Some C_1 is C_3.
∴ Some C_3 is not C_2.

25H. All C_1 is C_2.
Some C_3 is not C_1.
∴ Some C_3 is not C_2.

Write each argument in Exercises 26–31 using the forms of Definition A.2.1 for the categorical propositions. Use Venn diagrams to determine whether the arguments are valid.

26H. No Cubs are White Sox.
All White Sox are baseball players.
∴ Some baseball players are not Cubs.

27. Some voters are not Republicans.
Some people are not voters.
∴ Some people are not Republicans.

28. All trees are graphs.
Some structures are not graphs.
∴ Some structures are not trees.

29H. All dogs are animals.
No dogs are horses.
∴ No horses are animals.

30. Some integers are not perfect numbers.
All integers are real numbers.
∴ Some real numbers are not perfect numbers.

31. All movies are violent epics.
No violent epics are worth watching.
∴ Nothing worth watching is a movie.

Provide counterexamples to the propositions in Exercises 32–35.

32H. Every equation that has exactly two distinct solutions is a quadratic equation.

33. For every real number $x > 0$, $(x^2 + 1)/x < 100$.

34. Every personal computer costs less than $3000.

35H. For any prime p, $n^2 + n + p$ is prime for $n = 0, 1, \ldots,$
$p - 1$.

Give existence proofs of the propositions in Exercises 36 and 37.

36H. There is a solution in positive integers to

$$x^2 + y^2 = z^2.$$

*37. There is a polynomial $p(n)$ with $p(1) = p(2) = p(3) = 0$ and $p(4) = 162$.

38. Give an argument to show that the negation of

$$\text{No } C_1 \text{ is } C_2.$$

is

$$\text{Some } C_1 \text{ is } C_2.$$

39. Verify the counterexample to Euler's conjecture given in Example A.2.15.

40. Verify the counterexample to the claim that all Fermat numbers are prime given in Fermat's biography.

References

AHO, A., J. HOPCROFT, and J. ULLMAN, *The Design and Analysis of Computer Algorithms*, Addison-Wesley, Reading, Mass., 1974.

AINSLIE, T., *Ainslie's Complete Hoyle*, Simon and Schuster, New York, 1975.

BAASE, S., *Computer Algorithms: Introduction to Design and Analysis,* Addison-Wesley, Reading, Mass., 1978.

BARKER, S. F., *The Elements of Logic*, 2nd ed., McGraw-Hill, New York, 1984.

BELLMAN, R., K. L. COOKE, and J. A. LOCKETT, *Algorithms, Graphs, and Computers*, Academic Press, New York, 1970.

BELLMORE, M., and G. L. NEMHAUSER, ''The traveling salesman problem,'' *Oper. Res.*, 16 (1968), 538–558.

BERGE, C., *The Theory of Graphs and Its Applications*, Wiley, New York, 1962.

BERLEKAMP, E. R., J. H. CONWAY, and R. K. GUY, *Winning Ways,* Vols. 1 and 2, Academic Press, New York, 1982.

BERLINER, H., ''Computer backgammon,'' *Sci. Am.*, June 1980, 64–72.

BOGART, K. P., *Introductory Combinatorics*, Pitman, Marshfield, Mass., 1983.

BRUALDI, R. A., *Introductory Combinatorics*, North-Holland, New York, 1977.

BUSACKER, R. G., and T. L. SAATY, *Finite Graphs and Networks: An Introduction with Applications*, McGraw-Hill, New York, 1965.

CODD, E. F., "A relational model of data for large shared databanks," *Commun. ACM*, 13 (1970), 377–387.

COPI, I. M., *Introduction to Logic*, 7th ed., Macmillan, New York, 1986.

CULL, P., and E. F. ECKLUND, JR., "Towers of Hanoi and analysis of algorithms," *Am. Math. Mon.*, 92 (1985), 407–420.

D'ANGELO, H., *Microcomputer Structures*, McGraw-Hill, New York, 1981.

DATE, C. J., *An Introduction to Database Systems*, Vol. I, 4th ed., Addison-Wesley, Reading, Mass., 1986.

DEO, N., *Graph Theory with Applications to Engineering and Computer Science*, Prentice-Hall, Englewood Cliffs, N.J., 1974.

DUDA, R. O., and P. E. HART, *Pattern Classification and Scene Analysis*, Wiley, New York, 1973.

EULER, L., "Leonhard Euler and the Koenigsberg bridges," J. R. Newman, ed., *Sci. Am.*, July 1953, 66–70.

EVEN, S., *Algorithmic Combinatorics*, Macmillan, New York, 1973.

EVEN, S., *Graph Algorithms*, Computer Science Press, Rockville, Md., 1979.

FREEMAN, J., *The Playboy Winner's Guide to Board Games*, Playboy, Chicago, 1979.

FREY, P., "Machine-problem solving—Part 3: The alpha-beta procedure," *Byte*, 5 (November 1980), 244–264.

GARDNER, M., *Mathematical Puzzles & Diversions*, Simon and Schuster, New York, 1959.

GARDNER, M., *Mathematical Circus*, Knopf, New York, 1979.

GOLDBERG, S., *Introduction to Difference Equations*, Wiley, New York, 1958.

GOLDSTINE, H. H., *The Computer from Pascal to von Neumann*, Princeton University Press, Princeton, N.J., 1972.

GOSE, E. E., "Introduction to biological and mechanical pattern recognition," in *Methodologies of Pattern Recognition*, S. Watanabe, ed., Academic Press, New York, 1969, pp. 203–252.

GUSTASON, W., and D. E. ULRICH, *Elementary Symbolic Logic*, Waveland Press, Prospect Heights, Ill., 1973.

HALMOS, P. R., *Lectures on Boolean Algebras*, D. Van Nostrand, New York, 1967.

HALMOS, P. R., *Naive Set Theory*, Springer-Verlag, New York, 1974.

HARARY, F., *Graph Theory*, Addison-Wesley, Reading, Mass., 1969.

HILL, F. J., and G. R. PETERSON, *Switching Theory and Logical Design*, 2nd ed., Wiley, New York, 1974.

HOHN, F. E., *Applied Boolean Algebra*, 2nd ed., Macmillan, New York, 1966.

HOROWITZ, E., and S. SAHNI, *Fundamentals of Data Structures*, Computer Science Press, Rockville, Md., 1976.

HOROWITZ, E., and S. SAHNI, *Fundamentals of Computer Algorithms*, Computer Science Press, Rockville, Md., 1978.

HU, T. C., *Combinatorial Algorithms*, Addison-Wesley, Reading, Mass., 1982.

HUFFMAN, D. A., "A method for the construction of minimum-redundancy codes," *Proc. IRE*, 40 (1952), 1098–1101.

JACOBS, H. R., *Geometry*, W. H. Freeman, San Francisco, 1974.

JEFFREY, R. C., *Formal Logic: Its Scope and Limits*, McGraw-Hill, New York, 1967.

JOHNSONBAUGH, R., *Discrete Mathematics*, rev. ed., Macmillan, New York, 1986.

KLINE, M., *Mathematical Thought from Ancient to Modern Times*, Oxford University Press, New York, 1972.

KNUTH, D. E., *The Art of Computer Programming*, Vol. 1: *Fundamental Algorithms*, 2nd ed., Addison-Wesley, Reading, Mass., 1973.

KNUTH, D. E., *The Art of Computer Programming*, Vol. 3: *Sorting and Searching*, Addison-Wesley, Reading, Mass., 1973.

KNUTH, D. E., "Algorithms," *Sci. Am.,* April 1977, 63–80.

KNUTH, D. E., *The Art of Computer Programming*, Vol. 2: *Seminumerical Algorithms*, 2nd ed., Addison-Wesley, Reading, Mass., 1981.

KNUTH, D. E., "Algorithmic thinking and mathematical thinking," *Amer. Math. Mon.* 92 (1985), 170–181.

KOHAVI, Z., *Switching and Finite Automata Theory*, McGraw-Hill, New York, 1970.

KÖNIG D., *Theorie der endlichen und unendlichen Graphen,* Leipzig, 1936. (Reprinted in 1950 by Chelsea, New York.)

KROENKE, D., *Database Processing*, 2nd ed., Science Research Associates, Chicago, 1983.

KRUSE, R. L., *Data Structures and Program Design*, 2nd ed., Prentice-Hall, Englewood Cliffs, N.J., 1987.

LIPSCHUTZ, S., *Theory and Problems of Set Theory and Related Topics*, Schaum, New York, 1964.

LIPSCHUTZ, S., *Discrete Mathematics*, Schaum, New York, 1976.

LIPSCHUTZ, S., *Essential Computer Mathematics,* Schaum, New York, 1982.

LIU, C. L., *Introduction to Combinatorial Mathematics*, McGraw-Hill, New York, 1968.

LIU, C. L., *Elements of Discrete Mathematics*, 2nd ed., McGraw-Hill, New York, 1985.

McNAUGHTON, R., *Elementary Computability, Formal Languages, and Automata*, Prentice-Hall, Englewood Cliffs, N.J., 1982.

MENDELSON, E., *Boolean Algebra and Switching Circuits,* Schaum, New York, 1970.

NIEVERGELT, J., J. C. FARRAR, and E. M. REINGOLD, *Computer Approaches to Mathematical Problems*, Prentice-Hall, Englewood Cliffs, N.J., 1974.

NILSSON, N. J., *Problem-Solving Methods in Artificial Intelligence*, McGraw-Hill, New York, 1971.

NIVEN, I., *Mathematics of Choice*, Mathematical Association of America, Washington, D.C., 1965.

ORE, O., *Graphs and Their Uses*, Random House, New York, 1963.

PARRY, R. T., and H. PFEFFER, "The infamous traveling-salesman problem: A practical approach," *Byte*, 6 (July 1981), 252–290.

PEARL, J., "The solution for the branching factor of the alpha-beta pruning algorithm and its optimality," *Commun. ACM*, 25 (1982), 559–564.

REID, C., *Hilbert*, Springer-Verlag, New York, 1970.

REID, T. R., *The Chip: How Two Americans Invented the Microchip and Launched a Revolution*, Simon and Schuster, New York, 1984.

REINGOLD, E., J. NIEVERGELT, and N. DEO, *Combinatorial Algorithms*, Prentice-Hall, Englewood Cliffs, N.J., 1977.

RESNIK, M. D., *Elementary Logic*, McGraw-Hill, New York, 1970.

RIORDAN, J., *An Introduction to Combinatorial Analysis,* Wiley, New York, 1958.

RITTER, G. L., S. R. LOWRY, H. B. WOODRUFF, and T. L. ISENHOUR, "An aid to the superstitious," *Math. Teacher*, May 1977, 456–457.

ROBERTS, F. S., *Applied Combinatorics*, Prentice-Hall, Englewood Cliffs, N.J., 1984.

ROSS, K. A., and C. R. B. WRIGHT, *Discrete Mathematics*, Prentice-Hall, Englewood Cliffs, N.J., 1985.

SLAGLE, J. R., *Artificial Intelligence: The Heuristic Programming Approach*, McGraw-Hill, New York, 1971.

SOLOW, D., *How to Read and Do Proofs*, Wiley, New York, 1982.

STANDISH, T. A., *Data Structure Techniques*, Addison-Wesley, Reading, Mass., 1980.

STOLL, R. R., *Set Theory and Logic*, W. H. Freeman, San Francisco, 1963.

TSICHRITZIS, D. C., and F. H. LOCHOVSKY, *Data Base Management Systems*, Academic Press, New York, 1977.

TUCKER, A., *Applied Combinatorics*, 2nd ed., Wiley, New York, 1985.

ULLMAN, J. D., *Principles of Database Systems*, Computer Science Press, Rockville, Md., 1980.

VILENKIN, N. Y., *Combinatorics*, Academic Press, New York, 1971.

WILLIAMS, G., and R. MEYER, "The Panasonic and Quasar hand-held computers: beginning a new generation of consumer computers," *Byte*, 6 (January 1981), 34–45.

Hints and Solutions to Selected Exercises

SECTION 1.1

1. True   **4.** True

7.

p	q	$p \wedge \bar{q}$
T	T	F
T	F	T
F	T	F
F	F	F

10.

p	q	$(p \wedge q) \wedge \bar{p}$
T	T	F
T	F	F
F	T	F
F	F	F

13.

p	q	$(p \vee q) \wedge (\bar{p} \vee q) \wedge (p \vee \bar{q}) \wedge (\bar{p} \vee \bar{q})$
T	T	T
T	F	T
F	T	T
F	F	T

15. $p \wedge q$ **18.** $p \vee \overline{(q \wedge r)}$

21. Today is Monday or it is raining.

24. (Today is Monday and it is raining) and it is not the case that (it is hot or today is Monday).

27. 1   **30.** 3

SECTION 1.2

1. If a person is a Cub, then the person is a great baseball player.
4. If the function f is continuous, then f is integrable.
7. If the chairperson gives the lecture, then the audience will go to sleep.
9. (For Exercise 1) If the person is a great baseball player, then the person is a Cub.
11. True   **14.** False   **17.** False   **19.** $p \to \bar{q}$
22. $\overline{(r \wedge q)} \to \bar{r}$
25. If today is Monday, then it is raining.
28. It is not the case that today is Monday or it is raining if and only if it is hot.
31. Let p: $4 < 6$ and q: $9 > 12$.
    Given statement: $p \to q$; false.
    Converse: $q \to p$; if $9 > 12$, then $4 < 6$; true.
    Contrapositive: $\bar{q} \to \bar{p}$; if $9 \le 12$, then $4 \ge 6$; false.
34. Let p: $|4| < 3$ and q: $-3 < 4 < 3$.
    Given statement: $q \to p$; true.
    Converse: $p \to q$; if $|4| < 3$, then $-3 < 4 < 3$; true.
    Contrapositive: $\bar{p} \to \bar{q}$; if $|4| \ge 3$, then $-3 \ge 4$ or $4 \ge 3$; true.
35. $P \not\equiv Q$ **38.** $P \not\equiv Q$ **41.** $P \not\equiv Q$ **44.** $P \not\equiv Q$

45.

p	q	p impl q	p impl p
T	T	T	T
T	F	F	F
F	T	F	F
F	F	T	T

Since p impl q is true precisely when q impl p is true, p impl $q \equiv q$ impl p.

48.

p	q	$p \to q$	$\bar{p} \vee q$
T	T	T	T
T	F	F	F
F	T	T	T
F	F	T	T

Since $p \to q$ is true precisely when $\bar{p} \vee q$ is true, $p \to q \equiv \bar{p} \vee q$.

SECTION 1.3

1. $\overline{x_1 \wedge x_2}$

x_1	x_2	$\overline{x_1 \wedge x_2}$
1	1	0
1	0	1
0	1	1
0	0	1

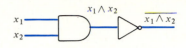

4.

x_1	x_2	x_3	$((x_1 \wedge x_2) \vee \overline{(x_1 \wedge x_3)}) \wedge \overline{x_3}$
1	1	1	0
1	1	0	1
1	0	1	0
1	0	0	1
0	1	1	0
0	1	0	1
0	0	1	0
0	0	0	1

7. 0   **10.** 1

13. Suppose that $x = 1$ and $y = 0$. Then the input to the AND gate is 1, 0. Thus the output of the AND gate is 0. Since this is then NOTed, $y = 1$. Contradiction. Similarly, if $x = 1$ and $y = 1$, we obtain a contradiction.

16.

19. $(A \wedge B) \vee (C \wedge \overline{A})$

A	B	C	$(A \wedge B) \vee (C \wedge \overline{A})$
1	1	1	1
1	1	0	1
1	0	1	0
1	0	0	0
0	1	1	1
0	1	0	0
0	0	1	1
0	0	0	0

22. $(A \wedge (C \vee (D \wedge C))) \vee (B \wedge (\overline{D} \vee (C \wedge A) \vee \overline{C}))$

25.

A	B	$(A \lor \bar{B}) \land A$
1	1	1
1	0	1
0	1	0
0	0	0

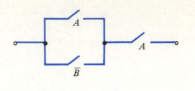

28.

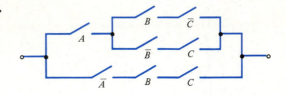

SECTION 1.4

In these hints, $a \land b$ is written ab.

1. $x_1 \bar{x}_2$

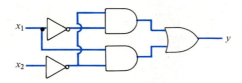

4. $x_1 \bar{x}_2 \lor \bar{x}_1 \bar{x}_2$

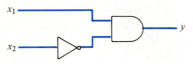

7. $x_1 x_2 x_3 \lor x_1 x_2 \bar{x}_3 \lor x_1 \bar{x}_2 x_3 \lor \bar{x}_1 x_2 \bar{x}_3 \lor \bar{x}_1 \bar{x}_2 x_3 \lor \bar{x}_1 \bar{x}_2 \bar{x}_3$

10. $x_1 \bar{x}_2 \bar{x}_3 \lor \bar{x}_1 x_2 x_3 \lor \bar{x}_1 x_2 \bar{x}_3 \lor \bar{x}_1 \bar{x}_2 x_3$

12. (For Exercise 1) $(\bar{x}_1 \lor \bar{x}_2) \land (x_1 \lor \bar{x}_2) \land (x_1 \lor x_2)$

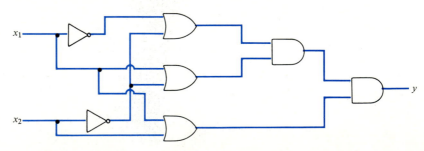

SECTION 1.5

1. $\bar{y}$ **4.** $\bar{x} \vee y$

7. (For Exercise 7)(This answer is not unique.)

$$(x \wedge y) \vee (x \wedge z) \vee (\bar{y} \wedge z) \vee (\bar{x} \wedge \bar{z})$$

8. $\bar{x}$ **11.** $\bar{x}$ **14.** $(\bar{x} \wedge \bar{y}) \vee z$

17.

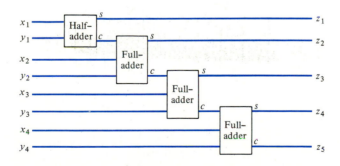

SECTION 1.6

1. $9 \cdot 10^3 + 8 \cdot 10^2 + 4 \cdot 10^1 + 1 \cdot 10^0$

4. $4 \cdot 10^4 + 8 \cdot 10^3 + 7 \cdot 10^2 + 0 \cdot 10^1 + 2 \cdot 10^0$

7. 9   **10.** 32   **13.** 100010   **16.** 110010000   **19.** 11000

22. 1001000   **25.** 58   **28.** 2563   **31.** (For Exercise 1) 2671

34. FE   **37.** 3DBF9

39. 2010 cannot represent a number in binary because 2 is an illegal symbol in binary. 2010 could represent a number in either decimal or hexadecimal.

41. 51   **44.** 4570   **47.** (For Exercise 13) 42

50. (For Exercise 41) 33

53. 9450 cannot represent a number in binary because 9, 4, and 5 are illegal symbols in binary. 9450 cannot represent a number in octal because 9 is an illegal symbol in octal. 9450 does represent a number in either decimal or hexadecimal.

54. $c = 0, s = 1$ **57.** $c = 0, s = 1$

59.

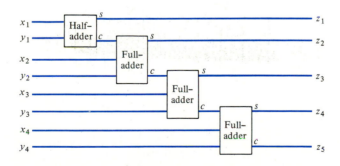

65. 1101   **68.** 00110101

71. The logic table is

b	FLAGIN	y	FLAGOUT
1	1	0	1
1	0	1	1
0	1	1	1
0	0	0	0

Thus $y = b \oplus$ FLAGIN and FLAGOUT $= b \lor$ FLAGIN. We obtain the circuit

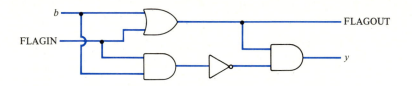

CHAPTER 1 SELF-TEST

1. False

2.

p	q	r	$\overline{(p \land q)} \lor (p \lor \bar{r})$
T	T	T	T
T	T	F	T
T	F	T	T
T	F	F	T
F	T	T	T
F	T	F	T
F	F	T	T
F	F	F	T

3. I take calculus and either I do not take philosophy or I take art.

4. $p \lor (q \land \bar{r})$

5. If Leah gets an A in discrete mathematics, then Leah studies hard.

6. Converse: If Leah studies hard, then Leah gets an A in discrete mathematics. Contrapositive: If Leah does not study hard, then Leah does not get an A in discrete mathematics.

7. True   **8.** $(\bar{r} \lor q) \to \bar{q}$

9.

x	y	z	$\overline{(x \wedge \bar{y})} \vee z$
1	1	1	1
1	1	0	1
1	0	1	1
1	0	0	0
0	1	1	1
0	1	0	1
0	0	1	1
0	0	0	1

10. 1

11.

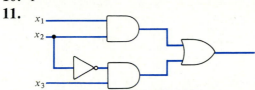

12. Suppose that x is 1. Then the upper input to the OR gate is 0. If y is 1, then the lower input to the OR gate is 0. Since both inputs to the OR gate are 0, the output y of the OR gate is 0 which is impossible. If y is 0, then the lower input to the OR gate is 1. Since an input to the OR gate is 1, the output y of the OR gate is 1 which is impossible. Therefore, if the input to the circuit is 1, the output is not uniquely determined. Thus the circuit is not a combinatorial circuit.

    In Exercises 13–16, $a \wedge b$ is written ab.

13. $x_1 \bar{x}_2 \bar{x}_3$

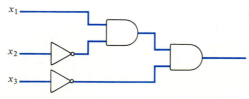

14. $x_1 x_2 \bar{x}_3 \vee x_1 \bar{x}_2 \bar{x}_3$

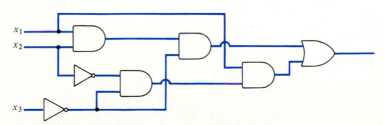

15. $x_1 x_2 x_3 \vee x_1 \bar{x}_2 \bar{x}_3 \vee \bar{x}_1 \bar{x}_2 \bar{x}_3$

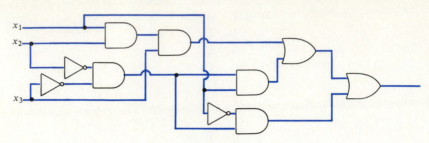

16. $x_1 x_2 \bar{x}_3 \vee x_1 \bar{x}_2 \bar{x}_3 \vee \bar{x}_1 x_2 x_3 \vee \bar{x}_1 \bar{x}_2 x_3$

17. x **18.** $\bar{x} \vee \bar{y}$ **19.** $(\bar{x} \wedge y) \vee z$ **20.** $(y \wedge z) \vee \bar{z}$.

21. 150   **22.** $1AE_{16}$, 110101110_2 **23.** 1000010   **24.** 3129

25.

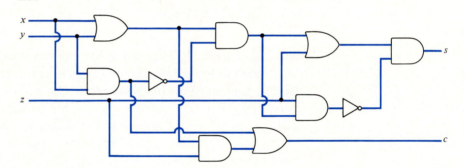

SECTION 2.1

1. $\{1, 2, 3, 4, 5, 7, 10\}$ **4.** $\{2, 3, 5\}$ **7.** $\varnothing$ **10.** U

13. $\{6, 8\}$ **16.** $\{1, 2, 3, 4, 5, 7, 10\}$ **17.** True   **20.** True

21. Equal   **24.** Equal

27. $\varnothing$, $\{a\}$, $\{b\}$, $\{a, b\}$. All but $\{a, b\}$ are proper subsets of $\{a, b\}$.

30. $2^n - 1$ **31.** $A \subseteq B$ **34.** $A \subseteq B$

35. $A \, \Delta \, B = \{1, 4, 5\}$

SECTION 2.2

1.   **4.** Same as Exercise 1

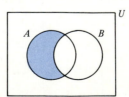

7.

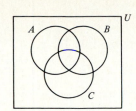

9. 10   **12.** 54   **15.** True   **18.** False

22. $\{(1, a), (1, b), (1, c), (2, a), (2, b), (2\ c)\}$

25. $\{(a, a), (a, b), (a, c), (b, a), (b, b), (b, c), (c, a), (c, b), (c, c)\}$

28. $\{(1, a, \alpha), (1, a, \beta), (2, a, \alpha), (2, a, \beta)\}$

31. $\{(a, 1, a, \alpha), (a, 1, a, \beta), (a, 2\ a, \alpha), (a, 2, a, \beta)\}$

35. $\{\{1\}, \{2\}, \{3\}, \{4, 5, 6, 7\}\}$ **38.** $\{\{1\}\}$

41. $\{\{a\}, \{b\}, \{c\}, \{d\}\}, \{\{a, b\}, \{c\}, \{d\}\}, \{\{a, c\}, \{b\}, \{d\}\}, \{\{a, d\},$
$\{b\}, \{c\}\}, \{\{b, c\}, \{a\}, \{d\}\}, \{\{b, d\}, \{a\}, \{c\}\}, \{\{c, d\}, \{a\}, \{b\}\},$
$\{\{a, b\}, \{c, d\}\}, \{\{a, c\}, \{b, d\}\}, \{\{a, d\}, \{b, c\}\}, \{\{a, b, c\}, \{d\}\},$
$\{\{a, b, d\}, \{c\}\}, \{\{a, c, d\}, \{b\}\}, \{\{b, c, d\}, \{a\}\}, \{\{a, b, c, d\}\}$

42. $|A| + |B|$ counts each element in regions 2 and 4 of Figure 2.2.1 once and each element in region 3 that represents $A \cap B$ twice. Thus $|A| + |B| - |A \cap B|$ counts each element in regions 2, 3, and 4, that represents $A \cup B$, once. The formula follows.

SECTION 2.3

1. (a) c (b) c (c) $cddcdc$

4. (a) 12   (b) 23   (c) 7   (d) 46   (e) 1   (f) 3
(g) 3   (h) 21

7. (a) 15   (b) 155   (c) $2n + 3(n - 1)n/2$ **10.** $3^n n!$

13. (a) $a_n = a_{n-1} + 2^{n-1}$, $n \geq 2$; $a_1 = 2$
(b) $a_n = a_{n-1} - (n - 1)$, $n \geq 2$; $a_1 = 21$

16. $b_1 = 2$, $b_2 = 3$, $b_3 = 5$, $b_4 = 8$, $b_5 = 12$, $b_6 = 257$

18. 00, 01, 10, 11

21. 000, 010, 001, 011, 100, 110, 101, 111, 00, 01, 11, 10, 0, 1, λ

24. $a_0 = 1$ because initially there is one pair of rabbits. After one month, there is still just one pair because a pair does not become productive until after one month. Therefore, $a_1 = 1$. The increase in pairs of rabbits $a_n - a_{n-1}$ from month $n - 1$ to month n is due to each pair alive in month $n - 2$ producing an additional pair. That is, $a_n - a_{n-1} = a_{n-2}$. Since $\{a_n\}$ satisfies the same recurrence relation as $\{f_n\}$ and the same initial conditions, $a_n = f_n$, $n \geq 0$.

SECTION 2.4

1. BASIS STEP. $1 = 1^2$
INDUCTIVE STEP. Assume true for n.

$$1 + \cdots + (2n - 1) + (2n + 1) = n^2 + 2n + 1 = (n + 1)^2$$

4. BASIS STEP. $1(1!) = 1 = 2! - 1$
INDUCTIVE STEP. Assume true for n.

$$1(1!) + \cdots + n(n!) + (n + 1)(n + 1)! = (n + 1)! - 1$$
$$+ (n + 1)(n + 1)!$$
$$= (n + 2)! - 1$$

7. BASIS STEP. $1^3 = 1 = \left[\dfrac{1 \cdot 2}{2}\right]^2$
INDUCTIVE STEP. Assume true for n.

$$1^3 + \cdots + n^3 + (n + 1)^3 = \left[\dfrac{n(n + 1)}{2}\right]^2 + (n + 1)^3$$
$$= \left[\dfrac{(n + 1)(n + 2)}{2}\right]^2$$

10. BASIS STEP. $\dfrac{1}{2^2 - 1} = \dfrac{1}{3} = \dfrac{3}{4} - \dfrac{1}{2 \cdot 2} - \dfrac{1}{2 \cdot 3}$
INDUCTIVE STEP. Assume true for n.

$$\dfrac{1}{2^2 - 1} + \cdots + \dfrac{1}{(n + 1)^2 - 1} + \dfrac{1}{(n + 2)^2 - 1}$$
$$= \dfrac{3}{4} - \dfrac{1}{2(n + 1)} - \dfrac{1}{2(n + 2)} + \dfrac{1}{(n + 2)^2 - 1}$$
$$= \dfrac{3}{4} - \dfrac{1}{2(n + 2)} - \dfrac{1}{2(n + 3)}$$

13. BASIS STEP. $\dfrac{1}{2 \cdot 1} = \dfrac{1}{2} \leq \dfrac{1}{2}$
INDUCTIVE STEP. Assume true for n.

$$\dfrac{1}{2(n + 1)} = \dfrac{n}{n + 1} \cdot \dfrac{2}{n} \leq \dfrac{n}{n + 1} \dfrac{1 \cdot 3 \cdot 5 \cdot \ldots \cdot (2n - 1)}{2 \cdot 4 \cdot 6 \cdot \ldots \cdot (2n)}$$
$$\leq \dfrac{1 \cdot 3 \cdot 5 \cdot \ldots \cdot (2n - 1)(2n + 1)}{2 \cdot 4 \cdot 6 \cdot \ldots \cdot (2n)(2n + 2)}$$

since

$$\frac{n}{n + 1} \leq \frac{2n + 1}{2n + 2}.$$

16. BASIS STEP. $(1 + x)^1 = 1 + x \geq 1 + 1 \cdot x$
INDUCTIVE STEP. Assume true for n.

$$(1 + x)^{n+1} = (1 + x)^n(1 + x) \geq (1 + nx)(1 + x)$$
$$= 1 + nx + x + nx^2 \geq 1 + (n + 1)x$$

19. BASIS STEP. $\overline{X_1} = \overline{X_1}$
INDUCTIVE STEP. Assume true for n.

$$\overline{X_1 \cup \cdots \cup X_n \cup X_{n+1}} = \overline{(X_1 \cup \cdots \cup X_n) \cup X_{n+1}}$$
$$= \overline{X_1 \cup \cdots \cup X_n} \cap \overline{X_{n+1}}$$

Theorem 2.2.3k

$$= \overline{X_1} \cap \cdots \cap \overline{X_n} \cap \overline{X_{n+1}}$$

Inductive assumption

22. $S_1 = 0 \neq 2; 2 + \cdots + 2n + 2(n + 1) = S_n + 2n + 2 = (n + 2)(n - 1) + 2n + 2 = (n + 3)n = S_{n+1}$.

SECTION 2.5

1. Is a function. Domain $= X$; range $= \{a, b, c\}$. $f(1) = a$, $f(2) = a$.

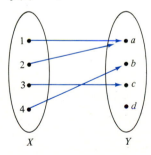

4. Not a function. The element 3 of the domain has not been assigned an element of the range.

7. 13   **10.** -13 **13.** 2   **16.** 1   **19.** $x = y = 2.5$
22. $x = y = 2.5$ **25.** $x = 2.5$ **28.** $P(x) = 0.01x$

30. If n is an odd integer, $n = 2k - 1$ for some integer k. Now

$$\frac{n^2}{4} = \frac{(2k - 1)^2}{4} = \frac{4k^2 - 4k + 1}{4} = k^2 - k + \frac{1}{4}.$$

Since $k^2 - k$ is an integer,

$$\left\lfloor \frac{n^2}{4} \right\rfloor = k^2 - k.$$

The result now follows because

$$\frac{n - 1}{2} \frac{n + 1}{2} = \frac{(2k - 1) - 1}{2} \frac{(2k - 1) + 1}{2} = \frac{(2k - 1)^2 - 1}{4}$$

$$= \frac{4k^2 - 4k}{4} = k^2 - k.$$

32.

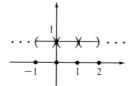

35.

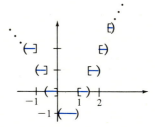

36. Friday   **39.** February, March, November

42. By looking at the calendar of a month whose first day is Sunday, it is apparent that a month has a Friday the 13th if and only if it begins on Sunday. Thus we determine which months begin on Sunday. Obviously, January itself begins on Sunday (leap year or not). Since January has 31 days and 31 *mod* 7 = 3, February begins on Wednesday. In a non-leap year, February has 28 days; and since 28 *mod* 7 = 0, March also begins on Wednesday. Arguing in this way, we see that in a non-leap year, April begins on Saturday; May begins on Monday; June begins on Thursday; July begins on Saturday; August begins on Tuesday; September begins on Friday; October begins on Sunday (and so has a Friday the 13th); November begins on Wednesday; and December begins on Friday.

In a leap year, each month excepting January and February begins one day later. Thus, in a leap year, January, April, and July have Friday the 13ths.

In the solutions to Exercises 45 and 48, *a:b* means store item *a* in cell *b*.

45. 10:0, 13:2, 377:3, 796:4, 281:6, 743:7, 53:9, 20:10

48. 714:0, 631:6, 26:5, 373:1, 775:8, 906:13, 509:2, 2032:7, 42:4, 4:3, 136:9, 1028:10

SECTION 2.6

1. Exercise 1—Neither one-to-one nor onto. Exercise 2—Not a function. Exercise 3—Both one-to-one and onto. Exercise 4—Not a function. Exercise 5—Neither one-to-one nor onto. Exercise 6—Both one-to-one and onto.

For Exercise 3: Inverse function = {(*c*, 1), (*d*, 2), (*a*, 3), (*b*, 4)}. Domain of inverse function = {*a*, *b*, *c*, *d*}. Range of inverse function = {1, 2, 3, 4}. Arrow diagram:

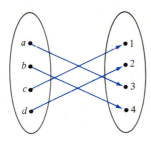

For Exercise 6: Inverse function = {(*d*, 1), (*a*, 2), (*c*, 3), (*b*, 4)}. Domain of inverse function = {*a*, *b*, *c*, *d*}. Range of inverse function = {1, 2, 3, 4}. Arrow diagram:

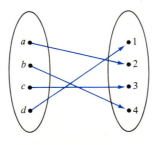

2. *f* is both one-to-one and onto.

5. *f* is both one-to-one and onto.

8. Define *f* from $X = \{1, 2, 3\}$ to $\{a, b, c, d\}$ as

$$f = \{(1, a), (2, b), (3, c)\}.$$

11. Strings that read the same from left to right as they do from right to left. (Such strings are called *palindromes*.)

14. $f = \{(0, 0), (1, 4), (2, 3), (3, 2), (4, 1)\}$; f is one-to-one and onto.

16. $f^{-1}(y) = (y - 2)/4$ **19.** $f^{-1}(y) = 1/(y - 3)$

22. $f \circ g = \{(1, x), (2, z), (3, x)\}$

25.

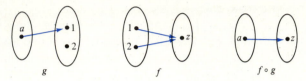

28. 4; one-to-one functions: $\{(1, a), (2, b)\}$ and $\{(1, b), (2, a)\}$. In this case, the onto and one-to-one functions are the same.

29. $f \circ g(x) = (2x + 3)^2$; $g \circ f(x) = 2x^2 + 3$

32. $f \circ g(x) = \sqrt{2(4x - 5)}$; $g \circ f(x) = 4\sqrt{2x} - 5$

35. Let $g(x) = \log_2 x$ and $h(x) = x^2 + 2$. Then $f(x) = g \circ h(x)$.

38. Let $g(x) = 2x$ and $h(x) = \sin x$. Then $f(x) = g \circ h(x)$.

SECTION 2.7

1. $a_{11} = 11$ **4.** $a_{33} = 9$

7. $\begin{pmatrix} 5 & 7 & 7 \\ -7 & 10 & -1 \end{pmatrix}$ **10.** $\begin{pmatrix} 3 & 18 & 27 \\ 0 & 12 & -6 \end{pmatrix}$ **13.** $\begin{pmatrix} 18 & 10 \\ 24 & -6 \\ 23 & 1 \end{pmatrix}$

16. $\begin{pmatrix} -11 & -6 \\ 18 & -8 \end{pmatrix}$

19. (a) $2 \times 3, 3 \times 3, 3 \times 2$

(b) AB, AC, CA, AB^2, and BC are defined.

$$AB = \begin{pmatrix} 33 & 18 & 47 \\ 8 & 9 & 43 \end{pmatrix}$$

22. BASIS STEP. $(n = 2)$

$$A^2 = \begin{pmatrix} 1 & 1 \\ 1 & 0 \end{pmatrix}\begin{pmatrix} 1 & 1 \\ 1 & 0 \end{pmatrix} = \begin{pmatrix} 2 & 1 \\ 1 & 1 \end{pmatrix} = \begin{pmatrix} f_2 & f_1 \\ f_1 & f_0 \end{pmatrix}$$

INDUCTIVE STEP. Assume that the identity holds for n. Now

$$A^{n+1} = A^n A = \begin{pmatrix} f_n & f_{n-1} \\ f_{n-1} & f_{n-2} \end{pmatrix}\begin{pmatrix} 1 & 1 \\ 1 & 0 \end{pmatrix} = \begin{pmatrix} f_n + f_{n-1} & f_n \\ f_{n-1} + f_{n-2} & f_{n-1} \end{pmatrix}$$

$$= \begin{pmatrix} f_{n+1} & f_n \\ f_n & f_{n-1} \end{pmatrix}.$$

25. If A is invertible, a solution is given by $X = A^{-1}C$.

28. Let $A = (a_{ij})$, $B = (b_{ij})$, $C = (c_{ij})$. Now

$$A + B = (a_{ij}) + (b_{ij}) = (a_{ij} + b_{ij}) = (b_{ij} + a_{ij})$$

$$= (b_{ij}) + (a_{ij}) = B + A.$$

31. Store the *ij*th element at position $2(j - 1) + i$.

CHAPTER 2 SELF-TEST

1. $\varnothing$ **2.** 256, 255   **3.** $A \subseteq B$ **4.** Yes

5.

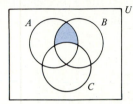

6. $\{(a, a), (a, c), (b, a), (b, c)\}$ **7.** $\{\{a, b\}, \{c, d\}\}$

8.

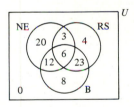

Six students read all three publications.

9. (a) 14   (b) 18   (c) 192   **10.** 89

11. (a) $b_5 = 35$, $b_{10} = 120$ (b) $(n + 1)^2 - 1^2$

12. (a) *ccddccccdd*   (b) *cccddccddc*   (c) 5   (d) 20

In Exercises 13–16, only the Inductive Step is given.

13. $2 + 4 + \cdots + 2n + 2(n + 1) = n(n + 1) + 2(n + 1)$
$$= (n + 1)(n + 2)$$

14. $2^2 + 4^2 + \cdots + (2n)^2 + [2(n + 1)]^2$

$$= \frac{2n(n + 1)(2n + 1)}{3} + [2(n + 1)]^2$$

$$= \frac{2(n + 1)(n + 2)[2(n + 1) + 1]}{3}$$

15. $1/2! + 2/3! + \cdots + n/(n + 1)! + (n + 1)/(n + 2)!$

$$= 1 - \frac{1}{(n + 1)!} + \frac{n + 1}{(n + 2)!}$$

$$= 1 - \frac{1}{(n + 2)!}$$

16. $2^{n+2} = 2 \cdot 2^{n+1} < 2[1 + (n + 1)2^n] = 2 + (n + 1)2^{n+1}$
$= 1 + [1 + (n + 1)2^{n+1}]$
$< 1 + [2^{n+1} + (n + 1)2^{n+1}]$
$= 1 + (n + 2)2^{n+1}$

17. f is a function from X into Y because each element in X is assigned a unique value in Y. Domain $f = X$. Range $f = \{\$, \&\}$.

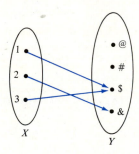

18. 2   **19.** $x = y = 2.3$
20. (a) X (b) Y
21. Domain $= \{1, 2, 3\}$; range $= \{\$, \&\}$. f is not onto Y.
22. $\{(s, 0), (t, 1)\}, \{(s, 1), (t, 0)\}, \{(s, 1), (t, 2)\}, \{(s, 2), (t, 1)\},$
$\{(s, 0), (t, 2)\}, \{(s, 2), (t, 0)\}$
23. (a) f is not one-to-one. For example, $f(aaba) = aab = f(aabb)$.
(b) Given a string α over $\{a, b\}$ of length 4, $g \circ f(\alpha)$ is equal to the first three symbols in α in reverse order.
24. $f \circ g(x) = 3(6 + \tan x)^2$; $g \circ f(x) = 6 + \tan 3x^2$
25. $a_{22} = -3$. AB is 2×8.
26. (a) $\begin{pmatrix} -6 & -12 \\ -9 & 9 \end{pmatrix}$ (b) $\begin{pmatrix} 12 & 12 \\ 5 & -7 \end{pmatrix}$ (c) $\begin{pmatrix} 6 & 12 \\ -1 & -11 \end{pmatrix}$

(d) $\begin{pmatrix} -2 & 4 \\ 15 & -3 \end{pmatrix}$

27. $A = \begin{pmatrix} 2 & 5 & 1 \\ 9 & 0 & 2 \end{pmatrix}$, $B = \begin{pmatrix} 2 & 7 \\ 3 & 0 \\ 5 & 2 \end{pmatrix}$

28. $x = 2, y = -5, z = -3$

SECTION 3.1

1. Since an odd number of edges touch some vertices (c and d), there is not path from a to a that passes through each edge exactly one time.

4. $(a, c, e, b, c, d, e, f, d, b, a)$

7. $V = \{v_1, v_2, v_3, v_4\}$. $E = \{e_1, e_2, e_3, e_4\}$. There are no parallel edges, loops, or isolated vertices. G is a simple graph. e_1 is incident on v_1 and v_2.

10. $V = \{v_1, v_2, v_3, v_4\}$. $E = \{e_1, e_2, e_3, e_4, e_5\}$. e_1 and e_2 are parallel edges. e_4 and e_5 are loops. v_4 is an isolated vertex. G is not a simple graph. e_1 is incident on v_1 and v_3.

13.

K_3

K_4

K_5

15. Bipartite. $V_1 = \{v_1, v_2, v_5\}$, $V_2 = \{v_3, v_4\}$.   **18.** Not bipartite

21. Bipartite. $V_1 = \{v_1\}$, $V_2 = \{v_2, v_3\}$.

22.

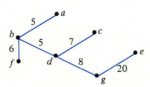
$K_{2,3}$ $K_{2,4}$ $K_{3,3}$

23. (b, c, a, d, e)

26. Two classes   **30.**

    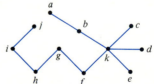

33.

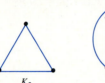

SECTION 3.2

1. Cycle, simple cycle   **4.** Cycle, simple cycle

7. Simple path

10. (a, a), (b, c, g, b), (b, c, d, f, g, b), (b, c, d, e, f, g, b), (c, g, f, d, c), (c, g, f, e, d, c), (d, f, e, d)

13. Every vertex has degree 4.

15.

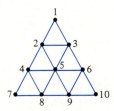

18. There are 17 subgraphs.

19. No Euler cycle   **22.** No Euler cycle

25. For

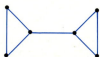

an Euler cycle is $(10, 9, 6, 5, 9, 8, 5, 4, 8, 7, 4, 2, 5, 3, 2, 1, 3, 6, 10)$. The method generalizes.

28. $m = n = 2$

31.

34. Let G be a simple, disconnected graph with n vertices having the maximum number of edges. Show that G has two components. If one component has i vertices, show that the components are K_i and K_{n-i}. Show that G has $f(i) = i(i - 1)/2 + (n - i)(n - i - 1)/2$ edges. Complete the square in i to show that

$$f(i) = \left(i - \frac{n}{2}\right)^2 + \frac{n^2}{4} - \frac{n}{2}.$$

Conclude that the maximum of $f(i)$ occurs when $i = 1$ or $i = n - 1$.

36.

38. Repeatedly taking three consecutive bits beginning with the first three, we obtain 000, 001, 011, 111, 110, 101, 010, 100. Since all possible three-bit strings are represented, 00011101 is a de Bruijn sequence for $n = 3$.

41. We first show that if G is a connected bipartite graph, then every closed path in G has even length.

Suppose that the disjoint vertex sets are V_1 and V_2. Let

$$P = (v_0, e_1, v_1, e_2, \ldots, v_{n-1}, e_n, v_n)$$

be a closed path from $v = v_0$ to $v = v_n$. Suppose that $v_0 \in V_1$. Then $v_1 \in V_2$, $v_2 \in V_1$, ... Notice that if i is odd, $v_i \in V_2$, and if i is even, $v_i \in V_1$. Since $v_n \in V_1$, it follows that n is even. Thus P has even length.

We conclude by showing that if every closed path in a connected graph G has even length, then G is bipartite.

Choose a vertex v in G. Let V_1 denote the set of vertices w in G that are reachable from v on a path of even length. Let V_2 denote the set of vertices w in G that are reachable from v on a path of odd length. First, notice that since G is connected, every vertex in G is in either V_1 or V_2. We claim that V_1 and V_2 are disjoint. To show this, we argue by contradiction. Suppose that some vertex w belongs to both V_1 and V_2. Then there is a path P_1 of even length from v to w and there is a path P_2 of odd length from v to w. Let P_3 be the path from w to v obtained from P_2 by reversing the direction of each edge in P_2. Then P_1 followed by P_3 is a closed path of odd length from v to v. This contradiction shows that V_1 and V_2 are disjoint. Now let e_1 be an edge incident on the vertices v_1 and v_2. Suppose that v_1 belongs to V_1. Then there is a path P of even length from v to v_1. Now P followed by e_1 followed by v_2 is a path of odd length from v to v_2. Thus v_2 is in V_2. It follows that G is a bipartite graph.

SECTION 3.3

1. $(d, a, e, b, c, h, g, f, j, i, d)$

3. We would have to eliminate two edges each at b, d, i, and k leaving $19 - 8 = 11$ edges. A Hamiltonian cycle would have 12 edges.

6. $(a, b, c, j, i, m, k, d, e, f, l, g, h, a)$

9.

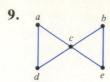

12. If n is even and $m > 1$ or if m is even and $n > 1$, there is a Hamiltonian cycle. The sketch shows the solution in case n is even.

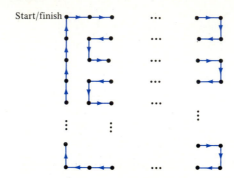

If $n = 1$ or if $m = 1$, there is no cycle and, in particular, there is no Hamiltonian cycle. Suppose that n and m are both odd and that the graph has a Hamiltonian cycle. Since there are nm vertices, this cycle has nm edges; therefore, the Hamiltonian cycle contains an odd number of edges. However, we note that in a Hamiltonian cycle, there must be as many "up" edges as "down" edges and as many "left" edges as "right" edges. Thus a Hamiltonian cycle must have an even number of edges. This contradiction shows that if n and m are both odd, the graph does not have a Hamiltonian cycle.

15. The five edges of smallest weight have weights 3, 4, 4, 5, 5. Thus the shortest Hamiltonian cycle has length at least $3 + 4 + 4 + 5 + 5 = 21$. However, three of these edges (having weights 3, 4, 4) are incident on vertex c. Thus the edges of weight 3, 4, 4 cannot all be in a Hamiltonian cycle. If we replace an edge of weight 4 with an edge of minimum replacement weight 6, we can conclude that the shortest Hamiltonian cycle has weight at least $3 + 6 + 4 + 5 + 5 = 23$. Since the given Hamiltonian cycle has length 23, we conclude that it is minimal.

18. An edge joins vertex i to vertex j if a knight can move from i to j or from j to i. A solution is a Hamiltonian cycle.

SECTION 3.4

1.

$$
\begin{array}{c c}
& \begin{array}{c c c c c} a & b & c & d & e \end{array} \\
\begin{array}{c} a \\ b \\ c \\ d \\ e \end{array} &
\left(\begin{array}{c c c c c}
0 & 1 & 1 & 1 & 1 \\
1 & 0 & 1 & 0 & 0 \\
1 & 1 & 0 & 1 & 1 \\
1 & 0 & 1 & 0 & 1 \\
1 & 0 & 1 & 1 & 0
\end{array}\right)
\end{array}
$$

4.

$$
\begin{array}{c c}
& \begin{array}{c c c c c c} v_1 & v_2 & v_3 & v_4 & v_5 & v_6 \end{array} \\
\begin{array}{c} v_1 \\ v_2 \\ v_3 \\ v_4 \\ v_5 \\ v_6 \end{array} &
\left(\begin{array}{c c c c c c}
0 & 1 & 1 & 0 & 0 & 0 \\
1 & 0 & 1 & 0 & 0 & 0 \\
1 & 1 & 0 & 0 & 0 & 0 \\
0 & 0 & 0 & 0 & 0 & 0 \\
0 & 0 & 0 & 0 & 0 & 1 \\
0 & 0 & 0 & 0 & 1 & 0
\end{array}\right)
\end{array}
$$

7.

$$
\begin{array}{c c}
& \begin{array}{c c c c c c c c} x_1 & x_2 & x_3 & x_4 & x_5 & x_6 & x_7 & x_8 \end{array} \\
\begin{array}{c} a \\ b \\ c \\ d \\ e \end{array} &
\left(\begin{array}{c c c c c c c c}
1 & 0 & 1 & 0 & 1 & 1 & 0 & 0 \\
1 & 1 & 0 & 0 & 0 & 0 & 0 & 0 \\
0 & 1 & 0 & 1 & 1 & 0 & 1 & 0 \\
0 & 0 & 0 & 1 & 0 & 1 & 0 & 1 \\
0 & 0 & 1 & 0 & 0 & 0 & 1 & 1
\end{array}\right)
\end{array}
$$

10.

$$
\begin{array}{c c}
& \begin{array}{c c c c c c c c} e_1 & e_2 & e_3 & e_4 & e_5 & e_6 & e_7 & e_8 \end{array} \\
\begin{array}{c} 1 \\ 2 \\ 3 \\ 4 \\ 5 \\ 6 \\ 7 \end{array} &
\left(\begin{array}{c c c c c c c c}
1 & 0 & 0 & 0 & 0 & 0 & 0 & 0 \\
1 & 1 & 0 & 1 & 1 & 1 & 0 & 0 \\
0 & 1 & 1 & 0 & 0 & 0 & 0 & 0 \\
0 & 0 & 1 & 1 & 0 & 0 & 0 & 0 \\
0 & 0 & 0 & 0 & 1 & 0 & 1 & 0 \\
0 & 0 & 0 & 0 & 0 & 1 & 1 & 1 \\
0 & 0 & 0 & 0 & 0 & 0 & 0 & 1
\end{array}\right)
\end{array}
$$

13.

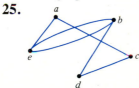

16.

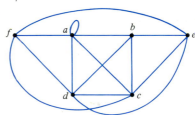

19. [For K_5]

$$
\begin{pmatrix}
4 & 3 & 3 & 3 & 3 \\
3 & 4 & 3 & 3 & 3 \\
3 & 3 & 4 & 3 & 3 \\
3 & 3 & 3 & 4 & 3 \\
3 & 3 & 3 & 3 & 4
\end{pmatrix}
$$

22. The graph is not connected.

25.

28. G is not connected.

29. Because of the symmetry of the graph, if v and w are vertices in K_5, there are the same number of paths of length n from v to v as there are from w to w. Thus all the diagonal elements of A^n coincide. Similarly, all the off-diagonal elements of A^n coincide.

32. If $n \geq 2$,

$$d_n = 4a_{n-1} \qquad \text{by Exercise 30}$$
$$= 4(\tfrac{1}{5})[4^{n-1} + (-1)^n] \qquad \text{by Exercise 31.}$$

The formula can be directly verified for $n = 1$.

SECTION 3.5

1. The graphs are not isomorphic since they do not have the same number of vertices.

4. The graphs are not isomorphic since G_2 has a vertex (c') of degree 4 but G_1 does not.

7. The graphs are not isomorphic since G_1 has a vertex (c) of degree 2 but G_2 does not.

10. The graphs are isomorphic. The orderings a, b, c, d, e, f, g, h, and e', c', a', g', d', f', b', h' produce identical adjacency matrices. The function F defined by

$$F(a) = e', \qquad F(b) = c', \qquad F(c) = a', \qquad F(d) = g',$$
$$F(e) = d', \qquad F(f) = f', \qquad F(g) = b', \qquad F(h) = h'$$

is an isomorphism.

11. Use the notation of Definition 3.5.7. If $(v_0, \ldots, v_n)$ is a simple cycle of length n in G_1, then $(f(v_0), \ldots, f(v_n))$ is a simple cycle of length n in G_n.

15. Is an invariant.

18.

21.

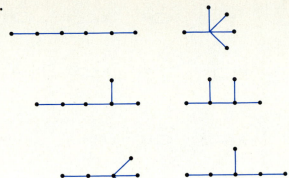

22.

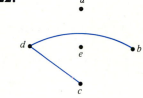

SECTION 3.6

1.

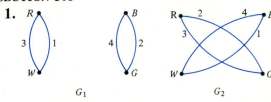

G_1 G_2

4.

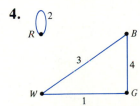

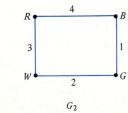

G_1 G_2

7. (a)

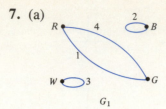
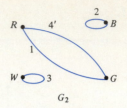

G_1 G_2

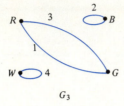
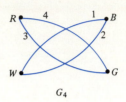

G_3 G_4

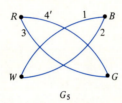
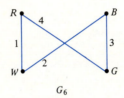

G_5 G_6

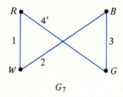

G_7

(b) Solutions are: G_1, G_5; G_1, G_7; G_2, G_4; G_2, G_6; G_3, G_6; and G_3, G_7.

11.

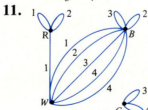

13.

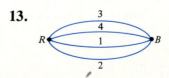
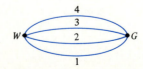

17. According to Exercise 12, not counting loops every vertex must have degree at least 4. In Figure 3.6.5, not counting loops vertex W has degree 3 and, therefore, Figure 3.6.5 does not have a solution to the modified version of Instant Insanity. Figure 3.6.3 gives a solution to regular Instant Insanity for Figure 3.6.5.

CHAPTER 3 SELF-TEST

1. $V = \{v_1, v_2, v_3, v_4\}$. $E = \{e_1, e_2, e_3\}$. e_1 and e_2 are parallel edges. There are no loops. v_1 is an isolated vertex. G is not a simple graph. e_3 is incident on v_2 and v_4. v_2 is incident on e_1, e_2, and e_3.

2. There are vertices (a and e) of odd degree.

3.

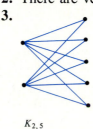

$K_{2,5}$

4. There are two classes.

$$\overset{\bullet}{v_2} \quad \overset{\bullet}{v_1} \quad \overset{\bullet}{v_4} \quad \overset{\bullet}{v_3} \quad \overset{\bullet}{v_5}$$

5. It is a cycle.

6.

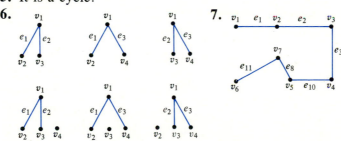

7.

8. No. There are vertices of odd degree.

9. $(v_1, v_2, v_3, v_4, v_5, v_7, v_6, v_1)$

10. A Hamiltonian cycle would have seven edges. Suppose that the graph has a Hamiltonian cycle. We would have to eliminate three edges at vertex b and one edge at vertex f. This leaves $10 - 4 = 6$ edges; not enough for a Hamiltonian cycle. Therefore, the graph does not have a Hamiltonian cycle.

11. $K_{2,3}$

12. In a minimal weight Hamiltonian cycle, every vertex must have degree 2. Therefore, edges (a, b), (a, j), (j, i), (i, h), (g, f), (f, e), and (e, d) must be included. We cannot include edge (b, h) or we will complete a cycle. This implies that we must include edges (h, g) and (b, c). Since vertex g now has degree 2, we cannot include edges (c, g) or (g, d). Thus we must include (c, d). This is a Hamiltonian cycle and the argument shows that it is unique. Therefore, it is minimal.

13.

	v_1	v_2	v_3	v_4	v_5	v_6	v_7
v_1	0	1	0	0	0	1	0
v_2	1	0	1	1	0	1	1
v_3	0	1	0	1	0	0	0
v_4	0	1	1	0	1	0	0
v_5	0	0	0	1	0	1	1
v_6	1	1	0	0	1	0	1
v_7	0	1	0	0	1	1	0

14.

	e_1	e_2	e_3	e_4	e_5	e_6	e_7	e_8	e_9	e_{10}	e_{11}
v_1	1	0	0	0	0	0	1	0	0	0	0
v_2	1	1	0	1	1	1	0	0	0	0	0
v_3	0	1	1	0	0	0	0	0	0	0	0
v_4	0	0	1	1	0	0	0	0	0	1	0
v_5	0	0	0	0	0	0	0	1	1	1	0
v_6	0	0	0	0	0	1	1	0	1	0	1
v_7	0	0	0	0	1	0	0	1	0	0	1

15. The number of paths of length 3 from v_2 to v_3.

16. No. Each edge is incident on at least one vertex.

17. The graphs are isomorphic. The orderings v_1, v_2, v_3, v_4, v_5 and w_3, w_1, w_4, w_2, w_5 produce equal adjacency matrices.

18. The graphs are isomorphic. The orderings v_1, v_2, v_3, v_4, v_5, v_6 and w_3, w_6, w_2, w_5, w_1, w_4 produce equal adjacency matrices.

19.   **20.**

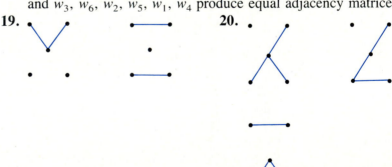

21.

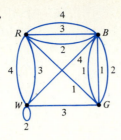

22. See the hints for Exercises 23 and 24.

23.

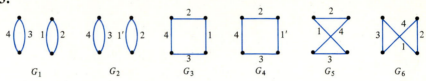

We denote the two edges incident on B and G labeled 1 in the graph of Exercise 21 as 1 and $1'$ here.

24. The puzzle of Exercise 23 has four solutions. Using the notation of Exercise 23, the solutions are G_1, G_5; G_2, G_5; G_3, G_6; and G_4, G_6.

SECTION 4.1

1. a-1; b-1; c-1; d-1; e-2; f-3; g-3; h-4; i-2; j-3; k-0

4. Height $= 4$

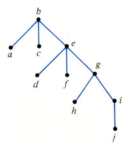

7. PEN   **10.** SALAD   **11.** 0111100010

14. 0110000100100001111

17. Let T be a tree. Root T at some arbitrary vertex. Let V be the set of vertices on even levels and let W be the set of vertices on odd levels. Since each edge is incident on a vertex in V and a vertex in W, T is a bipartite graph.

20. e, g

23.

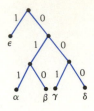

26. Another tree is shown in the hint for Exercise 23.

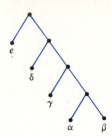

SECTION 4.2

1. Kronos   **4.** Apollo, Athena, Hermes, Heracles

7. (a) b; d (b) a; c

10. (a) $e, f, g, j; j$ (b) $g, h, i, m, n; k, l, q, r$

13. (a) a, b, c, d, e (b) $a, b, c, d, e, g, j, l, q$

16. They are siblings.   **21.**   **24.**

26. A single vertex is a ''cycle'' of length 0.

29. Each component of a forest is connected and acyclic and, therefore, a tree.

32. Since G^* is connected and has $n - 1$ edges, by Theorem 4.2.3, G^* is acyclic. But, adding an edge in parallel introduces a cycle. Contradiction.

SECTION 4.3

2. f

5.   **7.**

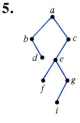

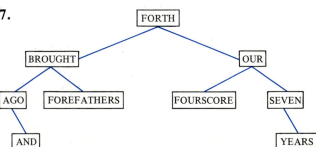

8. We first compare AGO with the word FORTH in the root. Since AGO is less than FORTH, we go to the left child. Next, we

compare AGO with BROUGHT. Since AGO is less than BROUGHT, we go to the left child. We find AGO at this vertex.

9. We first compare FOURTH with the word FORTH in the root. Since FOURTH is greater than FORTH, we go to the right child. Next, we compare FOURTH with OUR. Since FOURTH is less than OUR, we go to the left child. Since FOURTH is greater than FOURSCORE, we attempt to go to the right child. Since there is no right child, we conclude that FOURTH is not in the tree.

13. False. Consider

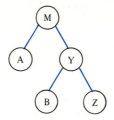

14. $i + 1$

17.

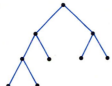

19. Balanced   22. Balanced

SECTION 4.4

1. Isomorphic. $f(v_1) = w_1$, $f(v_2) = w_5$, $f(v_3) = w_3$, $f(v_4) = w_4$, $f(v_5) = w_2$, $f(v_6) = w_6$.

4. Not isomorphic. T_2 has a simple path of length 2 from a vertex of degree 1 to a vertex of degree 1, but T_1 does not.

7. Isomorphic as rooted trees. $f(v_1) = w_1$, $f(v_2) = w_4$, $f(v_3) = w_3$, $f(v_4) = w_2$, $f(v_5) = w_6$, $f(v_6) = w_5$, $f(v_7) = w_7$, $f(v_8) = w_8$. Also isomorphic as free trees.

10. Not isomorphic as binary trees. The root of T_1 has a left child but the root of T_2 does not. Isomorphic as rooted trees and as free trees.

13. •———•———•   16.

19.

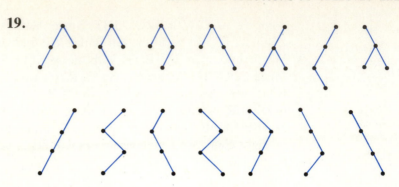

SECTION 4.5

1.

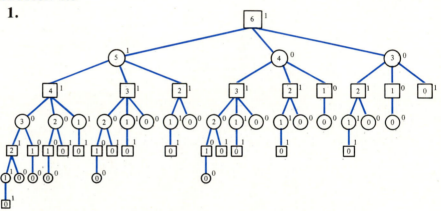

4. The second player always wins. If two piles remain, leave piles with equal numbers of tokens. If one pile remains, take it.

7.

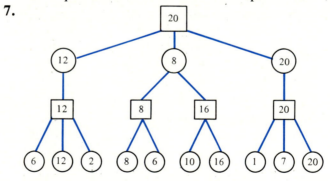

10. The value of the root is 3.

12. (For Exercise 9)

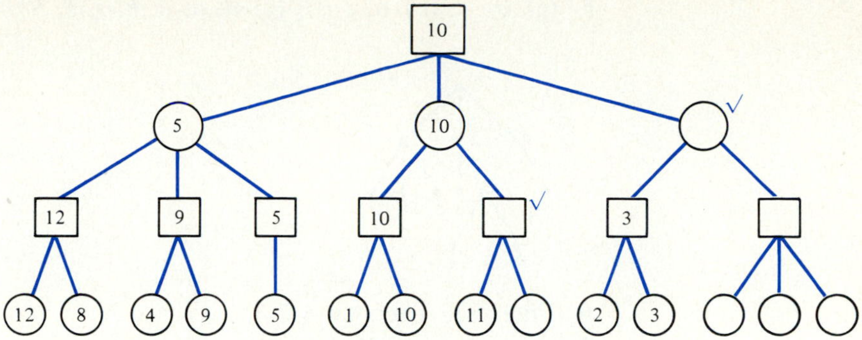

13. 3 − 2 = 1 **16.** 4 − 1 = 3

19. We first obtain the values 6, 6, 7 for the children of the root. Thus we order the children of the root with the rightmost child first and use the alpha-beta procedure to obtain

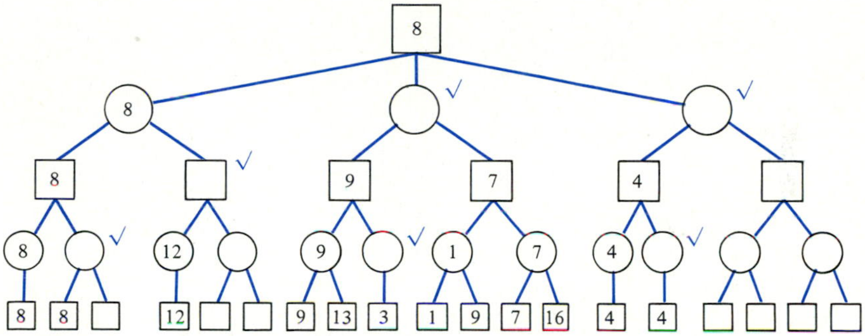

CHAPTER 4 SELF-TEST

1.

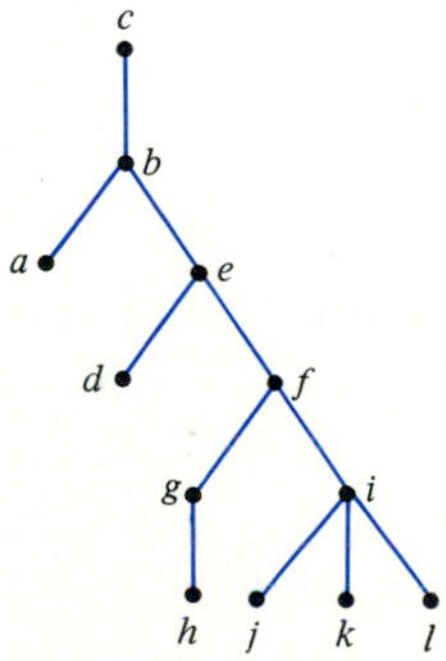

2. *a*-2, *b*-1, *c*-0, *d*-3, *e*-2, *f*-3, *g*-4, *h*-5, *i*-4, *j*-5, *k*-5, *l*-5

3. 5 **4.** *bade*

5. (a) *b*   (b) *a, c*   (c) *d, a, c, h, j, k, l*   (d)

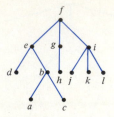 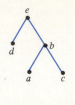

6. True. See Theorem 4.2.3.

7. True. A tree of height 6 or more must have seven or more vertices.

8. False.   **9.**

10.

WORD

PROCESSING

CLEAN   PRODUCES

BUT   MANUSCRIPTS   PROSE

CLEAR   NOT

NECESSARILY

11. We first compare NOT with the word WORD in the root. Since NOT is less than WORD, we go the the left child. Next, we compare NOT with PROCESSING. Since NOT is less than PROCESSING, we go to the left child. Since NOT is greater than CLEAN, we go to the right child. Since NOT is greater than MANUSCRIPTS, we go to the right child. We find NOT at this vertex.

12. We first compare MORE with the word WORD in the root. Since MORE is less than WORD, we go to the left child. Next, we compare MORE with PROCESSING. Since MORE is less than PROCESSING, we go to the left child. Since MORE is greater than CLEAN, we go to the right child. Since MORE is greater than MANUSCRIPTS, we go to the right child. Since MORE is less than NOT, we go to the left child. Since MORE is less than NECESSARILY, we attempt to go to the left child. Since there is no left child, we conclude that MORE is not in the tree.

13. True. If f is an isomorphism of T_1 and T_2 as rooted trees, f is also an isomorphism of T_1 and T_2 as free trees.

14. False.

T_1 T_2

15. Isomorphic. $f(v_1) = w_6$, $f(v_2) = w_2$, $f(v_3) = w_5$, $f(v_4) = w_7$, $f(v_5) = w_4$, $f(v_6) = w_1$, $f(v_7) = w_3$, $f(v_8) = w_8$.

16. Not isomorphic. T_1 has a vertex (v_3) on level 1 of degree 3, but T_2 does not.

17. $3 - 1 = 2$

18. Let each row, column, or diagonal that contains one X and two blanks count 1. Let each row, column, or diagonal that contains two X's and one blank count 5. Let each row, column, or diagonal that contains three X's count 100. Let each row, column, or diagonal that contains one 0 and two blanks count -1. Let each row, column, or diagonal that contains two 0's and one blank count -5. Let each row, column, or diagonal that contains three 0's count -100. Sum the values obtained.

19.

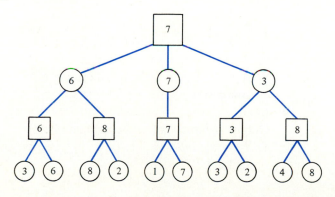

20.

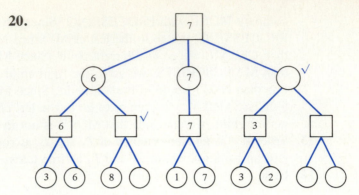

SECTION 5.1

  2. At line 1, since $a = 9 > 6 = b$ is true, we set x to 9. At line 2, since $c = 5 > 9 = x$ is false, we terminate the algorithm.

  5. At line 1, we set x to 5. At line 2, we set i to 2. At line 3, since $s_2 = 7 > 5$ is true, we set x to 7. At line 2, we set i to 3. At line 3, since $s_3 = 9 > 7$ is true, we set x to 9 and terminate the algorithm.

  8. Input: a, b, and c
Output: x
    1. [Find the smaller of a and b and call it x.] If $a < b$, then $x := a$; otherwise, $x := b$.
    2. [Find the smaller of x and c.] If $c < x$, then $x := c$.

  11. Input: a, b, c, d, e
Output: x (smallest), y (second smallest)
    1. If $a > b$, then swap a and b. (Swap x and y means: $temp := x$, $x := y$, $y := temp$.)
    2. If $c > d$, then swap c and d.
    3. If $a > c$, then swap a and c.
    4. If $a > e$, then swap a and e.
    5. $x := a$ (x is the smallest).
    6. If $b > c$, then swap b and c.
    7. If $d > e$, then swap d and e.
    8. If $b < d$, then $y := b$; otherwise $y := d$. (y is second smallest.)

  14. Input: $s_1, \ldots, s_n$
Output: x (smallest), y (second smallest)
    1. $x := s_1$, $index := 1$. ($x = s_{index}$ is the smallest element seen so far.)

2. For $i := 2$ to n, execute line 3.
3.   If $s_i < x$ then $x := s_i$, *index* $:= i$.
4. Swap s_n and s_{index}. (The smallest element is now at position n.)
5. $y := s_1$. (y is the smallest element seen so far.)
6. For $i := 2$ to $n - 1$, execute line 7.
7.   If $s_i < y$, then $y := s_i$.

17. Input: $s_1, \ldots, s_n$
Output: *index*
1. *index* $:= 1$, *large* $:= s_1$.
2. For $i := 2$ to n, execute line 3.
3.   If $s_i \geq$ *large*, then *index* $:= i$, *large* $:= s_i$.

20. Input: $s_1, \ldots, s_n$, *key*
Output: *index*
1. *index* $:= 0$.
2. For $i := 1$ to n, execute line 3.
3.   If *key* $= s_i$, then *index* $:= i$; stop.

23. Input: $s_1, \ldots, s_n$
Output: *index*
1. *index* $:= 0$.
2. For $i := 2$ to n, execute line 3.
3.   If $s_i > s_{i-1}$, then *index* $:= i$; stop.

26. Input: $s_1, \ldots, s_n$
Output: *max*
1. *max* $:= 0$, *sum* $:= 0$. (*max* is the maximum sum seen so far. After the ith iteration of the following loop, *sum* is the largest consecutive sum that ends at position i.)
2. For $i := 1$ to n, execute lines 3 and 4.
3.   If *sum* $+ s_i > 0$, then *sum* $:=$ *sum* $+ s_i$; otherwise, *sum* $:= 0$.
4.   If *sum* $>$ *max*, then *max* $:=$ *sum*.

29. Input: $b_n b_{n-1} \cdots b_1$
Output: $c_n c_{n-1} \cdots c_1$
1. *flag* $:= 0$. (*flag* $= 0$ means that we have not yet seen a 1 while scanning $b_n b_{n-1} \cdots b_1$ from the right. *flag* $= 1$ means that we have seen a 1 while scanning $b_n b_{n-1} \cdots b_1$ from the right.)
2. For $i := 1$ to n, execute lines 3 and 4.
3.   If *flag* $= 0$, then $c_i := b_i$; otherwise, $c_i := \bar{b}_i$.
4.   If $b_i = 1$, then *flag* $:= 1$.

SECTION 5.2

1. At line 2, since $a > b$ is true, we set x to 9. At line 6, since $c > x$ is false, we skip to line 8 where we return $x = 9$.

4. At line 1, we set *large* to $s(1) = 5$. At line 2, we set i to 2. At line 3, since $s(2) = 7 > 5 = large$ is true, we set *large* to 7. At line 2, we set i to 3. At line 3, since $s(3) = 9 > 7 = large$ is true, we set *large* to 9. At line 5, we return *large* $= 9$.

7. (For Exercise 4) At lines 1 and 2, we set *large* to 9 and *index* to 3. At line 3, we set i to 2. At line 4, since $s(2) = 7 > 9 = large$ is false, we simply return to line 3 where we set i to 1. At line 4, since $s(1) = 5 > 9 = large$ is false, we skip to line 7 where we return *index* $= 3$.

8. We first set i to 2. Since $m \bmod i = 9 \bmod 2 = 1$ is not 0, we continue the loop. We next set i to 3. Since $9 \bmod 3 = 0$, we return **false**.

11. We first set m to 9. Since **not** *is_prime*(9) is true, we set m to 10. Since **not** *is_prime*(10) is true, we set m to 11. Since **not** *is_prime*(11) is false, we return $m = 11$.

14. (For Exercise 8)

```
procedure min(a, b, c)
  if a < b then
    x := a
  else
    x := b
  if c < x then
    x := c
  return(x)
end min
```

SECTION 5.3

1. $55 = 11 \cdot 5$ **4.** $984 = 24 \cdot 41$ **7.** $-144 = -12 \cdot 12$

10. $71412 = -132(-541)$ **11.** $45 = 6 \cdot 7 + 3$

14. $221 = 17 \cdot 13 + 0$

17. $8 = 11 \cdot 0 + 8$ **20.** $490256 = 337 \cdot 1454 + 258$

21. Divide 60 by 90 to obtain $60 = 90 \cdot 0 + 60$. Thus gcd $(60, 90)$ $=$ gcd $(90, 60)$. Divide 90 by 60 to obtain $90 = 60 \cdot 1 + 30$. Thus gcd $(90, 60)$ $=$ gcd $(60, 30)$. Divide 60 by 30 to obtain $60 = 30 \cdot 2 + 0$. Thus gcd $(60, 30)$ $=$ gcd $(30, 0) = 30$. Therefore, gcd $(60, 90) = 30$.

24. 15   **27.** 1   **30.** 1

31. Since $c \mid m$, $m = cq_1$ for some integer q_1. Since $c \mid n$, $n = cq_2$ for some integer q_2. Now

$$m - n = cq_1 - cq_2 = c(q_1 - q_2).$$

Therefore, $c \mid (m - n)$.

34. In general,

$$r_i = r_{i+1}q_{i+2} + r_{i+2}. \qquad (*)$$

Taking $i = n - 3$ in $(*)$, we obtain

$$r_{n-3} = r_{n-2}q_{n-1} + r_{n-1},$$

which may be rewritten as

$$r_{n-1} = -q_{n-1}r_{n-2} + 1 \cdot r_{n-3}$$

We may take $s_{n-3} = -q_{n-1}$ and $t_{n-3} = 1$ to obtain

$$r_{n-1} = s_{n-3}r_{n-2} + t_{n-3}r_{n-3}. \qquad (**)$$

Taking $i = n - 4$ in $(*)$, we obtain

$$r_{n-4} = r_{n-3}q_{n-2} + r_{n-2}$$

or

$$r_{n-2} = -q_{n-2}r_{n-3} + r_{n-4}. \qquad (***)$$

Substituting $(***)$ into $(**)$, we obtain

$$r_{n-1} = s_{n-3}[-q_{n-2}r_{n-3} + r_{n-4}] + t_{n-3}r_{n-3}$$
$$= [-s_{n-3}q_{n-2} + t_{n-3}]r_{n-3} + s_{n-3}r_{n-4}.$$

Setting $s_{n-4} = -s_{n-3}q_{n-2} + t_{n-3}$ and $t_{n-4} = s_{n-3}$, we obtain

$$r_{n-1} = s_{n-4}r_{n-3} + t_{n-4}r_{n-4}.$$

Continuing in this way, we ultimately obtain

$$r_{n-1} = s_0r_1 + t_0r_0.$$

If we set $s = s_0$ and $t = t_0$, we have

$$\gcd(r_0, r_1) = r_{n-1} = sr_1 + tr_0.$$

37. $6 \mid 4 \cdot 3$ but $6 \nmid 4$ and $6 \nmid 3$.

SECTION 5.4

1. $O(n)$ **4.** $O(1)$ **7.** $O(n)$ **10.** $O(n^3)$ **13.** $O(n)$

16. $O(n^3)$ **18.** $O(n)$ **21.** $O(n^2)$ **24.** $O(n^3)$ **27.** $O(n)$

30. Input: The $n \times n$ matrices a and b

Output: $c = a + b$

1. For $i := 1$ to n, execute line 2.
2.   For $j := 1$ to n, execute line 3.
3. $c_{i,j} := a_{i,j} + b_{i,j}$.

The algorithm is $O(n^2)$.

33. $2^n = 2(2 \cdot 2 \cdot \ldots \cdot 2) \leq 2(2 \cdot 3 \cdot \ldots \cdot n) = 2n!$

CHAPTER 5 SELF-TEST

1. At line 1, we set x to 7. At line 2, we set i to 2. At line 3, since $s_2 = 9 > 7 = x$ is true, we set x to 9. Returning to line 2, we set i to 3. At line 3, since $s_3 = 17 > 9 = x$ is true, we set x to 17. Returning to line 2, we set i to 4. At line 3, since $s_4 = 7 > 17 = x$ is false, we terminate the algorithm.

2. Input: a, b, c

Output: x, y, z

1. $x := a$, $y := b$, $z := c$.
2. If $y < x$, swap(x, y).
3. If $z < x$, swap(x, z). (x is now smallest.)
4. If $y > z$, swap(y, z). (z is now largest.)

3. Input: $n \times n$ matrix A

Output: A^T

1. For $i := 1$ to n, execute line 2.
2.   For $j := i + 1$ to n, execute line 3.
3.   Swap(A_{ij}, A_{ji}).

4. Input: $s_1, \ldots, s_n$

1. $i := 1$, $j := 2$.
2. If $j > n$, then stop.
3. If $s_i = s_j$, then print s_i; otherwise, go to line 5.
4. If $j \leq n$ and $s_i = s_j$, then $j := j + 1$; go to line 3.
5. $i := j$, $j := j + 1$.
6. Go to line 2.

5. 5   **6.** 3

7. procedure $dup(s, n)$

$\quad i := 1$

$\quad j := 2$

$\quad$ **while** $j \leq n$ **do**

```
      begin
      if s(i) = s(j) then
         begin
         print s(i)
         while j ≤ n and s(i) = s(j) do
            j := j + 1
         end
      i := j
      j := j + 1
      end
end dup
```

8. **procedure** $numb\_occr(s, n, key)$

```
   count := 0
   for i := 1 to n do
     if s(i) = key then
        count := count + 1
   return(count)
end numb_occr
```

9. $333 = 24 \cdot 13 + 21$ **10.** 12   **11.** 2   **12.** gcd (b, r)

13. $O(n^3)$ **14.** $O(n^4)$ **15.** $O(n^2)$

16. Input: $n \times n$ matrices A and B

Output: X (X = True, in case $A = B$; X = False, in case $A \neq B$)

1. For $i := 1$ to n, execute line 2.
2.   For $j := 1$ to n, execute line 3.
3.   If $A_{ij} \neq B_{ij}$, then $X :=$ False; stop.
4. $X :=$ True.

Worst-case time is $O(n^2)$.

SECTION 6.1

1. $2 \cdot 4$ **4.** $3 \cdot 3 \cdot 4$ **7.** $8 \cdot 4 \cdot 5$ **9.** $6 \cdot 6$

12. $6 \cdot 6 - 5 \cdot 5$ (Total number of outcomes − outcomes with no 2)

14. $10 \cdot 5$ **17.** $10 \cdot 5 \cdot 5$ **18.** $26^3 10^2$ **21.** 2^4

24. $5 \cdot 4 \cdot 3$

27. $3 \cdot 4 \cdot 3$ (Assign Dolph, then choose someone for the highest remaining office, then choose someone for the remaining office.)

28. 5^3 **31.** $4 \cdot 3 \cdot 2 \cdot 1$

34. $5^3 - 4^3$ (Total number of strings − number of strings that do not contain A)

36. There are six choices for the top and, having chosen the top, there are four choices for the front, for a total of $6 \cdot 4$ choices.

39. $12 \cdot 11 \cdot 10 \cdot 9 \cdot 8 \cdot 7$

42. The first letter can be selected in 26 ways. The second alphanumeric can be selected or not selected in 37 ways. The last symbol can be selected or not selected in five ways. After subtracting the five disallowed possibilities, we obtain $(26 \cdot 37 - 5)5$ total possibilities.

45. Let $X = \{x_1, \ldots, x_n\}$. To construct a function from X to Y, we assign one element in Y to x_1. This can be done in m ways. We can assign one element in Y to x_2 in m ways, and so on. Thus the number of functions from X to Y is $m \cdot m \cdot \ldots \cdot m = m^n$.

48. Consider disjoint vertex sets $\{v_1, \ldots, v_m\}$ and $\{w_1, \ldots, w_n\}$. There are n edges from v_1 to the vertices $w_1, \ldots, w_n$. There are n edges from from v_2 to the vertices $w_1, \ldots, w_n$; and so on. Thus the total number of edges is $n + n + \cdots + n = mn$.

SECTION 6.2

1. $2^6 + 2^6$ **4.** $3 \; [(1, 3), (2, 2), (3, 1)]$

7. $6 \; [(1, 1), (2, 2), (3, 3), (4, 4), (5, 5), (6, 6)]$

10. $5 \cdot 3 + 5 \cdot 4 + 5 \cdot 2 + 3 \cdot 4 + 3 \cdot 2 + 4 \cdot 2$

13. $5 \cdot 4 + 5 \cdot 4 \cdot 3$

16. See the solution to Exercise 42, Section 2.2.

17. $2^5 + 2^7 - 2^4$ (According to Exercise 16, the total number of possibilities = number of strings that begin 100 + number of strings that have the fourth bit 1 − number of strings that begin 100 and have the fourth bit 1.)

20. $5 \cdot 4 + 3 \cdot 5 \cdot 4 - 2 \cdot 4$ (According to Exercise 16, the total number of possibilities = number in which Connie is chairperson + number in which Alice is an officer − number in which Connie is chairperson and Alice is an officer.)

SECTION 6.3

1. $4! = 24$

4. *abc, acb, bac, bca, cab, cba, abd, adb, bad, bda, dab, dba, acd, adc, cad, cda, dac, dca, bcd, bdc, cbd, cdb, dbc, dcb*

7. $P(11, 3) = 11 \cdot 10 \cdot 9$ **10.** $3!$

13. $4!$ contain the substring AE and $4!$ contain the substring EA; therefore, the total number is $2 \cdot 4!$.

16. We first count the number N of strings that contain either the substring AB or the substring BE. The answer to the exercise will be: Total number of strings $- N$ or $5! - N$.

According to Exercise 16, Section 6.2, the number of strings that contain *AB* or *BE* = number of strings that contain *AB* + number of strings that contain *BE* − number of strings that contain *AB* and *BE*. A string contains *AB* and *BE* if and only if it contains *ABE* and the number of such strings 3!. The number of strings that contain *AB* = number of strings that contain *BE* = 4!. Thus the number of strings that contain *AB* or *BE* is 4! + 4! − 3!. The solution to the exercise is

$$5! - (2 \cdot 4! - 3!).$$

17. $8!P(9, 5) = 8!(9 \cdot 8 \cdot 7 \cdot 6 \cdot 5)$ **20.** $5!/2$ **23.** $4!5!$

25. 10!

28. 3!5!3!2! (Pick the order the disciplines will appear, then order the books within each discipline.)

SECTION 6.4

1. $C(4, 3) = 4$ **2.** $\{a, b, c\}, \{a, b, d\}, \{a, c, d\}, \{b, c, d\}$

5. $C(12, 4)$ **6.** 7!   **9.** 7!/2!

12. $C(44, 6) = 44!/(6!38!) = 6{,}542{,}536; C(48, 6) = 48!/(6!42!) = 12{,}271{,}512$

15. $C(13, 5)$

18. A committee that has at most one man has exactly one man or no men. There are $C(6, 1)C(7, 3)$ committees with exactly one man. There are $C(7, 4)$ committees with no men. Thus the answer is $C(6, 1)C(7, 3) + C(7, 4)$.

21. $C(10, 4)C(12, 3)C(4, 2)$

24. $2^8 - 1$ (Total number of strings − number of strings that contain no 1's)

27. First, we count the number of eight-bit strings with no two 0's in a row. We divide this problem into counting the number of such strings with exactly eight 1's, with exactly seven 1's, and so on.

There is one eight-bit string with no two 0's in a row that has exactly eight 1's. Suppose that an eight-bit string with no two 0's in a row has exactly seven 1's. The 0 can go in any one of eight positions; thus there are eight such strings. Suppose that an eight-bit string with no two 0's in a row has exactly six 1's. The two 0's must go in two of the blanks shown:

$$\_ 1 \_ 1 \_ 1 \_ 1 \_ 1 \_ 1 \_ .$$

Thus the two 0's can be placed in $C(7, 2)$ ways. Thus there are

$C(7, 2)$ such strings. Similarly, there are $C(6, 3)$ eight-bit strings with no two 0's in a row that have exactly five 1's and there are $C(5, 4)$ eight-bit strings with no two 0's in a row that have exactly four 1's in a row. If a string has less than four 1's, it will have two 0's in a row. Therefore, the number of eight-bit strings with no two 0's in a row is

$$1 + 8 + C(7, 2) + C(6, 3) + C(5, 4).$$

Since there are 2^8 eight-bit strings, there are

$$2^8 - [1 + 8 + C(7, 2) + C(6, 3) + C(5, 4)]$$

eight-bit strings that contain at least two 0's in a row.

28. $1 \cdot 48$ (The four aces can be chosen in one way and the fifth card can be chosen in 48 ways.)

31. First, we count the number of hands containing cards in spades and hearts. Since there are 26 spades and hearts, there are $C(26, 5)$ ways to select five cards from among these 26. However, $C(13, 5)$ contain only spades and $C(13, 5)$ contain only hearts. Therefore, there are

$$C(26, 5) - 2C(13, 5)$$

ways to select five cards containing cards in spades and hearts.

Since there are $C(4, 2)$ ways to select two suits, the number of hands containing cards of exactly two suits is

$$C(4, 2)[C(26, 5) - 2C(13, 5)].$$

34. There are nine consecutive patterns: A2345, 23456, 34567, 45678, 56789, 6789T, 789TJ, 89TJQ, 9TJQK. Corresponding to the four possible suits, there are four ways for each pattern to occur. Thus there are $9 \cdot 4$ hands that are consecutive and of the same suit.

37. $C(52, 13)$

40. $1 \cdot C(48, 9)$ (Select the aces, then select the nine remaining cards.)

43. There are $C(13, 4)C(13, 4)C(13, 4)C(13, 1)$ hands that contain four spades, four hearts, four diamonds, and one club. Since there are four ways to select the three suits to have four cards each, there are $4C(13, 4)^3C(13, 1)$ hands that contain four cards of three suits and one card of the fourth suit.

45. 2^{10} **48.** 2^9 **50.** $C(50, 4)$

53. $C(50, 4) - C(46, 4)$ (Total number − number with no defectives)

58. Many partitions are not counted. For example, the partition

$$\{\{x_1, x_3\}, \{x_2\}, \{x_4\}, \{x_5\}, \{x_6\}, \{x_7\}, \{x_8\}, \{x_9, x_{10}\}\}$$

is not counted.

61. Each route can be described by a string of four R's (right) and four U's (up). For example, the route shown can be described by the string RRURUUUR. Count the number of such strings.

64. Look at the formula for $C(n, k)$.

SECTION 6.5

1. $(x + y)^4 = C(4, 0)x^4 + C(4, 1)x^3y + C(4, 2)x^2y^2 + C(4, 3)xy^3$
$\qquad\qquad + C(4, 4)y^4$
$\qquad\quad = x^4 + 4x^3y + 6x^2y^2 + 4xy^3 + y^4$

4. $(c + 3d)^3 = C(3, 0)c^3 + C(3, 1)c^2(3d) + C(3, 2)c(3d)^2$
$\qquad\qquad\quad + C(3, 3)(3d)^3$
$\qquad\qquad = c^3 + 3c^2(3d) + 3c(3d)^2 + (3d)^3$
$\qquad\qquad = c^3 + 9c^2d + 27cd^2 + 27d^3$

7. $(2c - 3d)^3 = (2c)^3 + 3(2c)^2(-3d) + 3(2c)(-3d)^2 + (-3d)^3$
$\qquad\qquad\quad = 8c^3 - 36c^2d + 54cd^2 - 27d^3$

10. $(3u - 2v)^6 = (3u)^6 + 6(3u)^5(-2v) + 15(3u)^4(-2v)^2$
$\qquad\qquad\quad + 20(3u)^3(-2v)^3 + 15(3u)^2(-2v)^4$
$\qquad\qquad\quad + 6(3u)(-2v)^5 + (-2v)^6$
$\qquad\qquad = 729u^6 - 2916u^5v + 4860u^4v^2 - 4320u^3v^3$
$\qquad\qquad\quad + 2160u^2v^4 - 576uv^5 + 64v^6$

11. $C(5, 2)x^3y^2 = 10x^3y^2$

14. $C(12, 6)(2s)^6(-t)^6 = 924(2s)^6(-t)^6 = 59136s^6t^6$

17. $[10!/(2!3!5!)](2x)^2y^3z^5 = 10080x^2y^3z^5$

20. $[12!/(2!3!2!5!)](2w)^2x^3(3y)^2(-z)^5 = -5987520w^2x^3y^2z^5$

21. 1 8 28 56 70 56 28 8 1

24. If we take $a = 1$ and $b = -1$ in the Binomial Theorem, we obtain

$$(1 - 1)^n = \sum_{k=0}^{n} C(n, k)1^{n-k}(-1)^k,$$

which gives the desired result.

27. Take $a = 1$ and $b = 2$ in the Binomial Theorem.

30. INDUCTIVE STEP

$$\sum_{k=1}^{n+1} kC(n+1, k) = \sum_{k=1}^{n} k[C(n, k-1) + C(n, k)]$$

$$+ (n+1)C(n+1, n+1)$$

$$= \sum_{k=1}^{n+1} kC(n, k-1) + \sum_{k=1}^{n} kC(n, k)$$

$$= \sum_{k=1}^{n+1} (k-1)C(n, k-1)$$

$$+ \sum_{k=1}^{n+1} C(n, k-1) + \sum_{k=1}^{n} kC(n, k)$$

$$= n2^{n-1} + 2^n + n2^{n-1} = (n+1)2^n$$

SECTION 6.6

1. 1357   **4.** 13678   **7.** 35789   **10.** 12435   **13.** 631245
16. 1523476
19. (For Exercise 1) At line 5, we find the rightmost s_m not at its maximum value. In this case, $m = 4$. At line 6, we increment s_m. This makes the last digit 7. Since m is the rightmost position, at lines 7 and 8, we do nothing. The next combination is 1357.
21. 123, 124, 125, 126, 134, 135, 136, 145, 146, 156, 234, 235, 236, 245, 246, 256, 345, 346, 356, 456
24. 12, 21
26. **procedure** *combination2*(*r*, *n*)
    {assume that $s(0)$ is defined}
    $s(0) := -1$
    for $i := 1$ **to** r **do**
      $s(i) := i$
    print $s(1), \ldots, s(r)$ {print the first r-combination}
    while true do
      begin
      $m := r$
      $max\_val := n$
      while $s(m) = max\_val$ **do**
        {find the rightmost element not at its maximum value}
        {this loop is guaranteed to terminate because the dummy value -1 is stored in $s(0)$}

```
      begin
      m := m − 1
      max_val := max_val − 1
      end
   {if m = 0 the algorithm terminates}
   if m = 0 then
      return
   {the rightmost element is incremented}
   s(m) := s(m) + 1
   {the rest of the elements are the successors of s(m)}
   for j := m + 1 to r do
      s(j) := s(j − 1) + 1
   print s(1), . . . , s(r) {print the ith combination}
   end
end combination2
```

SECTION 6.7

1. There are 12 possible names for the 13 persons. We can consider the assignment of names to people to be that of assigning pigeonholes to the pigeons. By the Pigeonhole Principle, some name is assigned to at least two persons.

4. Yes. Connect processors 1 and 2, 2 and 3, 2 and 4, 3 and 4. Processor 5 is not connected to any processors. Now only processors 3 and 4 are directly connected to the same number of processors.

7. Let a_i denote the position of the ith unavailable item. Consider

$$a_1, \ldots, a_{30}; a_1 + 3, \ldots, a_{30} + 3; a_1 + 6, \ldots, a_{30} + 6.$$

These 90 numbers range in value from 1 to 86. By the second form of the Pigeonhole Principle, two of these numbers are the same. If $a_i = a_j + 3$, two are three apart. If $a_i = a_j + 6$, two are six apart. If $a_i + 3 = a_j + 6$, two are three apart.

11. $n + 1$

12. Suppose that $k \leq m/2$. Clearly, $k \geq 1$. Since $m \leq 2n + 1$,

$$k \leq \frac{m}{2} \leq n + \frac{1}{2} < n + 1.$$

Suppose that $k > m/2$. Then

$$m - k < m - \frac{m}{2} = \frac{m}{2} < n + 1.$$

Because m is the largest element in X, $k < m$. Thus $k + 1 \leq m$ and so $1 \leq m - k$. Therefore, the range of a is contained in $\{1, \ldots, n\}$.

13. The second form of the Pigeonhole Principle applies.

14. Suppose that $a_i = a_j$. Then either $i \leq m/2$ and $j > m/2$ or $j \leq m/2$ and $i > m/2$. We may assume that $i \leq m/2$ and $j > m/2$. Now

$$i + j = a_i + m - a_j = m.$$

24. When we divide a by b, the possible remainders are $0, 1, \ldots, b - 1$. Consider what happens after b divisions.

CHAPTER 6 SELF-TEST

1. $7 \cdot 4 \cdot 5 \cdot 3$ 2. 2^4 3. $3 \cdot 4 \cdot 3$

4. $5 \cdot 5 \cdot 4 \cdot 3 \cdot 2$ (Assign Isaac an office; then fill the highest remaining office; then fill the next highest remaining office; etc.)

5. $5 + 5$ 6. $2^7 + 2^5$

7. $6 \cdot 5 \cdot 4 \cdot 3 + 6 \cdot 5 \cdot 4 \cdot 3 \cdot 2$

8. $6 \cdot 9 \cdot 7 + 6 \cdot 9 \cdot 4 + 6 \cdot 7 \cdot 4 + 9 \cdot 7 \cdot 4$

9. $8!$ 10. $4!$ 11. $4!4!$ 12. $9!P(10, 6)$

13. $6!/(3!3!) = 20$

14. $C(8, 3)C(6, 3)$

15. $4 \cdot 3C(13, 4)13$ (We can select the suit for the four cards and the suit for the fifth card in $4 \cdot 3$ ways. We can select four cards from the first suit in $C(13, 4)$ ways and we can select the fifth card from the second suit in 13 ways.)

16. $8!/(3!2!)$

17. $(s - r)^4 = C(4, 0)s^4 + C(4, 1)s^3(-r) + C(4, 2)s^2(-r)^2$
$$+ C(4, 3)s(-r)^3 + C(4, 4)(-r)^4$$
$$= s^4 - 4s^3r + 6s^2r^2 - 4sr^3 + r^4$$

18. $(2c + d)^5 = C(5, 0)(2c)^5 + C(5, 1)(2c)^4d + C(5, 2)(2c)^3d^2$
$$+ C(5, 3)(2c)^2d^3 + C(5, 4)2cd^4 + C(5, 5)d^5$$
$$= 32c^5 + 80c^4d + 80c^3d^2 + 40c^2d^3 + 10cd^4 + d^5$$

19. $C(12, 8)x^4(-2y)^8 = 126720x^4y^8$ 20. $C(n, 1) = n$

21. 12567   22. 234567   23. 6427153   24. 631245

25. Let the 15 individual socks be the pigeons and let the 14 types of pairs be the pigeonholes. Assign each sock (pigeon) to its type (pigeonhole). By the Pigeonhole Principle, some pigeonhole will contain at least two pigeons (the matched socks).

26. There are $3 \cdot 2 \cdot 3 = 18$ possible names for the 17 persons. We can consider the assignment of names to people to be that of assigning pigeonholes to the pigeons. By the Pigeonhole Principle, some name is assigned to at least two persons.

27. Let a_i denote the position of the ith available item. The 220 numbers

$$a_1, \ldots, a_{110}; a_1 + 19, \ldots, a_{110} + 19$$

range from 1 to 219. By the Pigeonhole Principle, two are the same.

28. Argue as in Example 6.7.5.

SECTION 7.1

1. $a_n = a_{n-1} + 4; a_1 = 3$

4. $a_n = a_{n-1} + a_{n-2}; a_1 = 3, a_2 = 6$

7. $a_n = 3a_{n-1}; a_1 = 2$ **10.** $a_n = 2a_{n-1}a_{n-2}; a_1 = a_2 = 1$

11. $A_n = (1.14)A_{n-1}$ **12.** $A_0 = 2000$

13. $A_1 = 2280, A_2 = 2599.20, A_3 = 2963.088$

14. $A_n = (1.14)^n 2000$

15. We must have $A_n = 4000$ or $(1.14)^n 2000 = 4000$ or $(1.14)^n = 2$. Taking the logarithm of both sides, we must have $n \log 1.14 = \log 2$. Thus

$$n = \frac{\log 2}{\log 1.14} = 5.29.$$

25. $A_n = (1.03)^4 A_{n-1}$ **26.** $A_0 = 3000$

27. $A_1 = 3376.53, A_2 = 3800.31, A_3 = 4277.28$

28. $A_n = (1.03)^{4n} 3000$ **29.** 5.86

35. $a_4 = a_0 a_3 + a_1 a_2 + a_2 a_1 + a_3 a_0 = 1 \cdot 5 + 1 \cdot 2 + 2 \cdot 1 + 5 \cdot 1 = 14$

38. $a_5 = 42, a_6 = 132, a_7 = 429$

41. (a) $a_2 = 1$ because two nodes must establish one link to share files.

Suppose that we have three nodes A, B, and C. If the following successive links are established—$A \leftrightarrow B, A \leftrightarrow C, A \leftrightarrow B$—then all nodes know all files. Since three links suffice, $a_3 \leq 3$. (Although it is true, we cannot write $a_3 = 3$ because we have not shown that it is impossible for three nodes to share files using fewer than three links.)

Suppose that we have four nodes A, B, C, and D. If the following successive links are established—$A \leftrightarrow B$, $C \leftrightarrow D$, $A \leftrightarrow C$, $B \leftrightarrow D$—then all nodes know all files. Since four links suffice, $a_4 \leq 4$.

(b) Suppose that we have $n \geq 3$ nodes. Let A and B be nodes. First, A and B share files. Next, all nodes except A share files (which requires a_{n-1} links). Finally, A and B again share files. At this point, all nodes know all files. Thus $a_n \leq a_{n-1} + 2$.

44. Suppose that we spend n dollars. If we buy tape the first day, there are C_{n-1} ways to spend the remaining money. If we buy paper the first day, there are C_{n-1} ways to spend the remaining money. If we buy pens the first day, there are C_{n-2} ways to spend the remaining money. If we buy pencils the first day, there are C_{n-2} ways to spend the remaining money. If we buy binders the first day, there are C_{n-3} ways to spend the remaining money. Thus

$$C_n = 2C_{n-1} + 2C_{n-2} + C_{n-3}.$$

46. $s_3 = 1/2$, $s_4 = 3/4$

48. $\{C_n\}$ satisfies $C_1 = 1$, $C_2 = 2$, and $C_n = C_{n-1} + C_{n-2}$; therefore, $C_n = f_n$, for $n = 1, 2, \ldots$.

50. There are S_{n-1} n-bit strings that begin 1 and do not contain the pattern 00 and there are S_{n-2} n-bit strings that begin 0 (since the second bit must be 1) and do not contain the pattern 00. Thus $S_n = S_{n-1} + S_{n-2}$. Initial conditions are $S_1 = 2$, $S_2 = 3$.

53. $S_1 = 2$, $S_2 = 4$, $S_3 = 7$, $S_4 = 12$

56. Let X be an n-element set and choose $x \in X$. Let k be a fixed integer, $0 \leq k \leq n - 1$. We can select a k-element set Y from $X - \{x\}$ in $C(n - 1, k)$ ways. Having done this, we can partition Y in P_k ways. This partition together with $X - Y$ partitions X. Since all partitions of X can be generated in this way, we obtain the desired recurrence relation.

58. $A(1, 1) = 3$, $A(1, 2) = 4$, $A(2, 2) = 7$, $A(2, 3) = 9$

59. BASIS STEP. $A(1, 0) = A(0, 1) = 2$
INDUCTIVE STEP. $A(1, n + 1) = A(0, A(1, n)) = A(0, n + 2) = n + 3$

SECTION 7.2

1. Linear, homogeneous of order 1   **4.** No   **7.** No
10. Linear, homogeneous of order 3   **11.** $a_n = 2(-3)^n$
14. $a_n = 2^{n+1} - 4^n$ **17.** $a_n = (2^{2-n} + 3^n)/5$

20. $a_n = 2(-4)^n + 3n(-4)^n$ **23.** $C_n = [(-1)^n + 2^{n+1}]/3$

26. $a_n = [2^{n+1} + (-1)^n]^2/9$

29. Show that $U_n - g(n)$ satisfies $a_n = c_1 a_{n-1} + c_2 a_{n-2}$.

30. $a_n = b2^n + d4^n + 1$ **33.** $a_n = b/2^n + d3^n - (4/3)2^n$

36. The argument is identical to that given in Theorem 7.2.11.

39. $a_n = [4^n + (-1)^{n+1}]/5$

SECTION 7.3

1. Input: n

Output: $x \ (= a_n)$

1. If $n = 1$, then $x := 2$; stop.

2. Invoke this algorithm with input $n - 1$ and denote the output by x. Execute $x := x + 3$.

4. Input: n

Output: $w \ (= a_n)$

1. If $n = 0$, then $w := 1$; stop.

2. If $n = 1$, then $w := 2$; stop.

3. If $n = 2$, then $w := 3$; stop.

4. Invoke this algorithm with input $n - 1$ and denote the output by x.

5. Invoke this algorithm with input $n - 2$ and denote the output by y.

6. Invoke this algorithm with input $n - 3$ and denote the output by z.

7. $w := 3x - 2y + 5z$.

7. Input: n

Output: $w \ (= L_n)$

1. If $n = 0$, then $w := 1$; stop.

2. If $n = 1$, then $w := 3$; stop.

3. Invoke this algorithm with input $n - 1$ and denote the output by x.

4. Invoke this algorithm with input $n - 2$ and denote the output by y.

5. $w := x + y$.

10. To print all the ways to spend n dollars, set α to the null string and then invoke this algorithm.

Input: n, α (a string)

Output: All the ways to spend n dollars. Each method of spending n dollars includes the extra string α in the list.

1. If $n = 1$, print α, "orange juice"; and stop.

2. If $n = 2$, print α, "orange juice   orange juice"; print α, "milk"; print α, "beer"; and stop.

3. Call this algorithm with input $n - 1$ and β, where β is "orange juice" followed by α.

4. Call this algorithm with input $n - 2$ and β, where β is "milk" followed by α.

5. Call this algorithm with input $n - 2$ and β, where β is "beer" followed by α.

13. To list all r-combinations of an n-element set, set α to the null string and then invoke this algorithm.

Input: r, n, α (a string)
Output: All r-combinations of $\{1, \ldots, n\}$. Each r-combination includes the extra string α.
1. If $r = n$, print "$12 \cdots n$"α and stop.
2. If $r = 0$, print α and stop.
3. Call this algorithm with input r, $n - 1$, and α.
4. Call this algorithm with input $r - 1$, $n - 1$, and β, where β is n followed by α.

16. **procedure** $recurs\_binary\_search(s, i, j, key)$
{Search for key among $s(i)$, $s(i+1)$, $\ldots$, $s(j)$.
We assume that $s(i) \leq s(i + 1) \leq \cdots \leq s(j)$.
We return the location of key if we find key; otherwise, we return 0.}
    if $i > j$ **then**
        return(0) {not found}
    $middle := \lfloor (i + j)/2 \rfloor$
    if $key = s(middle)$ **then** {found}
        return($middle$)
    if $key < s(middle)$ **then**
        return($recurs\_binary\_search(s, i, middle - 1, key)$)
    return($recurs\_binary\_search(s, middle + 1, j, key)$)
  end $recurs\_binary\_search$

SECTION 7.4

1. Algorithm 7.4.3 computes a^n by using the formula $a^n = a^m a^{n-m}$.
2. $b_n = b_{\lfloor n/2 \rfloor} + b_{\lfloor (n+1)/2 \rfloor} + 1$, $b_1 = 0$
3. $b_2 = 1$, $b_3 = 2$, $b_4 = 3$ **4.** $b_n = n - 1$
13. If $n = 1$, then $i = j$ and we return before reaching line 1a, 3, or 4. Therefore, $b_1 = 0$. If $n = 2$, then $j = i + 1$. There is one

comparison at line 1a and we return before reaching line 3 or 4. Therefore, $b_2 = 1$.

14. $b_3 = 3$, $b_4 = 4$

15. When $n > 2$, $b_{\lfloor(1+n)/2\rfloor}$ comparisons are required for the first recursive call and $b_{\lfloor n/2\rfloor}$ comparisons are required for the second recursive call. Two additional comparisons are required at lines 3 and 4. The recurrence relation now follows.

16. Suppose that $n = 2^k$. Then (7.4.4) becomes

$$b_{2^k} = 2b_{2^{k-1}} + 2.$$

Now

$$
\begin{aligned}
b_{2^k} = 2b_{2^{k-1}} + 2 &= 2[2b_{2^{k-2}} + 2] + 2 \\
&= 2^2 b_{2^{k-2}} + 2^2 + 2 = \cdots \\
&= 2^{k-1} b_{2^1} + 2^{k-1} + 2^{k-2} + \cdots + 2 \\
&= 2^{k-1} b_2 + 2^{k-1} + 2^{k-2} + \cdots + 2 \\
&= 2^{k-1} + 2^{k-1} + \cdots + 2 \\
&= 2^{k-1} + 2^k - 2 \\
&= n - 2 + \frac{n}{2} = \frac{3n}{2} - 2.
\end{aligned}
$$

22. $b_n = b_{\lfloor(1+n)/2\rfloor} + b_{\lfloor n/2\rfloor} + 3$

25. $b_n = 4n - 3$

28. BASIS STEP. $b_1 = 0 = 1 - 1$
INDUCTIVE STEP. $b_n = b_{\lfloor n/2\rfloor} + b_{\lfloor(n+1)/2\rfloor} + 1$
$\qquad\quad = \lfloor n/2\rfloor - 1 + \lfloor(n+1)/2\rfloor - 1 + 1$
$\qquad\quad = n - 1$

31. BASIS STEP. $b_1 = 1 \le 1 + 0 = 1 + \lg 1$
INDUCTIVE STEP. $b_n = 1 + b_{\lfloor n/2\rfloor} \le 1 + 1 + \lg\lfloor n/2\rfloor$
$\qquad\quad \le 2 + \lg n/2 = 2 + \lg n - \lg 2$
$\qquad\quad = 2 + \lg n - 1 = 1 + \lg n$

34. We will show that $b_n \le b_{n+1}$, $n = 1, 2, \ldots$. We have the recurrence relation

$$b_n = b_{\lfloor(1+n)/2\rfloor} + b_{\lfloor n/2\rfloor} + c_{\lfloor(1+n)/2\rfloor, \lfloor n/2\rfloor}.$$

BASIS STEP. $b_2 = 2b_1 + c_{1,1} \ge 2b_1 \ge b_1$.

INDUCTIVE STEP. Assume that the statement holds for $k < n$. In case n is even, we have $b_n = 2b_{n/2} + c_{n/2, n/2}$; so

$$b_{n+1} = b_{(n+2)/2} + b_{n/2} + c_{(n+2)/2, n/2}$$
$$\geq b_{n/2} + b_{n/2} + c_{n/2, n/2} = b_n.$$

The case n odd is similar.

CHAPTER 7 SELF-TEST

1. (a) 3, 5, 8, 12   (b) $a_1 = 3$ (c) $a_n = a_{n-1} + n$

2. $A_n = (1.17)A_{n-1}$, $A_0 = 4000$ **3.** 132

4. If the first domino is placed as shown, there are a_{n-1} ways to cover the $2 \times (n - 1)$ board that remains.

If the first two dominoes are placed as shown, there are a_{n-2} ways to cover the $2 \times (n - 2)$ board that remains.

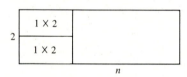

It follows that $a_n = a_{n-1} + a_{n-2}$.

By inspection, $a_1 = 1$ and $a_2 = 2$. Since $\{a_n\}$ satisfies the same recurrence relation as the Fibonacci sequence and $a_1 = f_1$ and $a_2 = f_2$, it follows that $a_i = f_i$ for $i = 1, 2, \ldots$.

5. Yes   **6.** $a_n = 2^{n+2} - 3$ **7.** $a_n = 2(-2)^n - 4n(-2)^n$

8. $a_n = 3 \cdot 5^n + (-2)^n$

9. Input: n

Output: w ($= a_n$)

1. If $n = 0$, then $w := 2$; stop.
2. If $n = 1$, then $w := 4$; stop.
3. Invoke this algorithm with input $n - 1$ and denote the output by x.
4. Invoke this algorithm with input $n - 2$ and denote the output by y.
5. $w := -4x - 4y$.

10. Input: n

Output: $w \; (= a_n)$

1. If $n = 0$, then $w := 1$; stop.
2. If $n = 1$, then $w := 1$; stop.
3. If $n = 2$, then $w := 0$; stop.
4. If $n = 3$, then $w := 0$; stop.
5. Invoke this algorithm with input $n - 1$ and denote the output by x.
6. Invoke this algorithm with input $n - 3$ and denote the output by y.
7. Invoke this algorithm with input $n - 4$ and denote the output by z.
8. $w := 6x - 4y + z$.

11. To print all the ways to spend n dollars, set α to the null string and then invoke this algorithm.

Input: n, α (a string)

Output: All the ways to spend n dollars. Each method of spending n dollars includes the extra string α in the list.

1. If $n = 1$, print α, "disk"; and stop.
2. If $n = 2$, print α, "disk   disk"; and stop.
3. If $n = 3$, print α, "disk   disk   disk"; print α, "printer cover"; and stop.
4. If $n = 4$, print α, "disk   disk   disk   disk"; print α, "disk   printer cover"; print α, "printer cover   disk"; and stop.
5. If $n = 5$, print α, "disk   disk   disk   disk   disk"; print α, "disk   disk   printer cover"; print α, "disk   printer cover   disk"; print α, "printer cover   disk   disk"; print α, "computer cover"; and stop.
6. Call this algorithm with input $n - 1$ and β, where β is "disk" followed by α.
7. Call this algorithm with input $n - 3$ and β, where β is "printer cover" followed by α.
8. Call this algorithm with input $n - 5$ and β, where β is "computer cover" followed by α.

12. Input: n

Output: $x \; (= \text{number of binary trees})$

1. If $n = 0$, $x := 1$; stop.

2. For $i := 0$ to $n - 1$, execute line 3.
3.   Call this algorithm with input i. Denote the output by s_i.
4. $x := 0$.
5. For $i := 0$ to $n - 1$, execute line 6.
6. $x := x + s_i s_{n-i-1}$.

13. $b_n = b_{n-1} + 1$, $b_0 = 0$ **14.** $b_1 = 1$, $b_2 = 2$, $b_3 = 3$
15. $b_n = n$
16. $n(n + 1)/2 = O(n^2)$. The given algorithm is faster than the straightforward technique and is, therefore, preferred.

SECTION 8.1

1. {(8840, Hammer), (9921, Pliers), (452, Paint), (2207, Carpet)}
4. {(a, a), (b, b)}

5.

a	6
b	2
a	1
c	1

8.

Maine	Augusta
Maryland	Annapolis
Massachusetts	Boston
Michigan	Lansing
Minnesota	St. Paul
Mississippi	Jackson
Missouri	Jefferson City
Montana	Helena

9.

12.

13. {(a, b), (a, c), (b, a), (b, d), (c, c), (c, d)}
16. {(b, c), (c, b), (d, d)}
17. (For Exercise 1) domain = {8840, 9921, 452, 2207}, range = {Hammer, Pliers, Paint, Carpet}
19. $R = \{(1, 1), (1, 4), (2, 2), (2, 5), (3, 3), (4, 1), (4, 4), (5, 2), (5, 5)\}$
22. {1, 2, 3, 4, 5}

25. $R = R^{-1} = \{(1, 1), (1, 2), (1, 3), (1, 4), (1, 5), (2, 1), (2, 2),$
$(2, 3), (2, 4), (3, 1), (3, 2), (3, 3), (4, 1), (4, 2),$
$(5, 1)\}$.
domain R = range R = domain R^{-1} = range R^{-1}
= $\{1, 2, 3, 4, 5\}$.

28. Antisymmetric   29. Antisymmetric

32. Reflexive, symmetric, antisymmetric, transitive, partial order

35. Reflexive, antisymmetric, transitive, partial order

37. $\{(1, 1), (2, 2), (3, 3), (4, 4), (1, 2), (2, 1), (2, 3), (3, 2)\}$

40. $\{(1, 1), (1, 2), (2, 1), (2, 2)\}$ 42. $R = \{(1, 2)\}, S = \{(2, 3)\}$

45. $R = \{(1, 2)\}, S = \{(2, 1)\}$

47. It may be the case that for some $x \in X$, there is *no* $y \in X$ such that
$(x, y) \in R$. Consider, for example,

$$X = \{1, 2, 3\}, R = \{(1, 1), (2, 2), (1, 2), (2, 1)\}$$

and $x = 3$.

SECTION 8.2

1. Is an equivalence relation. Equivalence classes are $\{1, 3\}, \{2\}, \{4\},$
$\{5\}$.

4. Is an equivalence relation. Equivalence classes are $\{1, 3, 5\}, \{2\},$
$\{4\}$.

7. Not an equivalence relation. Neither reflexive nor transitive.

9. $\{(1, 1), (1, 2), (2, 1), (2, 2), (3, 3), (3, 4), (4, 3), (4, 4)\}$. $[1] =$
$[2] = \{1, 2\}, [3] = [4] = \{3, 4\}$.

12. $\{(1, 1), (1, 2), (1, 3), (2, 1), (2, 2), (2, 3), (3, 1), (3, 2), (3, 3),$
$(4, 4)\}, [1] = [2] = [3] = \{1, 2, 3\}, [4] = \{4\}$.

15. R is reflexive because a city is in the same state as itself. If x and
y are cities in the same state, then y and x are cities in the same
state; therefore, R is symmetric. If x and y are cities in the same
state and y and z are cities in the same state, then x and z are
cities in the same state; therefore, R is transitive.

16. $\{$San Francisco, San Diego, Los Angeles$\}$
$\{$Pittsburgh, Philadelphia$\}$
$\{$Chicago$\}$

20. If the equivalence relation R on X has only one equivalence class,
then $R = \{(x, y) \mid x, y \in X\}$.

23. Five, corresponding to the partitions $\{\{1\}, \{2\}, \{3\}\}, \{\{1\}, \{2, 3\}\},$
$\{\{1, 2\}, \{3\}\}, \{\{1, 3\}, \{2\}\}, \{\{1, 2, 3\}\}$

24. (b) Cylinder

SECTION 8.3

1.
$$\begin{array}{c} \\ 1 \\ 2 \\ 3 \end{array} \begin{array}{cccc} \alpha & \beta & \Sigma & \delta \\ \begin{pmatrix} 0 & 0 & 0 & 1 \\ 1 & 0 & 0 & 0 \\ 0 & 1 & 1 & 0 \end{pmatrix} \end{array}$$

4.
$$\begin{array}{c} \\ 1 \\ 2 \\ 3 \\ 4 \\ 5 \end{array} \begin{array}{ccccc} 1 & 2 & 3 & 4 & 5 \\ \begin{pmatrix} 0 & 1 & 0 & 0 & 0 \\ 0 & 0 & 1 & 0 & 0 \\ 0 & 0 & 0 & 1 & 0 \\ 0 & 0 & 0 & 0 & 1 \\ 0 & 0 & 0 & 0 & 0 \end{pmatrix} \end{array}$$

7. $R = \{(a, w), (a, y), (c, y), (d, w), (d, x), (d, y), (d, z)\}$
10. $R = \{(w, w), (w, y), (y, w), (y, y), (z, z)\}$
13. $R = \{(1, 1), (2, 1), (2, 2), (3, 1), (3, 3)\}$
16. The test is, whenever the *ij*th entry is 1, $i \neq j$, then the *ji*th entry is *not* 1.
19. Take the transpose of the given matrix.

21. (a)
$$A_1 = \begin{pmatrix} 1 & 1 \\ 1 & 0 \\ 1 & 0 \end{pmatrix}$$
(b)
$$A_2 = \begin{pmatrix} 0 & 1 & 0 \\ 1 & 1 & 1 \end{pmatrix}$$

(c)
$$A_1 A_2 = \begin{pmatrix} 1 & 2 & 1 \\ 0 & 1 & 0 \\ 0 & 1 & 0 \end{pmatrix}$$

(d) We change each nonzero entry in part (c) to 1 to obtain

$$\begin{pmatrix} 1 & 1 & 1 \\ 0 & 1 & 0 \\ 0 & 1 & 0 \end{pmatrix}.$$

(e) $\{(1, b), (1, a), (1, c), (2, b), (3, b)\}$
24. Each column that contains 1 in row *x* corresponds to an element of the equivalence class containing *x*.
27. Input: *A*, the $n \times n$ matrix of the relation *R*.
   Output: Yes, if *R* is reflexive.
        No, if *R* is not reflexive.
   1. For $i := 1$ to *n*, execute line 2.
   2.   If $A_{ii} = 0$, then output ''No'' and stop.
   3. Output ''Yes.''
30. Input: *A*, the $m \times n$ matrix of the relation *R*.
   Output: Yes, if *R* is function.
        No, if *R* is not function.

1. For $i := 1$ to m, execute lines 2–5.
2. $sum := 0$.
3.   For $j := 1$ to n, execute line 4.
4. $sum := sum + A_{ij}$.
5.   If $sum \neq 1$, then output "No" and stop.
6. Output "Yes."

SECTION 8.4

1. {(1089, Suzuki, Zamora), (5620, Kaminski, Jones), (9354, Jones, Yu), (9551, Ryan, Washington), (3600, Beaulieu, Yu), (0285, Schmidt, Jones), (6684, Manacotti, Jones)}

5. EMPLOYEE [Name]
Suzuki, Kaminski, Jones, Ryan, Beaulieu, Schmidt, Manacotti

8. BUYER [Name]
United Supplies, ABC Unlimited, JCN Electronics, Danny's, Underhanded Sales, DePaul University

11. TEMP := BUYER [Part No = 20A8]
TEMP [Name]
Underhanded Sales, Danny's, ABC Unlimited

14. TEMP1 := BUYER [Name = Danny's]
TEMP2 := TEMP1 [Part No = Part No] SUPPLIER
TEMP2 [Dept]
04, 96

17. TEMP1 := BUYER [Name = JCN Electronics]
TEMP2 := TEMP1 [Part No = Part No] SUPPLIER
TEMP3 := TEMP2 [Dept = Dept] DEPARTMENT
TEMP4 := TEMP3 [Manager = Manager] EMPLOYEE
TEMP4 [Name]
Kaminski, Schmidt, Manacotti

22. Let R_1 and R_2 be two n-ary relations. Suppose that the set of elements in the ith column of R_1 and the set of elements in the ith column of R_2 come from a common domain for $i = 1, \ldots, n$. The *union* of R_1 and R_2 is the n-ary relation $R_1 \cup R_2$.

TEMP1 := DEPARTMENT [Dept = 23]
TEMP2 := DEPARTMENT [Dept = 96]
TEMP3 := TEMP1 *union* TEMP2
TEMP4 := TEMP3 [Manager = Manager] EMPLOYEE
TEMP4 [Name]

Kaminski, Schmidt, Manacotti, Suzuki

CHAPTER 8 SELF-TEST

1.

2. domain = range = {1, 2, 3, 4}

3. {(3, 1), (1, 3), (4, 2), (1, 1), (2, 2), (3, 3), (4, 4)}

4. Reflexive, transitive

5. Yes. It is reflexive, symmetric, and transitive.

6. [3] = {3, 4}. There are two equivalence classes.

7. {(a, a), (b, b), (b, d), (b, e), (d, b), (d, d), (d, e), (e, b), (e, d), (e, e), (c, c)}

8. Yes. Since R is reflexive, $(x, x) \in R$ for all $x \in X$. This implies that domain $R = X$.

9.
$$\begin{pmatrix} 1 & 0 & 1 & 0 \\ 0 & 1 & 0 & 1 \\ 1 & 0 & 1 & 0 \\ 0 & 0 & 0 & 1 \end{pmatrix}$$

10. Reflexive, symmetric, transitive, equivalence relation

11. (a) $\begin{pmatrix} 1 & 0 \\ 1 & 1 \\ 0 & 1 \end{pmatrix}$ (b) $\begin{pmatrix} 1 & 1 & 0 \\ 1 & 0 & 1 \end{pmatrix}$

(c) $\begin{pmatrix} 1 & 1 & 0 \\ 2 & 1 & 1 \\ 1 & 0 & 1 \end{pmatrix}$ (d) $\begin{pmatrix} 1 & 1 & 0 \\ 1 & 1 & 1 \\ 1 & 0 & 1 \end{pmatrix}$

(e) {(1, a), (1, b), (2, a), (2, b), (2, c), (3, a), (3, c)}

12. Input: A, the matrix of R
       n, the size of A
   Output: Each element $x \in X$ followed by the members of the
           equivalence class containing x
   1. For $i := 1$ to n, execute lines 2–4.
   2.   Print "Members of equivalence class" i.
   3.   For $j := 1$ to n, execute line 4.
   4.   If $A_{ij} = 1$, print j.

13. ASSIGNMENT [Team]
   Blue Sox, Mutts, Jackalopes

14. PLAYER [Name, Age]

Johnsonbaugh, 22; Glover, 24; Battey, 18; Cage, 30; Homer, 37; Score, 22; Johnsonbaugh, 30; Singleton, 31

15. TEMP1 := PLAYER [Position = p]

TEMP2 := TEMP1 [ID Number = PID] ASSIGNMENT

TEMP2 [Team]

Mutts, Jackalopes

16. TEMP1 := PLAYER [Age ≥ 30]

TEMP2 := TEMP1 [ID Number = PID] ASSIGNMENT

TEMP2 [Team]

Blue Sox, Mutts

SECTION A.1

1. If three points are not collinear, then there is exactly one plane that contains them.

4. If x is a nonnegative real number and n is a positive integer, $x^{1/n}$ is the nonnegative number y satisfying $y^n = x$.

7.

$x \cdot 0 + 0 = x \cdot 0$	because $b + 0 = b$ for all real numbers b
$= x \cdot (0 + 0)$	because $b + 0 = b$ for all real numbers b
$= x \cdot 0 + x \cdot 0$	because $a(b + c) = ab + ac$ for all real numbers a, b, c

Taking $a = b = x \cdot 0$ and $c = 0$, the equation above becomes $a + b = a + c$; therefore, $x \cdot 0 = b = c = 0$.

10. Valid $\quad p \to q$

$$\frac{p}{\therefore q}$$

13. Invalid $\quad (p \lor r) \to q$

$$\frac{q}{\therefore \bar{p} \to r}$$

15. Valid. If 64K is better than no memory at all, then we will buy a new computer. If 64K is better than no memory at all, then we will buy more memory. Therefore, if 64K is better than no memory at all, then we will buy a new computer and we will buy more memory.

18. Invalid. If we will not buy a new computer, then 64K is not better than no memory at all. We will buy a new computer. Therefore, 64K is better than no memory at all.

20. Invalid **23.** Invalid

26. An analysis of the argument must take into account the fact that "nothing" is being used in two very different ways.

SECTION A.2

2. Type (a) with: C_1 = class of all isosceles triangles.

C_2 = class of all equilateral triangles.

Negation: Some isosceles triangle is not equilateral.

5. Type (d) with: C_1 = class of all chips.

C_2 = class of all functional chips.

Negation: All chips are functional.

8. Type (a) with: C_1 = class of all violent movies.

C_2 = class of all R-rated movies.

Negation: Some violent movie is not R-rated.

11. Type (b) with: C_1 = class of all persons who can manage a crocodile.

C_2 = class of all despised persons.

Negation: Someone despised can manage a crocodile.

12. C_1 = class of all planar maps.

C_2 = class of all maps that can be colored using at most four colors.

All C_1 is C_2. Negation: Some planar map cannot be colored using at most four colors.

15. C_1 = class of all software with poor documentation.

C_2 = class of all objects not worth much.

All C_1 is C_2. Negation: Some software with poor documentation is worth much.

16. Valid   **19.** Valid   **22.** Valid   **25.** Invalid

26. C_1 = class of all Cubs; C_2 = class of all White Sox;

C_3 = class of all baseball players.

$$\frac{\begin{array}{l} \text{No } C_1 \text{ is } C_2. \\ \text{All } C_2 \text{ is } C_3. \end{array}}{\therefore \text{ Some } C_3 \text{ is not } C_1.}$$

Invalid

29. C_1 = class of all dogs; C_2 = class of all animals;

C_3 = class of all horses.

$$\frac{\begin{array}{l} \text{All } C_1 \text{ is } C_2. \\ \text{No } C_1 \text{ is } C_3. \end{array}}{\therefore \text{ No } C_3 \text{ is } C_2.}$$

Invalid

32. $(x - 2)^2(x - 1) = 0$ **35.** $p = 2$

36. $x = 3, y = 4, z = 5$

Index